Switching to Solar

Switching to Solar

What We Can Learn from Germany's Success in
HARNESSING CLEAN ENERGY

Bob Johnstone

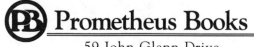
Prometheus Books

59 John Glenn Drive
Amherst, New York 14228-2119

Published 2011 by Prometheus Books

Inquiries should be addressed to
Prometheus Books
59 John Glenn Drive
Amherst, New York 14228–2119
VOICE: 716–691–0133
FAX: 716–691–0137
WWW.PROMETHEUSBOOKS.COM

15 14 13 12 11   5 4 3 2 1

Library of Congress Cataloging-in-Publication Data

Johnstone, Bob.
    Switching to solar : what we can learn from Germany's success in harnessing clean energy / by Bob Johnstone.
        p. cm.
    Includes bibliographical references and index.
    ISBN 978–1–61614–222–3 (pbk.)
    1. Photovoltaic power systems—Technological innovations. 2. Photovoltaic power generators—Government policy. 3. Building-integrated photovoltaic systems. 4. Solar power plants. 5. Photovoltaic power systems—Germany. I. Title.

GT3331.U6R67 2010
333.792/3—dc22
                                                                    2010030259

Printed in the United States of America

To Jack and Scott

CONTENTS

Part 1

THE FIRST COMING: HOW THE US INVENTED SOLAR TECHNOLOGY, BUILT AN INDUSTRY, THEN DROPPED THE BALL

Part 2

HUNDREDS, THOUSANDS, . . . MILLIONS? EARLY ATTEMPTS TO PUT SOLAR ON ROOFS IN THE UNITED STATES, SWITZERLAND, AND JAPAN

Part 3

HIER KOMMT DIE SONNE: HOW CLOUDY
GERMANY CAME FROM NOWHERE
TO LEAD THE SOLAR RACE

Part 4

THE SECOND COMING: PUSHING PV IN CALIFORNIA,
BOTTOM UP AND TOP DOWN

Part 5

JUST THE BEGINNING: UTILITY-SCALE SOLAR, CHINA,
AND THE SHAPE OF PV TO COME

INTRODUCTION

A man sitting at a New York sidewalk cafe is surprised when a battery splashes into his glass. From their windows office workers stare in bemusement as batteries begin raining down like bombs. In the street a car's windshield wipers struggle to clear accumulating precipitation, then the windshield shatters. Their drivers unable to see, cars collide; a trucker leaps from the cab of his tanker just before it crumples under the impact of a direct hit from a lead-acid battery. People panic, shielding their heads with their hands, running for cover as the hard rain falls. The scene shifts to London, where an explosion upends a double-decker bus onto a taxi; then to Paris, where a boy seeks refuge from the mayhem in a dress shop near the Eiffel Tower; to Africa, where a tribesman looks on dazed by the deluge as two terrified zebras gallop by; and to the Arctic, where Eskimos flee as their igloo is flattened.

Then comes the explanation:

970 trillion kilowatt-hours of energy falls from the sky every day.

The storm has passed. On the battery-strewn street people emerge from their hiding places.

Good we can't see it.

The people look up at the sky, where the sun is now shining brightly.

Bad we don't use it.

Cut to giant solar arrays deployed across green fields as far as the eye can see.

This apocalyptic TV commercial was made by a German solar panel manufacturer. An illustration by Richard Perez, a specialist in the measurement of solar radiation, reinforces the point. It depicts the world's energy resources as planets orbiting (appropriately enough) the sun. The other

11

renewables—wind, hydro, biomass—are minuscule, more like moons than planets; even the fossil fuels—coal, oil, gas—are merely a fraction of the size of the solar resource. Based on this comparison, Perez poses a rhetorical question: "Where should we invest for the long haul?"

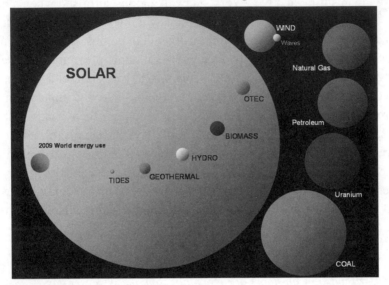

Courtesy of Richard R. and M. Perez, from "Making the Case for Solar Energy," *Daylight and Architecture Magazine* 9 (Fall 2008): 8–17. www.asrc.cestm.albany.edu/perez/.

The answer is obvious. We should choose solar because there is vastly more of it than any other type of energy. Also because solar energy is ubiquitous, environment friendly, and free.

Switching to solar has already begun. This book chronicles the first stages of what promises to be technological transformation on a grand scale. Some people call it a new industrial revolution; others, like the media mogul Ted Turner, declare it the greatest business opportunity in history. "The entire world is going to have to redo its energy regime," Turner explains. "Solar is going to be a very big part of it."

I began work on this book in 2006 because I was curious. Sales of solar systems were booming. In just a few years, the solar industry's output had leapt from millions of watts to billions. Something had obviously happened to trigger this boom. I wanted to know what it was. As a journalist who had covered the electronics industry in the eighties and nineties, I knew a little about solar, meaning photovoltaic cells that convert sunlight into electricity. The technology worked well enough, but solar had always been too expensive for most people to afford. Perhaps in the interim some breakthrough had occurred to bring down the cost.

The truth I soon discovered was that a breakthrough had indeed occurred, but not in the *technology* of solar cells. Rather, it was a breakthrough in *policy* that had made solar not just affordable for the average homeowner but an excellent investment. It turned out that this innovative, easy-to-understand policy mechanism had been developed and implemented not in solar's traditional strongholds—the US and Japan—but in cloudy Germany. Thanks to this mechanism, known as a feed-in tariff, Germans have already deployed more solar than the rest of the world put together, much of it on the roofs of their homes. Indeed, it is no exaggeration to say that Germany has made solar the boom industry it is today.

This book tells for the first time the story of that breakthrough, the people who made it, why they did it, what led up to it, and what has followed from it. The story is one of intrepid individuals who collectively have achieved—and continue to achieve—extraordinary outcomes. Unreasonable men, George Bernard Shaw would have called them. "The reasonable man adapts himself to the world; the unreasonable one persists in trying to adapt the world to himself. Therefore, all progress depends on the unreasonable man." (And, Shaw should have added, the unreasonable woman.) These unreasonable people are changing the way we generate our energy. They are responsible for the creation of tens of thousands of new jobs. They are helping to protect our planet from the rigors of climate change. There is much we can learn from them.

Switching to solar is just one part of a much larger, electronic revolution. It began more than sixty years ago with the invention of the transistor, a semiconductor device, at Bell Laboratories. The transistor together with its descendant the microchip have radically improved our world. Without semiconductors, personal computers, the Internet, cell phones, flat screens, and all the other electronic marvels that enrich our daily lives would be unthinkable. Solar cells are also semiconductors. They were invented at Bell Labs not long after the transistor. To a nonscientist such as myself, solar cells seem like magical devices. They somehow perform the alchemical trick of turning something free—sunlight—into something valuable—electricity. Like all semiconductor devices, solar cells are subject to the logic of the learning curve: the more you make of them, the cheaper they get.

In 1980, no one had a personal computer and the Internet was virtually unknown; in the developed world today, pretty much everyone has a PC and Internet connectivity. In 1990, few people had cell phones; now more than half the world's population does. These examples show that we are capable of taking giant steps in a relatively short period of time. Switching to solar can happen in the same way.

In this ongoing electronic revolution, the energy industry is lagging way behind. If you don't believe me, visit your nearest coal-fired power station. It's like traveling back in time, to the age of steam. To generate electricity, utilities burn fossilized trees dug out of huge holes in the ground. It's a primitive process that wastes vast amounts of energy (to transport the coal), heat (to create the steam), and water (to cool the steam down again). It is also one that discharges millions of tons of mercury-containing, asthma-inducing, atmosphere-warming greenhouse gases.

Generating electricity from solar is today more expensive than from coal and nuclear (unless you factor in the enormous cost of dealing with "negative externalities" like carbon pollution and radioactive waste). But the economics of solar are changing fast. For the past decade the industry has been growing by around 40 percent a year. Thanks to mass production, the price of panels is plummeting as the price of fossil fuels fluctuates upward. In some parts of the world, like Hawaii, Italy, and Japan, solar is already cheaper than electricity generated by conventional means. Optimistic Germans expect solar to reach price parity in their country by 2015.

Critics object that solar is an intermittent technology, meaning that electricity generation is not constant. As such it cannot replace coal and nuclear power plants, they say, which run 24/7. Not true—energy produced by solar thermal plants can be stored in the form of heat in molten salts during the day and used to generate electricity at night. Compact electricity storage is a solvable problem: advanced new batteries are under development. Meanwhile, the gap between sunset and sunrise can be filled by other renewable technologies like wind, biomass, and geothermal, along with relatively less-polluting fuels like natural gas.

Solar will not be the complete solution to our energy problem. We must also increase our energy efficiency and embrace every other form of alternative energy. But solar is different. Unlike wind turbines, solar panels are *silent*, having no moving parts. Unlike other renewables (but like the Internet), it is a *distributed* technology, meaning that you can put solar panels anywhere there is space for them—on a roof, in a field, along a road, above a parking lot. Unlike nuclear and fossil fuels, solar is *sustainable*: its fuel will never run out. Solar is virtually *maintenance free*—just set and forget. It is also uniquely *scalable*: solar can already power everything from a pocket calculator to an air force base, and even bigger applications beckon. Solar panels pump electricity into the grid precisely when it is most *valuable*, on hot summer afternoons when air conditioners are on full blast and demand spikes as a result. In keeping with the ethos of the age, solar is also a *personal* technology, one that delivers a degree of energy independence to users who are otherwise at the mercy of electricity monopolies. Finally, solar is a good strategy for individuals who want to generate power but still *contribute* to the reduction of greenhouse emissions.

This book tells the story of solar focusing on developments since the 1980s in Germany and California. Germany, because it leads the world in deploying solar; California, because it sets the pace in the US, potentially the world's largest market. My assumption is that what California does today, the other forty-nine states will do tomorrow. Other countries are already following Germany's lead. They include Spain, Italy, the United Kingdom, and the Czech Republic.

In both Germany and California, the switch to solar began with grassroots activists reacting to crises. In Germany, it was acid rain and nuclear accidents; in California, deregulation and its chaotic aftermath. The activists were determined to have solar, regardless of its high upfront cost. They understood that mass production was needed to bring the price down. But if manufacturers were to be persuaded to invest in new production lines, there had to be a guaranteed market. To create that market, to make solar affordable, a policy mechanism was required: in Germany, it was feed-in tariffs; in California, a mixture of rebates and tax breaks. Turning those mechanisms into law took the support of politicians who understood the significance of solar. In addition to providing clean air and energy independence, solar also meant new industry, green-collar jobs, and economic development. Once the policy mechanisms were in place, entrepreneurs and the market would take care of the rest.

Bad we don't use it. Studies show that a future powered entirely by renewables, solar prime among them, is eminently achievable. Moreover, so far from being a drag on the economy as critics claim, switching to solar will actually provide a gigantic boost.* In 2010 the solar industry reached a milestone, installing more than 10 gigawatts' worth of systems, the equivalent of five large-scale nuclear plants, up from just 1 gigawatt in 2004. In the US, sales were on track to reach 1 gigawatt, up by more than double in 2009. The switch has undoubtedly begun, but solar still has a very long way to go before reaching its potential. Today, in most places, solar accounts for only a fraction of a percent of our energy usage. More than half our electricity still comes from burning coal. The problem is that public understanding and political will are lacking. Solar is popular across the political spectrum, as opinion polls have repeatedly demonstrated. But people are confused. They think solar is too expensive. They do not realize that in many places, solar systems are already saving rate-payers money on their electricity bills.

Meanwhile, the fossil fuel industries and their proxies are doing their utmost to spread fear, uncertainty, and doubt. They babble about chimeras like "clean coal," an oxymoron if ever there was one, and a cynical excuse for business-as-usual. They complain about government subsidies for renewables, exploiting people's ignorance of the fact that such subsidies are peanuts compared to the vast amounts of public money lavished each year on coal, oil, gas, and nuclear. Armies of lobbyists defend investor-owned utilities bent on maintaining their monopoly on the supply of electricity. Vested interests spare no expense to ensure legislation that promotes individual ownership of the means of energy production is watered down or killed.

For three decades politicians in America have ignored the issue of energy security. Instead of keeping their eyes on the road ahead, lawmakers have been looking in the rear-view mirror. In Barack Obama the US finally has a leader who seems to understand the importance of energy. "We know the country that harnesses the power of clean, renewable energy will lead the twenty-first century," Obama told Congress in 2009. "We invented solar technology, but we've fallen behind countries like Germany and Japan in producing it. . . . It is time for America to lead again." Political leadership is

*See, for example, *Roadmap 2050,* a study published in April 2010 coauthored by the consultancy McKinsey for the European Climate Foundation, a not-for-profit policy advocacy group. The report concluded that not only is deriving 100 percent of electricity from renewables technically feasible, it is also economically compelling.

indeed vital. But for switching to solar to happen on a global scale, we must all get involved. We need to let our politicians know that we want solar. This book contains many examples of how grassroots activists have influenced the political debate. By telling their stories my intent is to inform that debate, and to inspire others to follow their example. Catastrophic climate change confronts us. We cannot delay any longer.

With their elegant feed-in tariff, the Germans have shown the world that switching to solar is feasible. They have mustered the political will and constructed the legal platform needed to support solar energy for the long haul. In addition to reducing their carbon emissions, the Germans have created a vibrant new solar industry that employs more workers than their coal mines and that, having survived the worst recession in living memory, continues to grow. Germany's goal is to become the world's first industrial power to source all its energy from clean technologies. If cloudy Germany can do it, so can we.

NOTE ON TERMINOLOGY

A **watt** (named after the eighteenth-century Scottish engineer James Watt) is a standard unit used to measure electrical power. A good way to think about watts is in terms of the work they can do. For example, 100 watts is enough to power a light bulb. To power a house, an average-size rooftop system (consisting of around twenty solar panels) outputs 3,000 watts, or 3 **kilowatts**. Using this rule of thumb, a million watts, 1 **megawatt**, is enough to power a large department store. A billion watts, 1 **gigawatt**, is equivalent to the output of a large coal-fired power station, enough juice to run a small city. In 2008 the planet's total electricity consumption was around 15 trillion watts, or 15 **terawatts**.

Part 1: The First Coming

HOW THE US INVENTED SOLAR TECHNOLOGY,
BUILT AN INDUSTRY, THEN DROPPED THE BALL

1

A MOST EXCITING ADVENTURE

*L*arry Kazmerski did not hesitate: he had to prevent the president of the United States from killing himself.

Jimmy Carter had designated May 3, 1978, as Sun Day, "to inform the general public, industry, and labor about solar technologies and to demonstrate the sun's potential in meeting America's energy needs." Now here he was, the chief executive himself, on the appointed day—a Wednesday, as it happened—visiting the brand-new US national Solar Energy Research Institute. The site chosen for SERI was a rocky mesa at the base of South Table Mountain just outside Golden, Colorado, home of Coors Beer and the last resting place of William Frederick "Buffalo Bill" Cody. Predictably, the sun was refusing to cooperate. In fact, as Kazmerski, then a young researcher who was the as-yet-unbuilt laboratory's first specialist in photovoltaic solar cells, recalled, "The day President Carter arrived, the weather changed, the sleet and snow were coming down—it was just miserable out there."

Carter had made it clear that he wanted to actually *see* things. Problem was, the lab was so new there was hardly anything to show him. Somehow at short notice the staff had rustled up a few bits and pieces. On his tour of the hastily assembled equipment, the raincoat-clad chief executive paused to inspect a solar array, borrowed at short notice from the Jet Propulsion Laboratory in Pasadena, California. "Back then, these things weren't sealed so well, and here's our president reaching out to wipe away some of the snow off this array," Kazmerski told me. "I grabbed his hand because I thought, My God—we might have the first president here electrocuted from photovoltaics!"

Not that Carter, a US Navy–trained nuclear engineer, was ignorant about solar energy or its potential. Like many people, the president had obviously been captivated by this space-age technology, which promised to help deliver the nation from its dependence on imported oil. "Nobody can embargo sunlight," Carter declared pointedly in his speech to the sodden audience that day. "No cartel controls the sun. Its energy will not run out."

Until the early seventies most Americans had taken energy for granted. Back in the sixties a joke had done the rounds of Washington wonks. It went, Q: What's our national energy policy? A: Pray for mild weather! That attitude changed forever on October 16, 1973, when the Organization of Petroleum Exporting Countries hiked the price of oil 70 percent. The next day the Arab states placed an oil embargo on the United States as punishment for resupplying Israel during the Yom Kippur War. The cartel quickly extended its ban to Western Europe and Japan. Almost immediately lines of irate motorists began forming outside gas stations.

Suddenly energy policy was a hot topic. In 1974 the Ford Foundation produced an influential report entitled *A Time to Choose: America's Energy Future*. Among its recommendations, the authors of the five-hundred-page study asserted that energy conservation and renewables were viable alternatives to finite fossil fuels like oil. They should be taken seriously. "Solar energy is the world's most abundant renewable energy resource," the report stated. Solar cells were "on the threshold of rapid technological advancement." A much stronger push was accordingly required: "the government solar energy R&D program should provide the funding to develop the requisite infrastructure in industry, the national labs, and the universities to create a base from which rapid development of the technology can take place."

The report's lead author was David Freeman, a feisty engineer-turned-lawyer from Tennessee who had been an advisor to Presidents Kennedy, Johnson, and Nixon. Freeman made a point of sending a copy of his report to the governor of each state. One went to Carter, who was then governor of Georgia. During the presidential campaign, Freeman would author statements on energy for the candidate. His report became the blueprint for the Carter administration's energy policy.

Another influential voice at this time was that of Amory Lovins. In October 1976 the twenty-nine-year-old physicist-turned-Friends-of-the-Earth-activist published a provocative essay entitled "Energy Strategy: The Road Not Taken?" in *Foreign Affairs* magazine. In it Lovins laid out a "soft" path based on energy conservation and renewable energy sources such as solar, as an alternative to the "hard" path of centralized power generation based on fossil fuels and nuclear. His ideas were widely embraced, not least by Jimmy Carter. In October 1977 the president invited Lovins to the White House to discuss energy policy.

Carter had arrived in Washington determined to make energy an urgent priority. In April 1977, during his first hundred days in office, the new chief executive outlined his proposals in a televised fireside chat. Wearing his trademark beige cardigan, Carter began by warning an anxious nation. "I want to have an unpleasant talk with you," he said. Unpleasant, because Carter knew that many of his proposals—especially those that involved personal sacrifice, people having to cut back on their energy usage—would be unpopular. "This difficult effort"—winning independence from imported energy—"will be 'the moral equivalent of war,'" the president asserted.

There had been two previous transitions in the way people used energy, Carter continued. The first, about two hundred years before, was from wood to coal, the motive force of the Industrial Revolution. The second, in the early part of the twentieth century, was from coal to oil, the enabler of the automobile age. Now, Carter argued, "because we are running out of oil, we must prepare for a third change." This would involve "strict conservation"—hence personal sacrifice—"the use of coal" (few concerns about global warming back then), and "permanent renewable energy sources, like solar power."

A new Department of Energy would be created (in October 1977) to coordinate the implementation of government policy in this area. Carter concluded by listing some fundamental principles and specific goals. Principles included the government taking responsibility for energy policy and ensuring that its policies were predictable. Without consistent government support—an idea that under his successor, Ronald Reagan, would suddenly and dramatically go out of fashion—nothing much would happen. Goals included the use of "solar energy in more than 2.5 million houses" by 1985.

In this context the term "solar energy" meant mostly the older technology of solar water heaters. In June 1979 Carter had a solar heater consisting of thirty-two panels installed on the roof of the West Wing of the White House. "A generation from now this solar heater could either be a curiosity, a museum piece, an example of a road not taken," Carter said at its dedication, referring to Lovins's tract. "Or, it can be just a small part of one of the greatest and most exciting adventures ever undertaken by the American people."

"In the year 2000 the solar water heater . . . will still be here, supplying cheap efficient energy," Carter predicted. He was mistaken. In 1986 during the Reagan administration the system would be dismantled and removed, supposedly to fix the roof underneath. The panels were never replaced.*

But "solar energy" also included the much newer technology of photo-voltaic cells for the direct conversion of sunlight into electricity. Such cells could be competitive with conventional sources like coal as early as 1990, Carter had told the staff of the solar institute in Colorado. "We know that [solar energy] works," he said, preaching to the converted. "The only question is how to cut costs so that solar power can be used more widely."

This was not the first time that the thorny question of how to reduce the high cost of photovoltaics had been posed, nor would it be the last. For the moment, however, the solution that Carter proposed was the logical one. "The government will speed this program by increasing demand for solar hardware, so that mass production can help to bring prices down." In addition, to expedite the spread of residential solar systems, over the next seven years the federal government would offer more than a billion dollars in tax credits, up to $2,000 per home-owner. A national Solar Energy Development Bank would be established with annual funding of $100 million to make available financing, at reasonable terms, for solar investments in residential and commercial buildings.

To get his radical energy legislation through Congress, Carter had to fight what he later described as "bloody battles." Many of his hard-won outcomes would not last past the mideighties. During the Reagan Revolution, government procurement of photovoltaics would be slashed, the tax credits for renewables phased out, and the Solar Bank demolished. For many years Carter would be regarded as a head-in-the-clouds idealist whose energy policies had failed. But with the passage of time other, more favorable views would increasingly be voiced. "I don't think there's been a president like Carter before or after him," an admiring David Freeman told an interviewer in 2003. "[H]e was no politician and didn't pretend to be one, but he did what he thought was right. And he was right a lot of the time."

"If we'd done what Jimmy Carter wanted us to do," Freeman told me, "we wouldn't have climate change now."

---

*In 1991 the panels were acquired by Unity College, Maine, where they continued to operate until their boiler burst in 2005. In March 2007 one panel was donated to the Jimmy Carter Library and Museum in Atlanta. In October 2010, US Energy Secretary Steven Chu announced plans to install PV panels on the roof of the White House residence. "This project reflects President Obama's strong commitment to US leadership in solar energy," Chu said.

In June 1979 President Carter delivered a message to Congress, his last major policy pronouncement on solar energy. In it Carter recommended that the United States should commit to an ambitious national goal of meeting 20 percent of its energy needs with solar and other renewable resources (including hydropower) by the turn of the century. At this time the US economy was reeling from the impact of the interruptions to the oil supply that followed the Iranian Revolution of Ayatollah Khomeini. Just a few weeks earlier the partial core meltdown of a reactor at a supposedly "accident-proof" nuclear plant, Three Mile Island in Pennsylvania, had thrown the future of nuclear energy into doubt.

Crises tend to galvanize people. During the 1980s forest dieback from acid rain caused by emissions from coal-fired power stations and the nuclear catastrophe of Chernobyl would lead directly to the rise of photovoltaics in Germany. In California the chaotic aftermath of the botched deregulation in the electricity industry ensured that energy would remain a key issue in that state. It would thus be reasonable to suppose that the US solar industry emerged in response to challenges posed by the oil crises. Reasonable, perhaps, but that is not what actually happened: the dates don't match.

In fact, while the nascent PV industry derived considerable, albeit short-lived, benefits from federal and state governments during the 1970s and early '80s, its origins derive from an entirely different dynamic. To wit, as Jimmy Carter put it on that miserable day in Colorado, from "private enterprise, individual initiative, and the inventive genius of the United States." In particular, the drive and persistence of four doughty entrepreneurs—Elliot Berman, Joseph Lindmayer and Peter Varadi, and Bill Yerkes—who would found the solar industry's three most seminal firms. These were determined men who made gutsy calls, took big risks with their careers, their families, their homes, in short, with everything they had, to bring forth something new and different. Before we meet these exceptional individuals, first let us take a brisk stroll through the invention and early history of the solar cell.

Tapping the limitless energy of the sun was an idea that had long beguiled scientists and inventors. Thomas Edison, the man who had almost single-handedly invented electric power generation, transmission, and distribution, was disgusted by the profligate practice of burning finite oil and coal to produce electricity, "this old, absurd Prometheus scheme of fire," in his scathing words. "We are like tenant farmers chopping down the fence around our house for fuel," Edison grumbled, "when we should be using nature's inexhaustible sources of energy—sun, wind, and tide." For the future, Edison's money was on the sun: "What a source of power! I hope we don't have to wait till oil and coal run out before we tackle that."

Jimmy Carter was correct to say that solar cells were a proven technology. When the first oil crisis struck in 1973, the solar cell was already nearly twenty years old. The photovoltaic effect—electricity generated by exposing a material to light—was discovered in 1839 by a French prodigy, nineteen-year-old Edmund Becquerel, while doing experiments at the Musée d'Histoire Naturelle in Paris. The reason for the transformation remained a mystery until 1905. In that year the twenty-five-year-old über-prodigy Albert Einstein published a paper that explained light consisted of energetic particles, what we now call photons. It was for his explanation of the photovoltaic effect (and not for his much harder-to-understand work on relativity) that Einstein won the Nobel Prize for Physics in 1921.

Upon striking a suitably prepared material, photons dislodge electrons, which can then be marshaled to form an electric current (see appendix). Early photovoltaic devices employed a light-sensitive metallic element called selenium. But selenium was inefficient, able at best to convert only 1 percent of incident light. This piddling amount sufficed for photosensors, such as the switches used to turn street lamps on and off, and the light meters used to gauge exposure in cameras. For broader applications, photovoltaics would have to become much better convertors.

The breakthrough that would catapult the solar cell from niche to mainstream was made at the Bell Telephone Laboratories in Murray Hill, New Jersey. Bell Labs was one of the great research powerhouses that made American industry the envy of the postwar world. The labs' mission was long term: to develop technologies that would be useful to the telephone system in twenty years. To this end Bell hired the best scientists then let them loose to follow their instincts. Researchers were encouraged to work in small, casually formed interdisciplinary teams of physicists, chemists, and engineers. In 1947 one such trio had invented the transistor. First-generation transistors were made of germanium. By the early fifties the limitations of this material were apparent. The focus of research switched to developing transistors made of silicon, an abundant element commonly found in sand.

Though it depended on a fundamental understanding of materials science, the transistor was developed to solve a practical problem—namely, to replace the unreliable vacuum tubes in the amplifiers that Bell positioned along its lines every few miles to boost telephone signals. Another item the phone company wanted replaced was the dry-cell batteries it used to power amplifiers located in tropical regions like Florida. Batteries were unreliable. Humidity caused their terminals to corrode, shortening their working life.

In early 1953 Daryl Chapin, a forty-three-year-old electrical engineer, was tasked with finding a replacement source of power. One possibility Chapin considered was solar cells. He bought a commercial selenium cell, but even in bright sunlight the cell could only generate a measly half a volt. Then, in one of the serendipitous comings-together for which Bell Labs was famous, Chapin happened to bump into Gerald Pearson, an old friend whom he knew from Willamette University in his hometown of Salem, Oregon. Pearson, a physicist, had been working on silicon transistors, devices made up of layers of material that had been doctored to alter the number of electrons they could muster. Pearson happened to shine a lamp on a sample of silicon he was testing. To his surprise light from the lamp caused the needle on his meter to jump. Advising his buddy not to waste any more time on selenium, Pearson handed Chapin his precious piece of silicon. In sunlight the silicon cell proved to be five times more efficient than selenium at turning photons into electrons. This was good, but not good enough.

After several frustrating months of experiments, Chapin turned for help to Pearson's colleague Calvin Fuller, a chemist who had grown the crystalline silicon for the original sample. Fuller rejiggered his recipe to grow the active layer—in which incoming photons jolt electrons out of their orbits around the atomic nucleus to form a current—as close to the surface of the sample as possible. There, it would be exposed to the maximum amount of sunlight. The new devices, small strips of silicon wafer about the size of a razor blade, worked wonderfully well. They were able to convert sunlight into electricity with an efficiency of 6 percent, the target Chapin had set himself.

On April 25, 1954—a Sunday, appropriately enough—Bell Labs presented the team's work to the world. The demonstration showed "how the sun's rays could be used to power the transmission of voices over telephone wires." This was sensational news. "Vast Power of the Sun Is Tapped by Battery Using Sand Ingredient," ran a headline on the front page of the *New York Times* the following day. It was the beginning, the paper speculated, of a new era "leading eventually to one of mankind's most cherished dreams—the harnessing of the almost limitless energy of the sun for the uses of civilization."

"Since it has no moving parts and nothing is consumed or destroyed, the Bell Solar Battery should theoretically last forever," a company press release promised. In 2003 an original Bell solar cell was exhibited at a conference in Osaka. After almost fifty years it had lost just 15 percent of its efficiency.

The phone company immediately began field tests on its new invention. The world's first solar panels—"frames" the Bell people called them—were rigged up on poles to power rural telephone lines outside the small southwestern Georgia town of Americus. This, as it happens, is just ten miles from the tiny hamlet of Plains, George, the birthplace of Jimmy Carter. In 1954 the future president had just been discharged from the navy. He returned to Plains to take charge of the family peanut farm. Carter used to say that he never forgot the major change to his lifestyle as a boy growing up in Plains brought about by the rural electrification program of that other great energy policy innovator, Franklin D. Roosevelt. It is thus entirely plausible that the first president to endorse solar power could have seen with his own eyes this original terrestrial application of photovoltaic technology.

The cells performed well, despite Bell's concerns about what *Time* magazine delicately described as "indifferent birds" (a weekly cleaning soon solved the bird-shit problem) and, a more serious menace, "stone-throwing boys." But the application soon became redundant, ironically because the same silicon transistors that had led to the invention of the solar cell were adopted as amplifiers. Compared to the vacuum tubes they replaced, transistors could be driven by the minuscule amounts of power that phone lines carried. There was thus no longer any need to boost the signal every few miles. Photovoltaics would eventually find applications in long-distance communications, but it would be in microwave wireless telephony, not land lines.

What else could solar cells be used for? A few novelty items, like solar-powered radios, were produced. From the start one application obvious to many was to power homes. "[A]bout 100 square feet of the silicon panel wafers mounted on a roof in a sunny area may one day furnish electricity for household use," a GE researcher speculated, adding "if much cheaper manufacturing methods are found." In fact, it wasn't so much the manufacturing as the material itself that would have to get cheaper. Silicon was said to be the second most common element in the earth's crust, after oxygen. It was, as the *New York Times* had pointed out, obtained from sand. But high-purity silicon could cost as much as five hundred dollars a pound. Even that was nowhere near pure enough for device-grade uses.

To produce ultra-pure "single-crystal" silicon, you began by melting chunks of pure silicon in a graphite crucible. Into the melt you dipped a tiny seed crystal stuck on the tip of a rotating rod. Then, ever so gradually, you withdrew the rod. The method was named after its Polish inventor, Jan Czochralski. In 1916 Czochralski had absent-mindedly dipped his pen into a vat of molten tin instead of his inkwell. He was surprised to note a thin thread of solidified metal dangling from its nib. The product of the process was a salami-shaped ingot of silvery semiconductor material that in the midfifties was worth its weight in gold. In 1956 Chapin calculated that a 5-kilowatt solar array, enough to power an average house, would cost almost $1.5 million. It would be like covering your roof with gold leaf.*

To most people the idea that crystalline silicon would ever be cheap enough to compete with conventional sources of power seemed inconceivable. Two other approaches were seen as more likely. One was more bang for the buck. You could take a few super-efficient cells and concentrate sunlight on them using lenses or mirrors to extract the maximum juice. The other approach might be characterized as less bang, but far fewer bucks. You could spray thin films of photovoltaic material onto a panel of glass or metal. (We shall consider the development of both these approaches in part 5.) What nobody imagined was that thanks to applied human smarts and— eventually—advanced manufacturing techniques, crystalline cells would get cheaper and cheaper. The result is that today, silicon cells still account for 80 percent of all solar panels made. But we are getting ahead of our story.

Curiously enough in the light of subsequent developments, it was a German researcher who came up with the application that would transform solar cells from little more than a laboratory curiosity into an essential component of modern communications. Hans Ziegler was one of the select group of scientific talent the US Military scooped up from occupied Germany

---

*In fact device-grade silicon was *more* expensive than gold. In 1958 gold cost about one dollar a gram, whereas ultra-pure crystalline silicon cost five dollars a gram, according to Werner Freiesleben, a German chemist who at Munich-based Wacker helped pioneer its commercialization. By 1985 the price of gold had leapt to fifteen dollars a gram while the price of silicon had dropped to just twenty cents.

many after World War II. An electrical engineer, Ziegler was assigned along with twenty-three of his fellow countrymen to the US Army Signal Corps base at Fort Monmouth, New Jersey. In 1954, the year he became a US citizen, Ziegler went to Bell Labs—located just thirty miles up the road from Fort Monmouth—where he was shown the brand-new silicon solar cell. Ziegler came away deeply impressed. "Future development may well render [the solar cell] into an important source of electrical power," he wrote in a bulletin to his colleagues, which John Perlin quotes in his book *From Space to Earth: The Story of Solar Electricity.* "[T]he roofs of all our buildings in cities and towns equipped with solar [cells] would be sufficient to produce this country's entire demand for electrical power."

Roofs would come later, much later. In the short term, however, there was one obvious application crying out for photovoltaics: in space, powering communications from satellites. Obvious because science fiction writers like Arthur C. Clarke had long predicted that satellites would be solar powered. The timing was right. In 1954 the US satellite program had just begun. The US Army and Navy had both submitted proposals; the navy's Vanguard program won. This was unfortunate for Ziegler: the navy had nixed the use of PV because it was an unproven technology. Then in October 1957, the Soviet Union launched *Sputnik* and the space race began in earnest.

Sputnik and the first US satellites were powered by chemical batteries, an unsatisfactory solution. Batteries were heavy and bulky. They took up almost half of the available payload space—in those days, not much bigger than a grapefruit—and ran out within a couple of weeks. Solar cells, Ziegler argued, were much better suited for the job. They were lightweight, when attached to the outside of the satellite they would not take up precious instrument space, and they would in theory last forever.

His persistence paid off. The first solar-powered satellite, *Vanguard 1*, was launched on March 17, 1958. Eight little solar panels containing 108 cells were fixed symmetrically around the satellite. They ensured that power would continue to flow to the beacon signal generator during the spacecraft's random tumbling. The longevity of the signal surpassed all expectations. Designed to last for twelve months, the satellite kept on transmitting for six and a half years. There was no way to turn the bloody thing off.

Four years later, Bell Labs' parent company, AT&T, launched *Telstar*, the first commercial communications satellite, to provide transatlantic telephone and television broadcasting services. Despite the success of *Telstar*, NASA remained unconvinced that solar cells would provide enough oomph to power the next, much larger generation of missions. The skeptical space agency was still betting on nuclear. That is, until April 1964, when a plutonium-powered navigation satellite failed to achieve orbit, disintegrating as it fell back through the atmosphere to Earth. A subsequent soil sampling program found, embarrassingly, that radioactive debris from the satellite was present "in all continents and at all latitudes." From then on, all space vehicles would be solar powered.

2

TO HELL WITH THE GENERALS

*T*he space race, the informal Cold War competition between the United States and the Soviet Union, was largely about national prestige. That meant, in essence, getting there first. The contest kicked off in 1957 with a surprise win to the Russians, who launched *Sputnik*, the first artificial satellite to orbit Earth. It ended in 1969 with a triumphant victory to the Americans via *Apollo 11*, the craft that landed the first men on the moon. The superpower rivalry had the happy side effect of turbocharging the development of certain key technologies, most notably that of integrated circuits, better known as microchips. The Americans, whose rockets were not as powerful as their Russian rivals, were always looking for ways to reduce payload weight. Photovoltaics also benefitted, with the US government pouring some $50 million into their development during the decade of the space race.

But with the government space program the sole customer, solar cells were just a tiny niche market. By the early 1970s only two commercial firms were in the PV business, Spectrolab and Centralab, both based in Los Angeles. In addition, there was also a hybrid entity called Communications Satellite Corporation, or COMSAT, which was headquartered in Maryland. It tested satellite-use solar cells and did R&D. From these roots would spring two entrepreneur-driven start-ups—one on the East Coast, the other on the West—which between them would develop most of the technology and pioneer most of the markets for terrestrial applications. In so doing they would lay the foundations for today's multibillion-dollar solar industry.

COMSAT was established by Congress in 1963 to prevent AT&T, which had launched *Telstar* the previous year, from gaining a monopoly on satellite communications. Run by a bunch of retired air force generals, it was a peculiar outfit. In 1969 COMSAT opened a fancy-looking Cesar Pelli–designed R&D center in rural Clarksburg, Maryland, halfway between Washington and Frederick. To establish and direct its physics and chemistry labs, the corporation hired two expatriate Hungarians.

Joseph Lindmayer and Peter Varadi had fled their native land following the failure of the 1956 uprising against its Stalinist government. Along with the likes of Intel's Andy Grove (born András Gróf), they were members of the extraordinary Hungarian diaspora that had such a disproportionately large impact on twentieth-century science and technology. Lindmayer (1929–1995) was a solid-state physicist who gained his PhD at the Technical University of Aachen (a city that coincidentally, as we shall see in part 3, would play a crucial role in the spread of solar). Varadi is a chemist who graduated from the University of Szeged, also known as the City of Sunshine.

At COMSAT Lindmayer discovered that in the decade since their invention solar cells had barely improved. They were made, as he contemptuously put it, using "ancient technology." Lindmayer quickly set about improving the conversion efficiency of cells from 10 percent to around 14 percent, an enormous step forward. As a salaried employee, Lindmayer did not receive any financial reward for his efforts. Unimpressed, he determined that the next time he made a major breakthrough he would do it in his own company, so as to be properly recompensed for his bright ideas. After all, this was America.

By 1972 Lindmayer and Varadi had been at COMSAT for four years. Their labs were established and running smoothly. Demand for space-use solar cells was steady but flat. "So by then we didn't have much challenge," Varadi told me. Toward the end of that year the two friends began kicking around ideas about solar cells that could be used for terrestrial purposes. "It was in the air at that time that one should do something about it," Varadi recalled. Obviously the technology for terrestrial cells would have to be different; in particular, much less expensive. The pair wrote a proposal to do exploratory research into developing such technology and submitted it to COMSAT's management. The generals showed no interest in funding such work.

At a New Year's Eve party, following a couple of glasses of champagne, Lindmayer and Varadi decided that, "to hell with the generals, we were going to do it ourselves." In classic entrepreneurial style, they grabbed a piece of paper (history does not record, but in accordance with tradition, it would probably have been a cocktail napkin) and began scribbling a business plan. The first question was what to call their company. Lindmayer said he didn't care what the name was, so long as it ended with an *x*. That made the designation easy: their start-up would be known as Solarex.

The next issue was the division of roles. Lindmayer was obviously an excellent solid-state physicist, so it was agreed that he should handle technology and R&D. Varadi modestly allowed that "since as of that time I had not received the Nobel Prize in chemistry, I should get out of chemistry and go into some other field, maybe I should try to run the business. It sounded an excellent idea to run the operations such as the production, the sales and marketing, accounting, administration, etc., since I had never done it before. But on the other hand, why should I not be able to do it?"

On they went, point by point, sorting out what to do and when. Finally the pair agreed that they would quit their jobs and open the doors of the company for business the following August. Next morning Lindmayer and Varadi woke up shaking their heads and asking themselves "why well-paid scientists would get into such a harebrained idea when there is no technology, no product, no market, and we have no money." Nonetheless, they decided to "damn the torpedoes and go full steam ahead."

The first step was to find some funding. "It sounded very simple," Varadi recalled. "We will send the excellent business plan to venture capitalists and they will love the idea because it is totally new." Varadi ended up making presentations to around twenty money men. "It was an absolute success," he remembered wryly three decades later. "All of them learned how to spell P-H-O-T-O-V-O-L-T-A-I-C-S, [but] we didn't get the money." The VCs were leery of solar technology and dubious that two white-coats with no previous business experience could run a company.

After a couple of frustrating weeks Varadi changed tack, raising $250,000 in seed capital from friends and acquaintances. He and Lindmayer resigned their jobs at COMSAT. "Until then we got a salary every two weeks," Varadi said. "Now we had to provide our own somehow. It was a very strange feeling." They rented space in a small two-story office building in Rockville, Maryland, on the outskirts of Washington. The factory was in the basement, the R&D department upstairs. Operation began on schedule, on August 1, 1973. A few weeks later, the first oil crisis erupted. "I could tell you that we foresaw the oil crisis and that was the reason we planned to open the company on the first of August," Varadi later confessed. "Well, that would be a lie. We had absolutely no inkling that there would be an oil crisis a couple of months later. But on the other hand, I have to admit that the oil crisis had a very profound effect on us. We realized what an incredible business we had got ourselves involved in."

From the very beginning, there was never any doubt about the intense public interest in photovoltaics. Especially after the first oil crisis, when gasoline prices shot up. "One of our sales people had the idea of putting a bingo card—you know, one of those prepaid postage things you tear out and send in—into a magazine," Varadi told me. "Well, a big truck came delivering those to us, from housewives, because everybody wanted to use solar for their house."

Those were heady days. When an order came in everyone would work nights and weekends to fill it, without overtime. It wasn't just a job at this point: employees were on a mission to save the planet. "Everybody became almost religious about alternative energy," Lindmayer said. But initial sales were disappointing. Every time payday came around, the founders worried about how they were going to pay their employees. "It was always a horrible feeling," Lindmayer recalled. "You've got to have guts to live like this." It is sometimes said that the only true entrepreneurs are those who have had to take out a second mortgage on their home in order to meet the payroll.

Lack of cash forced Solarex to invent cheaper solutions. One problem was the supply of high-purity single-crystalline silicon, the starting material for solar cells. During boom times in the semiconductor industry, which used the same material to make its microchips, the supply would dry up. Solarex could not afford its own Czochralski-style crystal pullers. Lindmayer came up with a cheaper way to produce silicon: by casting ingots in a mold. This produced polycrystalline material.

Telling single-crystalline silicon from poly is easy. The former is evenly colored, whereas the latter has a visible grain, a "metal flake" effect with highlights. Cells made from poly were not as efficient, but they were good enough for commercial purposes. Previous attempts to make solar cell–grade polysilicon had failed: by trial and error, Lindmayer figured out that the key was to slow down the speed at which the cast material cooled, letting it sit for a couple of days or so. Today polysilicon is used in around half of all silicon solar cells.

Lindmayer and Varadi started their company because they were bored. Bill Yerkes started his because he was fired. Though he embarked on his venture two years later than the Hungarians, Yerkes's influence on the nascent industry would be more profound. "Bill's pretty much a classic entrepreneur start-up guy," said Yerkes's former colleague Charlie Gay. "A very bright man, very enthusiastic and committed to the business, a gregarious soul who brought a lot of the innovations to the early stages of cost reduction in solar." "He's very determined," added Peter McKenzie, another former colleague. "Bill will not stop, he just keeps going, no matter what. That was what you needed in the early days of the solar business—you never say die, you kept chasing customers no matter how many times they say no."

Bill Yerkes had an unusual childhood. He was born in Kansas in 1934 but grew up in the Panama Canal Zone, where his father was an architect. A mechanical engineering student at Stanford, Yerkes was infatuated with cars and car design. He loved watching the auto races held in those far-off days in San Francisco's Golden Gate Park. Italian cars were Yerkes's passion: upon graduating the following year he bought himself an Alpha Romeo Giulietta Sprint and drove it back east to Detroit, where he had landed a job at the Chrysler Institute. His exotic choice of wheels did not go down well in the Motor City. In supermarket parking lots the bright red coupe was a provocation: "People would sometimes bang into me just to show they didn't like those foreign cars."

After three years working in Chrysler's sound and vibration lab, Yerkes became restless. He joined Boeing, where ultimately he was assigned to a satellite lunar mapping program. There, for the first time, he encountered solar cells and met Alfred Mann, the remarkable serial entrepreneur who had founded Spectrolab in 1956. That year Mann sold the business to Textron, a diversified conglomerate. One of the only two solar cell companies in the world, Spectrolab had forged a reputation for reliability that enabled the company to grab the bulk of the space business. In 1967 Mann hired Yerkes as the company's vice president and general manager so that he could move on to new developments. Yerkes moved down from Seattle to Sylmar, at the northern tip of Los Angeles, where Spectrolab was located. Three years later he was promoted to president.

The space program was Spectrolab's only customer. The most exciting project Yerkes worked on was building a small solar array that would be left behind on the moon to power a seismic activity experiment. It was a rush job. The original idea had been to use a nuclear reactor, which would be started by astronauts using tongs to insert radioactive pellets into the generator. Then NASA got nervous. The astronauts might accidentally drop some of this hot material, then use their hands to pick it up and brush off the moon dust, with potentially cancer-inducing consequences. To the space agency this seemed like the very stuff of which bad publicity is made.

With just three months left before liftoff, Spectrolab built a little solar panel that attached to the leg of the lunar module. A grateful NASA included Yerkes and two of his colleagues among 1,300 people who signed a sheet that was photographically reduced to a microdot and bonded to the lander. Since those days, however, Yerkes had become disenchanted with the space market, which wasn't very large, wasn't growing, and did not need much salesmanship. "We just waited for government requests to come through for various satellites," he told me. "That wasn't very satisfying."

Yerkes was running a company that employed between two and three hundred people. It was profitable, but bigger opportunities beckoned. The government had sent out feelers to Spectrolab and COMSAT to determine whether solar cells could provide electricity for terrestrial uses. During his fruitless hunt for backers, Peter Varadi had actually approached Yerkes for help. Yerkes took the Hungarian to Providence, Rhode Island, where Textron was based. "Bill tried to convince the people there to get into terrestrial," Varadi recalled. "His idea was that we should start the terrestrial business and then somehow merge" with Spectrolab. But Yerkes could not persuade the suits to fund the venture. Lindmayer and Varadi would end up "friendly competitors," along with Yerkes.

In 1975 Spectrolab was acquired by its largest customer, Hughes Aircraft. Hughes's only interest was the space communications business. Three months after the takeover, Yerkes was informed that he was "surplus to requirements." The new owners wanted one of their own people in charge. Yerkes got a year's salary and an hour to clear out his desk. Being sacked was a blow to the forty-one-year-old's ego. "I *was* the company," Yerkes said later, "I had created all the things that were working well."

Three days later, not one to take things lying down, Yerkes decided that he would start up his own solar cell company and grow it ten times bigger than Spectrolab. He would show those dumb aerospacers. From the outset the target market for his company was terrestrial applications. "The object was to make photovoltaic panels that would be reliable and as inexpensive as possible." To do that, to become a large-scale manufacturer of a standard solar panel, to offer a better product at a lower price than anybody else, he would have to completely reinvent the process for making photovoltaics.

Though solar cells were the epitome of the space age, the way they were manufactured was very old-fashioned. In fact the cells were like pieces of jewelry, individually handcrafted by artisans. Each satellite had different power requirements, calling for a unique design. There was thus no chance of volume production to bring prices down. Manufacturing was done in small batches. Masked workers carried cells around the shop on cafeteria trays, transferring them from one process to another using tweezers. Worse, the production processes were frozen: NASA insisted that every step had to be carefully documented so that if a problem occurred it could be traced back to its origin. No changes were allowed, effectively ruling out the possibility of innovation. Every cell was exhaustively tested for resistance to vibration, to humidity, to whatever. *Reliability* was the watchword, along with *efficiency* and *light weight*. NASA didn't care how much the cells cost: that was insignificant compared to the cost of getting satellites into space. With solar cells for terrestrial applications, it would be the opposite: cost would be far and away the most important consideration. The issue was, How many watts per dollar? The cheaper, the better.

Bill Yerkes did not even attempt to get venture capital for his new firm, grandly named Solar Technology International (STI). He knew from his experience trying to help Peter Varadi that funding from the money men would not be forthcoming. Varadi and Lindmayer had started with $250,000; Yerkes made do with even less. His mother, two aunts, and an old friend from Boeing each chipped in $10,000. Yerkes himself contributed $50,000 from selling his stock in Textron, plus his year's salary from Hughes.

Lack of funds meant that everything had to be done on a shoestring. By late summer 1975 Yerkes had found a suitable industrial building in Chatsworth, about ten miles west of Sylmar, which he rented for $12,000 a month. From Texas Instruments' used equipment warehouse in Dallas he bought the basic machinery he needed. He had to get the installed machines certified. The official the city sent discovered that the belt furnace he had bought didn't seem to have the requisite Underwriters Laboratories safety label. Undaunted, the resourceful Yerkes obtained a label from the manufacturer then stuck it under the box containing the circuit breakers. When the official returned, Yerkes casually told him, "Look, there's the label, the reason you couldn't find it was the dummies put it underneath the box!" By that September his company was in business.

At STI Yerkes was responsible for at least two key innovations that transformed the economics of silicon solar cell manufacture, laying down a template that would be used by almost all subsequent makers. One was a simplified method for printing the silver metal contacts used to interconnect the cells and extract electric current from them. The conventional method employed a vacuum chamber in which evaporated metal was deposited onto the silicon through a pattern mask. This was expensive, time-consuming, and wasteful because it left much of the metal on the mask. Yerkes opted instead for the much simpler technology of screen printing. This technique had recently migrated to electronics from the art world, where Andy Warhol had used it to make his celebrity portraits and soup cans. A rubber squeegee pushing silver paste through the pattern in the screen could coat 1,000 to 1,500 cells an hour, with little precious metal wasted.

The other major innovation was how to protect the precious silicon from the elements. Rival companies like Solarex stuck their cells onto circuit boards then encapsulated them by pouring silicone—a transparent rubberlike polymer—over them, like syrup on pancakes. Silicone was less than ideal. It was expensive and dirt stuck to it outdoors, necessitating regular scrubbings with soap and water. Exposed to the elements, silicone tended to degrade within a few years. Plus, it was unpleasant to work with: the stuff had a way of getting all over everything, permeating the workplace with a distinctive, plastic smell.

Casting around for a replacement, Yerkes drew inspiration from his time in the auto industry. He remembered noticing that in wrecking yards while the steel bodies of old cars were all rusted up, their glass windshields were usually still in relatively good condition despite having been left out in the open for years. Instead of sticking the cells on top of a circuit board, Yerkes thought, why not just glue them onto the back of a sheet of tempered glass? Glass was cheap and, once outdoors, did not need cleaning—any dirt that landed on it would be washed off by rain. The glue would be the standard transparent stuff the auto industry used to laminate its windshields. In addition to being weather proof, the resultant tempered glass panels would also be robust.

The annals of entrepreneurship contain many instances of evangelists attempting to demonstrate to skeptical potential customers that their new gizmo is indestructible. If the product is small enough, you hit it with a hammer; if not so small, you jump on it. In 1976 the federal government tasked the Jet Propulsion Laboratory with reducing the cost of silicon solar cells. In the preliminary phase of this project the laboratory asked makers to submit samples of their panels for testing. The folks at JPL had lots of experience with solar panels for space use. These were light and not especially strong.

Yerkes devised his own simple strength test to prove that, though covered with glass, his products were not fragile. He placed a solar panel between two chairs. Then, in front of an astonished JPL committee, he climbed onto the panel and jumped up and down. Next, he stood the panel up on a chair and threw a coffee cup at it. The cup shattered, but the glass was unscathed. "We wanted kids to be able to throw rocks at these panels and not have our customers calling us all the time upset," Yerkes explained. It turned out that his panels were more or less indestructible. Even when kids did manage to break the glass—as happened to an array of forty panels that STI sold to a Navaho tribe in New Mexico—it merely cracked, just like an automotive windshield, while the cells went on producing power.

Yerkes built things to last. On the Internet today you can buy used solar panels made by his company that, after thirty-plus years exposed to all weathers, are still producing 80 percent of their rated output.

For his workforce Yerkes hired five girls, friends of his daughter Kathy, who had previously worked for Taco Bell. "Everybody thought that that was kind of funny," Yerkes said, "because it rhymed—solar cells and Taco Bells." A weekly production cycle was soon established. On Mondays the girls would start by taking a thousand round silicon wafers out of their boxes. They would etch the surface of the wafers, coat them with the chemical needed to modify the electrical characteristics of the silicon, then diffuse it into the material by baking the wafers in the furnace. On Tuesdays they screen-printed the contacts on the front and back of the cells, then fired them. On Wednesdays they soldered the cells together, thirty-six in a row, three rows to a panel. On Thursdays they laminated the panels. On Fridays they placed the finished panels outside in the Southern California sunshine for testing, then packed them in boxes ready for shipping. "So that was like $10,000 worth of product and my silicon wafers cost me $1,000," Yerkes said. "We went on like this, week after week, actually making money."

In December 1975, just three months after his company began production, Yerkes personally delivered his first consignment of panels, to Explorer Motor Home of San Diego. The lightweight panels were intended to replace the heavy diesel generators that recreational vehicles carry to charge batteries that run refrigerators, televisions, and lights. "I actually drove down with the shipment and it was cash-on-delivery. They wanted to argue with me, saying that they were a big company and they didn't need to pay COD." Yerkes told them that he was going to wait right there: they would get their panels when he got his check. "So I waited two more hours, they called around various people, until finally they agreed, 'Oh all right, we'll write you a check.' So then I said, 'Where's your bank?' I took their check over to the bank, and I cashed it right away. I drove back to Los Angeles with $10,000 in my pocket and we had a very nice Christmas."

Within a year STI had racked up more orders than the little company could handle. The energetic Yerkes was having great fun doing almost everything himself, up to and including writing out invoices for the panels they shipped. At the same time, however, he also had a vision for the future. Yerkes knew that if the company was to grow it needed capital investment. This would fund the construction and equipping of a much bigger factory in which to make more powerful panels in higher volume, thus significantly lowering manufacturing costs.

Over the next couple of years Yerkes reckoned he must have visited seventy-five venture capitalists, without luck. He did manage to arrange some additional financing through old Stanford connections, but not enough to underwrite the kind of large-scale investment he had in mind to beat other, better-heeled rivals. Attracted by the lucrative contracts that the government was dishing out, large firms like Motorola were entering the solar cell business. "A number of these big companies had come in," Yerkes said. "They kind of scared me because they all had fancy sales departments and everything. They could get on a plane and fly here and there." How could a tiny outfit like STI compete?

Then in 1978 Yerkes was approached by Atlantic Richfield, usually known as ARCO, one of America's largest oil companies. We've decided to get into the solar energy business as a hedge against future shifts in the energy market, they told him, and we've decided to buy you. It was a good match. ARCO had considered acquiring Solarex, but plumped for STI instead. STI was conveniently located not far from ARCO's corporate headquarters in a twin-tower building on Wiltshire Boulevard in downtown LA. Yerkes was attractive because he had worked for a large company. Under the deal he struck, the new company—named ARCO Solar—would build a state-of-the-art automated factory to produce solar cells and panels. In addition, a one-hundred-person research department would be set up, which would report to Yerkes. This was not a trivial undertaking. Over the next decade, much to Yerkes's astonishment, ARCO would pour at least $200 million into its solar business.

ARCO was not the first oil company to get into photovoltaics, nor the last. Solarex would also be acquired five years later, by Amoco in 1983. But the first involvement by an oil company in solar dates back to 1970. The company in question, the one whose muscle most frightened Yerkes, was Exxon.

3

CELLS AND GASOLINE

*O*il companies—understandably, given the seemingly suspicious way the price of their product fluctuates—attract conspiracy theorists. When word got out that the "seven sisters," as the oligopoly was known back in the paranoia-prone 1970s, were buying solar start-ups, theorists jumped to the conclusion that Big Oil was up to no good. The gas giants were clearly out to gain control over this attractive new alternative energy source, which might conceivably, one day, pose a threat to their core business.

Like most conspiracy theories, this one does not stand up to scrutiny, in particular because, during the twenty-five years between 1975 and the turn of the century, oil companies like Exxon, Mobil, ARCO, Amoco, BP, and Shell would between them pour hundreds of millions of dollars into funding the development of solar energy. If the oil companies had wanted to kill off photovoltaics, the best way would have been to ignore the technology, not invest a king's ransom in nurturing it.

In fact solar was not actually a direct competitor to the sisters' core business, selling gasoline for automobiles. It turns out that the oil companies had other motives for wanting to get into PV, most of them quite innocuous. Not the least of these was that petroleum producers quickly discovered that solar provided a solution to an in-house problem particular to them. This application was first identified by Elliot Berman, an ebullient figure who deserves credit for being one of the first people to realize that terrestrial solar cells would find a market.

Berman was born and brought up in Quincy, Massachusetts. His father was a salesman whose customers included drugstores. Young Elliot thus had access to a ready supply of chemicals, which he used to create interesting pyrotechnic effects in the family cellar. Fortunately the cellar had a concrete floor and granite walls. Berman went on to enjoy a successful career as an industrial chemist. After he developed and commercialized a new photographic film process, a grateful employer gave him time off to think about what he wanted to do next. "I'm a photochemist, and I decided, Gee, solar energy's kind of a nice thing," Berman recalled. "Because you have sunlight everyplace and if you could make electricity out of sunlight, you could do some good things." Like improving the lives of the 2 billion people in developing countries in places like Africa who had no access to an electric grid.

You didn't have to invent a device that converted sunlight to electricity. That already existed in the shape of the silicon solar cell. So why wasn't the device being used to improve people's lot? The problem, Berman soon saw, was that its economics were all wrong. He therefore proposed to develop a new, much cheaper type of solar cell based on organic materials, such as dyes. Having worked on photographic films, he knew a good deal about such materials. When his company rejected his proposal, Berman looked elsewhere for support. He found an enthusiastic audience at Jersey Enterprises, the recently formed venture capital arm of Standard Oil of New Jersey, which in 1973 would change its name to Exxon. Jersey Enterprises had been set up specifically to diversify the company's portfolio out of oil and into promising new areas, like computer peripherals.

Berman hired some scientists and set up a group, dubbed the Solar Energy Conversion Unit, based at the oil company's laboratories in Linden, New Jersey. He reported to a senior manager named Ben Sykes. One day, Sykes came to see Berman with a question. "I know solar cells have been used in space, but space is space, there's no rain, no dirt, no night. How do you know solar cells are going to work on Earth?"

The organic cells Berman was proposing would take years to develop. Meantime, to keep his boss happy and to gauge what the terrestrial market would be, Berman set off to buy some conventional silicon solar cells with which to cobble together an interim product. His first stop was California, where he tried to buy some cells from Centralab, Spectrolab's smaller rival. But all that Centralab had for sale were rejects from the space program, not nearly enough to meet Berman's needs.

Berman next flew to Japan to visit the consumer electronics maker Sharp. In the United States companies had talked about making solar cells to power transistor radios and the flashing hazard lights and foghorns on navigational buoys. In Japan such pronouncements were taken seriously, especially by Sharp, which was one of the smaller manufacturers. In the fierce competition that characterizes the Japanese domestic electronics market, such companies must try harder to survive. They seize on any promising-looking technology.

Japan would not launch its first satellite until 1976. Sharp began mass-producing solar cells for terrestrial uses in 1963. By the time Berman arrived the company had already equipped buoys in Yokohama and other harbors with solar. (There they had solved the tricky issue of "seagull fouling" by the simple expedient of surrounding the panels with long spikes.) Sharp had also built what the company claimed was the world's largest solar system. It powered a full-scale lighthouse located on a breakwater at Nagasaki. At the firm's headquarters in Nara, Berman met Sharp's canny chief of engineering, Tadashi "Dr. Rocket" Sasaki. But the two could not agree on a price and the American went home empty-handed.

That left no option for Berman and his colleagues. They would have to go into the business of making silicon cells themselves. Back in 1968 Berman had registered a company name, Solar Power Corporation. Solar Power Corporation, a wholly owned subsidiary of Exxon, opened its doors for business on April Fools' Day 1973, four months before Solarex. The Japanese had shown that the obvious initial market was navigational aids. It occurred to Berman that his parent company might itself have a use for such devices on its numerous offshore oil production platforms in the Gulf of Mexico. The idea was sound enough: to prevent ships from running into them and consequent liability suits, the platforms had to be equipped with lights that could be seen five miles away and horns that could be heard two miles away. So could these platforms use some PV to power the warning devices? "Oh no," Berman was told, "they're *loaded* with power!"

The truth, as Berman discovered on a trip to Grand Isle, Louisiana, was very different. "We went and visited some Exxon platforms in the gulf. What we learned, which everybody down there knew, but nobody at headquarters knew, is that the way you do oil platforms is, you have one platform that's loaded with power, and that's where all the crews live. However, most of the platforms are unmanned and have no power."

The system for powering platforms relied on huge lead-acid batteries, each weighing several hundred pounds. The batteries were delivered by wooden barges equipped with winches to hoist them onto the platforms. Initially, rechargeable batteries were used. The barges would come out periodically with fresh ones, taking the depleted batteries back to shore for recharging. Then some bright spark decided it would be cheaper to use primary batteries. When done, crews could simply dispose of the batteries by tossing them over the side. Being highly toxic, such batteries had a devastating effect on the local marine life. In 1978 the Environmental Protection Agency outlawed the dumping of batteries in the ocean. That merely added an element of urgency to what was already a compelling economic case.

The nonrechargeable batteries cost $2,100 each. To rent the barge needed to ferry them out to the platforms cost $3,500 a day. The solar-powered batteries were much smaller, much cheaper, and could be transported by speedboat or by helicopter. "A helicopter could carry four boxes of solar panels with four panels in each box and some batteries that were the size of car batteries," Bill Yerkes explained. "They could fly out to a platform, install a solar array in an hour or two, then fly on to the next one. They could do three platforms a day then fly back to Houston." This represented a huge saving in time and money.

Oil platforms played a big part of the making of the infant solar industry. The major producers bought hundreds of panels from Solar Power, Solarex, and STI. Money from the oil companies helped keep the little firms afloat during the crucial start-up stage. By 1980, solar-powered navigation aids had become standard equipment for production platforms in the gulf. By then, however, Elliot Berman had left the company, frustrated with Exxon's unreasonable demand that it should break even. In 1975, he returned to Boston University, his alma mater, where he founded the Center for Energy and Environmental Studies. Three years later, Berman joined ARCO Solar as chief scientist, heading up the company's newly formed research department. In 1983, following the fall in oil prices, Exxon abandoned all attempts at diversification. Though there were potential buyers, company management chose to shut down Solar Power Corporation, auctioning off its equipment as scrap.

In the late 1970s, flush with cash from windfall profits garnered from the first oil crisis, the oil companies embarked on a corporate shopping spree. They could afford to make any acquisition they wished. Solar energy was well within their purview. ARCO picked up Solar Technology International for a song, paying less than $1 million, hardly big bucks for a company that in 1978 had revenues of $12.7 billion. There was, however, also a personal reason why Robert Orville Anderson, chief executive officer of ARCO, was interested in Bill Yerkes's company.

Even by the exotic standards of the oil industry, Bob Anderson was quite a character. He was a wildcatter, a prospector for black gold in unlikely places. His most famous find was Alaska's North Slope—the largest oil field ever discovered in North America—where, after a series of failures, he insisted on drilling one more exploratory well. Oil began to flow from this field in 1977, through an eight-hundred-mile pipeline all the way to Valdez, the nearest ice-free port. With his trademark Stetson and cowboy boots, Anderson looked the part of a Texan roughneck. But appearances can deceive. He was actually born in Chicago in 1917, the son of a Swedish banker. Though a lifelong Republican, Anderson was also a man of pronounced liberal sympathies. He chaired the Aspen Institute for Humanistic Studies and helped found the Worldwatch Institute in Washington to monitor global environmental trends.

A shrewd businessman who liked to think big, Anderson had parlayed an interest in refineries via a series of mergers into what was by the late 1970s the nation's seventh-largest oil company. By then he was also the largest private landowner in America, raising 30,000 cattle on over a million acres of ranch land in New Mexico. On a hot day a cow can drink up to 150 liters of water. New Mexico is arid. It gets very little rain. Groundwater must be pumped to the surface. Pumps are traditionally powered by windmills. But in midsummer, the hottest time of the year, the wind drops. Anderson thus had a strong interest in photovoltaics as a way of pumping water for his herds. He commuted by helicopter from his home in Roswell, New Mexico, to his office at the ARCO twin towers in downtown LA. Elliot Berman remembered bumping into Anderson there one day in his cowboy hat and boots. "He asked about what we were doing with solar, and told us not to forget about water pumping."

It so happened that when ARCO came shopping for STI, Yerkes had just installed his first solar-powered water pump. The customer was the Bureau of Indian Affairs in Albuquerque. "I didn't realize that the government was in the job of raising cattle," Yerkes said, "but basically they had all these Indian reservations out there that had cattle which needed water." The bureau's officers hated it when the wind dropped. They would have to jump in a pickup truck equipped with a petrol-driven generator, dash off to a well where all these thirsty cattle would be waiting around mooing at them, start the generator, pump up the water, then drive off and do the same at the next well. Solar offered a neat alternative to this onerous extra work.

A friend in the swimming pool business set Yerkes up with a pump and a DC motor. "We ran a little test on it, then I got a guy who was off from UC Davis for the summer to load this system onto my Volkswagen van, drive over to Albuquerque, and install it. That was our first system job: a 500-watt solar panel powering a pump for filling a tank with water on an Indian reservation. That was a big deal." Water pumping for irrigation would subsequently become one of ARCO Solar's main businesses.

For the first two years after the acquisition, however, things did not go well at ARCO Solar. The problem was the plant managers delegated by the parent company to run the subsidiary were fossil fuel men. They didn't know the first thing about semiconductors, glass, bonding, or, worse, manufacturing. The first manager's previous job had been planning the huge Black Thunder coal mine in Wyoming. After a year of wasting money on the wrong things, he was sent to the showers. His replacement was hell-bent on empire building. This new manager's goal was to hire hundreds of new employees. The more people you had working for you, the greater the status you enjoyed within the corporate pecking order.

Yerkes had succeeded in convincing ARCO's management to let him build a large, new automated plant. This would produce the first solar panel to be designed as a standard product. Standardization coupled with automation would drive down the cost of manufacturing. Also, since cells would be based on four-inch wafers rather than the conventional three-inch ones, the new panels would be more powerful, capable of producing 35 watts instead of the usual 10 or 20. The facility was built outside at Camarillo, a small farming community located on Highway 101 between Los Angeles and Santa Barbara.

The second manager couldn't wait for the plant to be completed. Rashly he decided to go ahead and try to make the new product in the company's existing, unautomated plant. It was a disaster. Yield collapsed. Unfortunately for the manager, in late 1979 a reporter from *Forbes* magazine dropped in to see how ARCO was bringing automation to the solar industry. The article he wrote began with a description of the plant's assembly line. To him it seemed like something out of Charlie Chaplin's *Modern Times*.

Silicon wafers emerge from a large oven ready for the next chemical process. The next conveyor belt is located across the room. Enter a disheveled teenager with acne wielding a Teflon pancake flipper and pan.

He scoops the wafers into the pan and runs them across the room to the belt. One can easily imagine the belt speeding up and, despite frantic flipping, expensive silicon chips [*sic*] spewing out onto the factory floor. One can also imagine, alas, how far away a rational process for the economical production of solar energy equipment may be.

The corporate types at the twin towers were not amused by this portrayal of their company as a laughingstock. The second manager was duly relieved of his position. The parent company dispatched Ron Arnault, a financial whiz who was also its director of strategic planning, to find a replacement. For three days Arnault interviewed candidates until finally he got around to seeing the company's founder. "It looks to me like you were president of a division of Textron, you know how to add, subtract, and close the books," Arnault told Yerkes "So tell me—why aren't *you* running this company? Then he said, OK, I'm going to give you the company for six months, you'll be general manager, and I'll be back in six months to see what happens."

The new plant had all sorts of teething troubles. Yerkes shut down the production line then sent everyone off on a two-week vacation. During that time he and three other managers set about fixing the problems. "We just went ahead and moved to Camarillo. We gritted our teeth and worked like hell until we got the plant running. Then two weeks later everyone came back and started making panels, and the plant has never stopped since. I was able to double shipments every month for six months in a row. When that six months was up, Ron Arnault came back, it turned out that that month we had our first million-dollar shipment of panels. He said, 'Congratulations, the board has agreed to make you president of ARCO Solar—you got your own company, see you next year.' So basically things worked real well, that year we actually shipped 1.2 megawatts of solar panels, it was the first time in the world that any company had shipped a million watts. That was 1980."

By 1983 ARCO Solar, this little pipsqueak outfit ARCO had bought, was the largest photovoltaic company in the world. That year Solarex was acquired by Amoco. Solar Power, the Exxon subsidiary, folded. To have beaten mighty Exxon in any business was a big deal for the suits at ARCO headquarters. As a reward, Yerkes was promoted to group vice president of new ventures and given an office in the towers. "It was a big payoff from their little investment," he concluded. "They had stumbled onto something, it turned out to be a good thing, and as a result they looked good."

The glory days did not last. In 1985, unhappy with his new job and fed up playing the game of corporate musical chairs, Yerkes resigned. Ron Arnault was promoted to the parent corporation's CFO. The following year, Bob Anderson, who had thrown his support behind solar, retired as chairman of ARCO. In 1989, having invested at least $200 million in PV and never making a profit, ARCO would sell its solar subsidiary to the giant German firm Siemens. ARCO told reporters it wanted to concentrate on its core business, oil. It had been a big gamble, everybody had understood that, but ARCO was making $2 billion a year and had money to burn.

In 1998, five years after acquiring Solarex, Amoco merged with BP, which had entered the solar business in 1983. As a result of the merger BP acquired the Solarex facilities in Maryland. For a couple of years, the British oil company would be the world's largest solar panel producer, then it too would scale back its involvement.

Tim Bruton is a former director of R&D at BP Solar (based at the corporation's research laboratories located, appropriately enough, in the London suburb of Sunbury). He worked at the company for over twenty years, from 1983 to 2004. Looking back, Bruton explained how solar meshed with BP's ever-changing corporate image: "BP has redefined itself many times: in the early eighties we were a technology-driven company, and photovoltaics fitted that. Then we became a natural resources company, which solar still fitted. In the early 1990s when we became an environmentally friendly hydrocarbon company, solar got slightly marginalized but it still fitted the environmentally friendly bit. For most of the next ten years, BP has seen itself as an energy solutions company. So in its redefining, BP saw solar as being a key part of its ongoing strategy. And, you know, it was good publicity: BP saw Asia as the emerging market. All the local management in the different Asia-Pacific countries said, Solar's very good—when I talk to the local energy minister, he wants to talk solar, not oil. So there was a kind of feel-good factor within the company, and having a solar arm definitely opened doors, particularly within the developing world."

But with the Amoco merger and the rapid growth in the solar market after the turn of the century, the emphasis shifted from BP Solar being a technology leader to being a financially viable company. "Through the early eighties when BP was cash rich, the annual expenditure on solar was around $20 million," Bruton said. This amount, according to a friend of his who worked for BP's financial arm, was less than the uncertainty on any given day as to how much cash BP had. No one bothered much about such a paltry outlay. But through the 1990s the price of oil stayed low, at one point dropping below ten dollars a barrel. "Money became relatively tight in BP and operational efficiency became the watchword. Solar, though it wasn't costing the company anything in real terms—at one point it had a turnover of $100 million and losses of $100 million—was seen as a financial liability. So BP kept saying, Right, we've got to tighten up."

Just when a new generation of start-ups, many of them German, were aggressively investing in solar production facilities, BP started cutting back. In 2006 BP's great rival, Shell, quietly sold off most of its renewable energy assets (including ARCO Solar, which it had bought from Siemens, to Solar-World, one of the new German upstarts). Solar, Bruton recalled a former CEO of Shell's PV subsidiary sneering, was "a dysfunctional business."

While the oil companies might ultimately have been disappointed in solar, they were never against it. As the hundreds of millions of dollars they invested demonstrate, they remained true believers for far longer than made financial sense. Solar's real enemy was not the oil companies, but the corporations that felt PV actually did threaten their business—the utilities. The electric power generators were the ones who wanted to get control of solar. If the utilities could not control solar, then they would do everything in their power to stop it.

4

CAMEL WITH THE FRIDGE ON TOP

*P*eter McKenzie well remembered the first day he laid eyes on one of Bill Yerkes's products. "It was July 22, 1978, that we opened up these containers with long boxes in them," he said. The boxes contained 20-watt STI panels. McKenzie had no idea what they were. "I had never seen or read about anything about PV."

The place was Port Moresby, capital of Papua New Guinea, where McKenzie, an Australian, was working under contract with Telikom PNG. His job was to fuel and maintain the diesel generators that powered the microwave repeaters used to transmit long-distance phone calls across the nation's impenetrable terrain. His current mission was to install power at five microwave repeater sites that would connect the capital, Port Moresby, with Lae, the country's second city. This hot, green, wet outpost of civilization is best known as the departure point for Amelia Earhart, whence she flew off into oblivion while attempting to cross the Pacific in 1937.

To reach Lae, signals had to be sent up and over the Owen Stanley range, the lofty chain of densely forested mountains that form the backbone of the island nation. Installing repeaters on isolated mountaintops, some of which are over 13,000 feet high, is a tricky business. The only way to reach such sites is by helicopter. To keep the diesels fueled and maintained meant making an expensive helicopter trip every few weeks. Someone at Telikom PNG headquarters had come up with the idea of using solar instead to make the sites self-sufficient. The problem was nobody had told McKenzie. "The guy in the planning division who had ordered the panels took a six-month sabbatical the week before they arrived because he knew I would be chasing his blood."

The first attempt to cobble together the solar panels into functioning systems was a failure. Improperly configured, they didn't work. McKenzie called Bill Yerkes. "I told him I was going to send them all back if he didn't send somebody who knew something about this stuff." The sale to Papua New Guinea had been ARCO Solar's first big order. It was too important to lose. Yerkes came all the way out to PNG himself. He flew around the Owen Stanley range, making sure the systems were properly set up. By the end of the trip, McKenzie—a man who by his own admission had diesel running through his veins—was a convert to solar. Yerkes hired him to be ARCO Solar's Asia-Pacific representative.

In this early, off-grid period, PNG would be one of the best markets for solar precisely because there was no practical way of putting in conventional power there. (It would be the same several decades later with mobile phones.) "There's probably more PV per head of the population in Papua New Guinea [about 6 million] than in any other country," McKenzie reckoned. "Many villages have got their lights powered by solar, their pumping systems for potable water, their refrigeration for vaccines in their clinics, their telecoms systems, many many applications. It's a common thing—even in the bush people know about PV."

Among the technical support people Yerkes sent to PNG was a young man named Mark Mrohs. His first job while a junior at the Harvey Mudd College in Los Angeles in 1973 had been at Spectrolab making solar cells for satellites. There he met Yerkes. Upon graduating Mrohs went to work at the company full-time until he had saved enough money to subsidize his other love, acting. Mrohs decamped for Europe for a couple of years. A self-confessed bohemian, he studied mime and theater in Paris. On his return to the United States in 1978, he joined ARCO Solar. The fledgling firm then had perhaps fifty employees. Mrohs would draw on his twin passions—physics and drama—to become a one-man training department for the company, running classes and making training videos. With so little known about photovoltaics, even how to spell the name, there was a desperate need to educate people, as McKenzie knew from his own recent experience. "We had so many things happen you wouldn't believe," he said. "People would hook their panels up back to front, they would put them facing the wrong direction, away from the sun. The worst case was probably in a ski resort, where they had installed the panels under the eaves of a chalet's roof, to protect the panels from the snow, on the north side of the building, so they never got any sun at all!"

Mrohs was a born teacher, passionate about his subject. Over the next three decades he would train hundreds of technicians and dealers for the fledgling industry. "I was able to give competence to these guys so they knew enough about a solar cell to be able to say that it was more than magic, yet feel confident enough that they kind of understood how it worked. They could go on from there to do the real work, which of course was designing and selling systems.

"The ARCO Solar period was one of explosive growth," Mrohs recalled. "It was very exciting. We were the leaders, there was nothing we couldn't do; we had some competition, the US companies were strong, but there was not much else in the world. The domestic market was zilch really, so the main markets were international—remote applications, both for industrial applications like telecom and rural electrification applications like water pumping and village lighting."

During his first fifteen years in the solar business Mrohs reckoned he visited something like twenty-eight countries. But he would never forget the adventures he had in the early days. In particular, what fascinated Mrohs was the bizarre coming together of the preindustrial primitive with the ultra-modern. "Here was this effectively twenty-first-century technology. In the early years, the focus was on remote applications. We were dealing with inherently undeveloped areas, either on a mountaintop, or in a village, both of them remote. So you'd have this juxtaposition of a photovoltaic system sitting next to a half-naked farmer who's smiling away because it's running his water pump, and he no longer has to walk so far, or pump it by hand.

"We did a lot of work in Papua New Guinea in the early days of ARCO Solar. I remember one time with Peter, we set up a water-pumping system in this village, and the village did a celebration for us, a 'sing-sing,' a sort of musical festival. The singing in PNG is really good, there's lots of missionary influence there, so there's a real mixture of their indigenous songs with Christian hymns. During the day these people would be office workers in the city, but they also had their traditional village outfits. They didn't have proper tattoos anymore, so they'd draw tattoo markings all over their arms and legs and faces using magic markers. The women would be naked from the waist up, just wearing these grass skirts. The guys would be out there too with feathers, pearls, and animal skins representing birds, trees, and mountain spirits. The whole village came out to watch the dancers doing their thing. They had carved a couple of boats out of local wood and they'd strung them between two poles. They'd pull the string to move the boats as they sang, reenacting the legendary story of these boats making the voyage from island to island. So here's Papua New Guineans, who are still reaching back to the past and have cannibals living a hundred miles away, with a photovoltaic-powered radio telephone and a water-pumping system that needs no batteries. That juxtaposition always just blew my mind—I had one foot in space and satellites, and one foot back maybe four hundred years."

Terry Jester was another who looked back fondly on the early days ARCO Solar. In her final year as a mechanical engineering student at California State, Jester was doing an internship at a Los Angeles aerospace company. One day she came out of her lab and was taken aback to see a missile being wheeled by. "I guess somebody's got to do this work," Jester thought to herself, "but it's not going to be me!" She quit her job forthwith.

A friend suggested that he contact Bill Yerkes on her behalf. "I was still a college student then, Bill called me that same day, and I went in and interviewed at ARCO Solar. I'd studied some solar energy in college and liked it, but I didn't know there was this start-up until this friend of mine told me, so it was really quite fortuitous. I don't know if I would have come across it in any other way, in the early days there wasn't much advertising, it wasn't something that people knew about.

"ARCO Solar was literally two miles from the place where I went to college [Northridge]. I would go there after school until I graduated, then started working there full-time. It was just really exciting and new, the environment was one of creativity and a lot of passion. There was a feeling that what we were doing could truly change the world. It was a really young organization, most of us were just out of school and willing to work ninety hours a week for something that we really cared about. It was just kind of anything goes. We could pretty much do anything we wanted, design anything we wanted, build anything we wanted—we really had a lot of fun.

"One of my fondest memories is this little skunk works Bill started. The organization was growing and he was getting a little frustrated with some of the bureaucracy that goes with a company getting bigger, the rules and regulations you have to follow. He created this small design group that he directed himself. We rented a separate building, and Bill put five or six of us over there to develop these products that he knew weren't being done anywhere. We were doing all kinds of stuff, for example, solar-powered refrigerators. We ended up selling refrigerators to Africa, to keep vaccines cool on their way to remote health clinics in places like Chad and Djibouti. One of the most famous photographs from the ARCO Solar days was the one of a camel with a refrigerator on its back. In a way that photo sort of epitomizes what we were working on at the time, you know, we could actually start something, even with that little application, that could do some good."

The image of the camel with the fridge on its hump represented the positive side of solar power in the public mind. At the same time, however, it was something remote, far removed from everyday life. But in those early days solar also acquired another, negative image, one that subsequently would often be used to denigrate it. That was of photovoltaics as something of interest to only long-haired, sandal-wearing environmentalists, tree-huggers and, above all, hippies. To understand how this image originated, and also how the retail solar business got started, it is necessary to drive about five hours up Highway 101 from San Francisco to the tiny Northern California town of Willits, billed as the "Gateway to the Redwoods."

In Mendocino County, where Willits is located, and neighboring Humboldt County, the economy has long been heavily dependent on a single cash crop: "contemporary tobacco," as marijuana is sometimes euphemistically known. While the town has around five thousand inhabitants, there are probably twice that many people living out in the hills in remote homesteads, the largest such community anywhere in the United States. Not many of them were connected to the electric grid, and even those who were didn't necessarily want strangers coming by to read their meters.

In addition to the marijuana growers, there were also the back-to-the-landers. These were people disillusioned with life in the big city. Disgusted by the smog and other forms of pollution associated with an urban lifestyle, they felt out of touch with nature. They wanted to move to the country, where they could grow their own food and live a better, more wholesome, alternative existence. Their bible was the *Whole Earth Catalog*, which told them how to access the tools they needed to survive sustainably.

Nowhere was the back-to-the-land phenomenon more prevalent than in California. There, during the late sixties and seventies, a steady stream of people from the Bay Area and Los Angeles migrated up to the unspoiled mountains and forests of its rural north. One such migrant was Wayne Robertson. A self-described "Southern California beach kid," in 1978 he had just gotten married and he and his wife were looking for a safe place to settle and start a family. He found what he was looking for outside Willits, a bare piece of land that he bought with the proceeds from selling his house in the big city.

On his ranch Robertson built a homestead, putting in the plumbing himself. There was, however, one thing lacking: electricity. The nearest utility poles were twelve miles away. Life without power was a common problem that homesteaders faced. No television they could tolerate, but having to do without a sound system was hard. For those not prepared to embrace the full macrobiotic-style diet of red beans and brown rice, no refrigeration meant eating melted ice cream and drinking warm beer. Plus, it was difficult to read the kids bedtime stories by the flickering light of a smoky kerosene lamp. Renting a diesel generator would have done the trick, but not many people could afford that. Anyway, it would have run counter to their principles to depend on something whose exhaust fumes polluted the atmosphere. One solution, as suggested in DIY books like *How to Be Your Own Power Company,* was to run jumper cables from the battery in your car and hook them up to run your sound and lights. But it was a drag waking up the next morning to find the damn thing wouldn't start because its battery was stone dead. There had to be a better way.

Wayne Robertson discovered solar electricity while working in a Willits lumber yard, selling solar hot water systems. A lumber yard was a good place to locate an outlet, because folks from the hills were always coming down to the yard to buy materials for their houses. One day a sales rep from ARCO Solar dropped in. Though Robertson didn't normally care for salespeople, he said, "I was just so fascinated by the technology and how it worked so well." He persuaded his boss to set up a distributorship, Solar Electric Specialties. Within a short time SES was the largest retailer of solar systems in the country. Many of the company's initial customers were marijuana growers. They came down in pickup trucks. They would pay for panels with cash wads of hundred-dollar bills, no purchase orders necessary, then disappear to whence they came. Back then in Willits grateful retailers printed bumper stickers that read "Growers are Good Customers." Robertson remembered the first time a group of law enforcement officers brought in a photovoltaic module they had confiscated during a raid on a marijuana farm. "It had one of my Solar Electric Specialties stickers on the back: they wanted to know what it was."

Wayne Robertson was one of the first distributors Mark Mrohs trained. "The kinds of people that were involved early on were definitely true believers," Mrohs said. "The concern back in the 1970s and '80s was pollution—Earth Day had just been started—and photovoltaics were seen as one of these possibilities to generate power without causing pollution. They wanted to save the Earth, and these guys [like Robertson] were willing to talk to customers over and over again, and to make long trips to visit remote customers, because all the customers were remote.

"The dealers had to be businessmen, but they also had to be a little bit crazy, willing to try something new, believe in it, and stick to it. They also had to find customers that were similar-minded. I think that's what made it fun for them, because it wasn't just another sale, or another couple of hundred sales. You found a like-minded spirit that made you feel that you were all part of saving the world. So I think that was contagious, and that [business] kept us going for something like ten years."

Those were heady days for Robertson and other solar installers like Real Goods that sprung up around this time. Located in nearby Hopland, Real Goods started like SES in 1978, with Robertson's wife its second employee. The company has since grown to become one of the largest retailers of solar products in the country. "Solar was alternative energy and a lot of us loved the alternative lifestyle thing," Robertson said. "At that time there was a lot of experimentation, and great camaraderie in the industry. Someone would discover something new, and we'd all apply ourselves to it."

Over thirty years on, Robertson remains in the solar business. While no longer with SES, he still loves the solar industry, though he grumbles that it is not as much fun as it used to be. These days solar is becoming a hardcore business like any other. But the thrill is by no means entirely gone. "I mean, for somebody that started off doing one and two modules at a time, a few tens of watts, to be able to quote and sell megawatts is very exciting."

For many years, despite the best efforts of industry executives, solar would continue to be associated with alternative lifestyles. They desperately tried to portray solar, as an ARCO Solar marketing man told the *New York Times* in 1981, "more as a business now, not a hobby, not a ban-the-nukes/save-the-whales bunch of guys who stand around airports and pass out literature." Solar was not seen as a serious alternative, something that could usefully augment the steady 24/7 "base-load" power that the utilities generated in their massive central plants. But as we shall see in the next chapter, the utilities soon learned how solar power could be useful—so long as it was they who remained in charge of its production.

5

SKUNK THE GOVERNMENT

*F*or most of the twentieth century, generating energy was a more or less predictable business. In the United States, as elsewhere, electric utilities came in two varieties. Dominating the sector were the big, investor-owned utilities, or IOUs. Given the colossal amounts of money they had to borrow in order to build their huge power stations, transmission lines, and distribution channels, this was an entirely appropriate acronym.

IOUs accounted for about three-quarters of generation, with the much smaller municipal utility districts—known as munis or MUDs—making up the rest. Thus California, for example, has three IOUs: Pacific Gas & Electric (PG&E), which holds sway over the northern portion of the state, and Southern California Edison and San Diego Gas & Electric, which between them dominate the south. In addition, there are also dozens of munis—like Los Angeles Department of Water & Power, the nation's largest, and Sacramento Municipal Utility District—which serve individual cities.

IOUs are regulated monopolies, meaning that they have the exclusive right to sell electricity to customers within a given area. In return, they must submit to oversight from a state government body—a public utilities commission, or PUC—which sets the rates that utilities can charge and the "fair return"—profit—to which they are entitled. The rates charged by munis, which are self-regulated entities run by locally elected officials, are invariably lower than those of the IOUs.

California's PUC was jocularly known as the "Public Futility Commission." Like many regulatory agencies, the commission was dominated by the entities it regulated. "The revolving door was unending as the (nominally elected) commissioners came and went from the utility companies," Bill Clark, a former *Wall Street Journal* reporter who covered the power industry during the 1950s, told me. "You very seldom had a public representative on the public utility commission." For many years the CPUC entertained almost continuous rate increases from PG&E and its Southern California counterparts.

Not surprisingly, given this tidy, well-defined regulatory environment, utilities tended to be somewhat stodgy organizations. All they had to do was keep increasing their output in line with the nonstop economic growth of the postwar period. Innovation was not part of their mission statement. "The generation and transmission of electrical energy, once a forefront engineering domain, now had all the excitement of accountancy," writes Phillip Schewe in *The Grid*. "The best engineering students, as if smelling the complacency, seldom went into the power business anymore."

Then, beginning in the 1970s, this cosy world was turned upside down. Two new factors caused disruptive changes in the federal and state regulatory framework, on top of which came the rapid development of new technologies for the generation of electricity. The first new factor was growing concern about pollution. The National Environmental Policy Act, which President Richard Nixon signed into law in 1970, required plant builders to conduct an environmental assessment before proceeding with construction. Public awareness of the damage resulting from acid rain caused by sulphuric emissions from coal-burning power stations increased during the 1970s. Such concerns ruled out the construction of any new coal-fired plants in states like California, where the environmental movement was strong.

In the 1970s the Golden State still depended on fossil fuels for more than 80 percent of its electricity. To meet the expected growth in demand, California's utilities planned to build dozens of new nuclear power plants all along the Pacific coast. But finding suitable sites for such plants in the earthquake-prone state was fraught with difficulties and delays. Back in 1961 PG&E had begun construction of a nuclear plant at a rocky inlet called Botega Head, about forty miles northwest of San Francisco. A large hole had been dug before it was discovered that the location was right on top of the San Andreas fault. Local opponents dubbed the site "Hole in the Head." It was abandoned.

To get its Diablo Canyon nuclear plant approved, PG&E had to undergo six years of hearings, referendums, and litigation. There were also huge cost overruns. When construction started in 1968 the site—at Avila Beach, near San Louis Obispo—was deemed safe. By the time the plant was completed in 1973, however, a seismic fault had been identified a couple of miles out to sea. Embarrassingly, it turned out that blueprints for the plant's earthquake safety features had been read backward.

In 1981 during two turbulent weeks at Diablo Canyon, 1,900 protesters were arrested, the largest number in the history of the US anti-nuclear movement. By then it was apparent that the expected growth in demand for electricity had failed to materialize. There was therefore less need for new plants, especially not pricey nuclear ones. (Diablo Canyon ended up costing more than $5 billion.) Until very recently no US utility had ordered a new nuclear plant since 1973. In the sixties nuclear power had been hyped as "too cheap to meter." By the eighties it had become too expensive to build.

The other new factor in the wake of the oil crisis was the impetus for the United States to become energy independent. Reducing energy consumption was a key part of the strategy. Especially in California, where in May 1974 the state legislature passed the Warren-Alquist Act, which established the California Energy Commission. The new body was given the authority to approve new power plants. Before doing so the commission would first consider anticipated energy demand, and the potential of efficiency efforts to reduce that demand. The CEC would also consider the environmental consequences of new plants and whether there were less environmentally damaging alternatives.

For utilities, promoting energy efficiency was an unnatural act. Thus far, their entire rationale had been to encourage customers to use *more* energy, not less. Their rate structure was designed to reward consumption, by charging customers less for the additional kilowatt-hours they used. Their domestic sales departments came up with promotional gimmicks to encourage usage. Utility salespeople gleefully handed out free hair dryers to their customers. They vigorously endorsed electricity-guzzling tumble dryers. "Clotheslines are for the birds," scoffed a 1950s utility slogan. The assertion was taken seriously by California's upwardly mobile homeowners' associations. Hanging out clothes to dry came to be perceived as something only poor people did.

Now with the coming of energy efficiency, these same utility salespeople were supposed to sally forth offering inducements to customers to unplug inefficient appliances, like the ancient refrigerators they used as beer coolers in their garages. Promoting conservation was, understandably, something the utilities had to be prodded into doing.

The CEC came up with a solution to this problem in 1982, passing legislation that enabled "decoupling," severing the link between revenue and sales. This made it possible for generators to sell less power and not lose money. Rates would be set high enough for utilities to recover their costs. Incentives would reward them for any drop in sales. Decoupling works: per capita electricity consumption in California has remained flat, while increasing by 60 percent in the United States as a whole. The Golden State would continue to lead the nation in energy policy innovations. It was already, and would remain, by far the largest solar market in the United States, accounting in 2008 for more than two-thirds of the nation's installations.

The biggest challenge to business-as-usual in utility-land came as a consequence of the Public Utility Regulatory Policies Act (PURPA), part of Jimmy Carter's National Energy Plan. The main intention of the legislation, enacted in November 1978, was to eliminate rate structures that promoted energy consumption. But the act also contained another provision, little noticed at the time, that would have huge unintended consequences. Its ostensible aim was to obtain more electricity without burning more fuel. Utilities were required to purchase power generated by private companies that produced electricity as a by-product of their activities. For instance, a paper maker that needed steam to heat the drums on which it dried its paper might use the excess steam to produce electricity. This simultaneous production of heat and power was called co-generation.

Previously, any electricity generated in excess of the firm's own internal needs would be wasted. Now, the company could sell it to the local utility, and at a favorable rate since obtaining electricity in this way would be much cheaper than building new power stations. According to Phillip Schewe, "The utilities were not thrilled. Their business was making and selling electricity, not buying it from other companies. Furthermore, since the scheduling of electricity—the perpetual balancing act between load and generation—is a tricky thing, it would be an imposition to have to buy orphan power in small amounts and at odd hours. The utilities didn't like being forced to accept this mongrel electricity, and they fought the whole thing in the courts for years."

In California a political battle ensued, which the utilities ended up losing. The upshot was the establishment of a standard no-hassle contract. If the supplier could deliver power at the given price, then the utilities were legally obliged to accept it. PURPA provided a tremendous boost to innovation. It spurred the development of nontraditional technologies for generating electricity, turbines driven by gas and wind. The former were a spin-off from engines used by airliners. They were much more efficient than the heavy-duty industrial steam turbines on which the utilities had long relied. "The economy of scale for the electric power industry had been turned on its head," Schewe writes. "Smaller was cheaper. Moreover, [gas] turbines could be built quickly, they better matched load growth, and they caused less pollution."

A gas turbine–based plant could dispatch power at peak times and sell it, under contract, to the local utility. Here was an opportunity for energy entrepreneurs. Now new, unregulated companies—the so-called nonutility generators, or NUGs—could enter the power generation business. And enter they did. By 1990 NUGs accounted for over half of new additions to electricity generation capacity.

In California the financial and environmental circumstances were especially propitious. State tax write-offs were available. Wind blew nonstop in the Altamont Pass, across the bay from the load center of San Francisco: no need to build expensive new transmission lines. When the standard contract was announced, people thought that wind might account for a couple of megawatts of additional generation capacity. In fact, wind turbines capable on paper of delivering around *5,000* megawatts were built. Not all of this capacity was productive, however. Authorities learned the hard way that tax breaks should be granted only for *production* of electricity, not for the *construction* of facilities. Nonetheless, by breaking down the barriers to entry, PURPA ended the utilities' monopoly on power generation. It was the thin end of a wedge that would lead twenty-odd years later to deregulation and the opening up of the electricity market to competition.

As we saw in chapter 1, the Carter administration was gung-ho about the prospects for solar as an alternative source of energy. It passed the Solar Energy Research, Development and Demonstration Act as part of the same package as PURPA. The explicit long-term goal of this legislation was "to make electricity from photovoltaic systems economically competitive with electricity from conventional sources." Under this aggressive R&D program, the government hoped to drive down the cost of silicon solar technology to make large-scale application economically viable as soon as possible. Once low-cost technology that could compete with conventional methods of electricity generation had been developed, the government assumed that "the market will enter an explosive self-sustaining growth phase." Dale Myers, the undersecretary at the Department of Energy responsible for overseeing the development of the technology, said that his goal was "to move it all out into the industry and get the hell out of the business."

That the main users of solar would be utilities was taken for granted. The conventional wisdom was that solar would be deployed in much the same way as nuclear had been, in the form of large central stations that would provide economies of scale. Thus the bulk of the money from the Department of Energy's solar R&D budgets of the late 1970s and early '80s went to support research aimed at reducing the manufacturing cost of silicon panels. Jet Propulsion Laboratory in the San Gabriel Valley district of Los Angeles, where Bill Yerkes had performed his dramatic solar panel reliability test, was given primary responsibility for this research. Other federally funded facilities were also involved. They included Aerospace Corporation, a US Air Force think-tank, located just a few miles south of JPL in El Segundo. Primarily a space program actor, the corporation was tasked with performing technical and financial analyses of photovoltaics in residential, commercial, and central station applications.

On a tour of Aerospace Corporation's facilities in 1981, Bill Yerkes was horrified to discover that the government was planning to build a megawatt-scale solar power plant. Contracts for its design and construction had already been drafted. He dashed back to see Ron Arnault, his boss at ARCO Solar. "This is a disaster," Yerkes complained to Arnault. "Our government is going to put us out of business!" From his experience working for the space program, Yerkes knew how the feds operated. As with any government project, there would be schedule slippages and cost overruns. Building a large-scale solar plant would probably take Aerospace Corporation and its subcontractor, Martin Marietta, five years. Meantime, everybody else would be held up, because the government would refuse to green-light other projects until it had proved that its own baby worked. "This is absolutely the worst thing that could happen to us," Yerkes moaned.

"How long would it take us to build a megawatt plant?" Arnault wanted to know. "In another couple of months we'll probably have a megawatt's worth of solar panels in inventory," Yerkes told him. "We could just go out there and put them up." His boss nodded: "OK—let's do it." A few phone calls later, Arnault had identified a willing partner in Southern California Edison. "When a senior vice president of the Atlantic Richfield company calls up somebody at Southern California Edison or PG&E, they get a different reception than somebody at some independent outlier would," Arnault told me. "We weren't asking them to do very much of anything other than pay us half of the price. I mean, we took all the responsibility, all the risk."

The unmanned solar plant would be built on a twenty-acre site belonging to the utility at Hesperia Plains, in the Mojave Desert northeast of Los Angeles. The whole thing—from design to construction to getting the plant into operation and online—was done in a matter of months. "It was kind of fantastic to watch," Yerkes recalled. "A guy comes into this field and lays it out with a post-hole digger. A truck comes along and drops the pipes onto the ground. Then a cement truck comes in and pours a yard of cement in each of the holes. They stick the pipes in the holes. We put a simple gearbox on top of each pipe to enable the arrays to track the sun."

Trackers were crucial because they allowed the system to harvest the evening sun all the way to 8 PM in the summertime. Evening was when the utility really needed the extra juice, to cope with the load from domestic air conditioners. If the arrays had been fixed at a 45-degree angle, the setting sun would have dropped below their bottom edge at around 4 in the afternoon, cutting power output to zero.

The resultant plant, seen in the late afternoon sun with the low sierra in the background, looked eerily beautiful. Lined up facing west in rows, like giant playing cards tilted at an angle of about eighty degrees, the 108 arrays bore a slight but unmistakable resemblance to the stone heads on Easter Island. Capable of supplying enough electricity to power up to four hundred homes, Hesperia Plains was three times larger than any other PV system in the world at that point. "Our objective was to skunk the government," Yerkes recalled. "That was not difficult to do, because we were a private company, we just made decisions then went off and executed them."

Inspired by this success, ARCO Solar went on to build a much larger, 6.5-megawatt power plant at Carrisa Plains, in California Valley, west of Bakersfield. This time PG&E provided a 160-acre site in return for the right to monitor the plant's performance. By then Yerkes had been promoted to vice president of new ventures at ARCO headquarters. His replacement, whose previous experience had been running an oil refinery, embarked on what turned out to be an overambitious plan. Mirrors would be attached to the sides of the panels to concentrate sunlight impinging on the cells, reducing the cost. Yerkes warned that this approach was fraught with problems. Tests he had done showed that the mirrors would get dirty within a couple of weeks and, as a result, the efficiency of the panels would drop. The mirrors would need to be washed, but there was no water out there in the desert. Plus, the cells were not designed to run at elevated temperatures. Sure enough, come the summer, the cells overheated, the panels turned brown, and the voltage dropped. This in turn tripped the inverters—which convert DC power into AC—to turn off.

It was too late: the go-ahead had already been given. ARCO was racing to take advantage of federal tax credits that were about to expire. Completed in 1985, by which time Yerkes had left the company, the plant was dismantled just ten years later. "Basically it was kind of a tragedy," he said. Fiasco might be a better word. Once some of the technical problems had been fixed, however, Carrisa Plains did at least provide PG&E with a lot of data about the performance of PV. In particular, the almost perfect correlation during hot summer afternoons and evenings between good sunlight, hence power generated (supply), and peak load (demand). PV reached its maximum output at precisely the time when electricity was at its most valuable.

Ultimately, a total of four large photovoltaic central power stations would be built in the United States during the eighties. All four were in California. In addition to Hesperia and Carrisa Plains, beginning in 1984 ARCO Solar also installed a 2-megawatt plant right in front of the giant cooling towers of the Sacramento Municipal Utility District's troubled nuclear power plant at Rancho Seco (see chapter 9). The last of the first-generation PV central stations, begun in 1989, was a 650-kilowatt plant built by PG&E outside the city of Davis, near the state capital. This facility was originally named Utility Scale Systems Research. Then someone realized that the acronym read USSR, not the most propitious of names when asking the Department of Energy for funding. A more patriotic alternative was concocted: PVUSA, which stood for Photovoltaics for Utility Scale Applications. The objective of this plant was to test not so much the panels but rather the surrounding components, like inverters, what utility people call the "balance of system," which made up half the cost.

Skeptical utilities had long argued that it would be disruptive, not to mention unsafe, to hook up this newfangled PV stuff to their precious grid. As a result of experience gained from operating these four large plants, by 1989 it was clear that solar electric power stations were actually easy to install and operate. The fact that PG&E had been involved in the trials made the conclusions credible. Though PV plants took up a lot of space, their environmental impact was otherwise minimal. They were, however, still far too expensive to be competitive with conventional forms of power generation. By the late eighties, however, some people were beginning to wonder whether large-scale plants were really the best way to exploit PV's unique and fundamentally different characteristics. Prime among them were Carl Weinberg and his group of pioneering young researchers at PG&E.

A civil engineer by training, Carl Weinberg spent the early part of his career in the US Air Force. There he worked on biomedical projects, like studying the effects of acceleration on a pilot ejecting from an airplane. He ended up a desk-jockey at the Pentagon. In 1974 Weinberg retired from the military. A friend at PG&E offered him a job. Weinberg was assigned to PG&E's Department of Engineering Research. This was based at Crow Canyon Road, San Ramon, on the eastern side of the hills behind Oakland, about fifty miles' distance from the utility's headquarters in San Francisco.

The seventies were, as we have seen, a turbulent time for utilities. PG&E was beset on the one hand by criticism from environmentalists, and on the other by regulatory pressure to investigate energy efficiency and alternative forms of generation. Previously the utility's researchers had mostly been construction engineers. Now the department was also taking on biologists and physicists and electronics people. In 1979 Weinberg was made head of the department with a mandate to think outside the box. "Carl was terribly important," said Don Aitken, an architect who directed one of the regional solar energy centers set up under the Carter administration around the same time. "He realized that [solar] was an important way to go. At that time PG&E didn't pay much attention to the R&D group, so Carl could do whatever he wanted to do—and he did."

"Oil and coal and gas are all forms of stored energy," Weinberg told me. "We got to looking around and finally concluded that what you really needed to do was intercept the energy flux before it gets stored. So we began to look at, What are the technologies out there that would deal with the flux?" Wind turbines were one obvious form of flux-interceptor. Photovoltaics were another. The researchers noted that big oil companies like ARCO were getting involved with solar cells. The technology was beguiling. "It looks like a simple device," Weinberg said, "it doesn't require water, it doesn't emit anything—it's almost like a magic crystal!" To find out whether PV was good enough for utility-type applications, first it was necessary to test its performance. Testing quickly became the research department's main mission.

"Our idea was, Why don't we buy one of every type of photovoltaic panel made and stick them on our roof to get long-term data on them?" That way Weinberg's researchers could determine whether the ratings provided by manufacturers were credible, which they needed to be if the ratings were to be used as the basis for calculating energy output under actual utility conditions. By the mid-1980s, after a lengthy program of daily and hourly measurements, the researchers had accumulated prodigious amounts of data. While manufacturers might have been inclined to overstate the output of their products, there was no question that, overall, photovoltaic technology was extremely reliable. The question now was, Where to apply it?

As head of the department, Weinberg's philosophy was to surround himself with what he described as "a wonderful, innovative group of individuals." One of his bright young crew of engineers was Kay Firor. She had been "a total PV addict," she confessed, since her first encounter with photovoltaics in the engineering library at the University of Colorado in 1977. "I had come across the term 'solar cell' and was looking it up in a couple of books to find out what on earth it might mean. I found a single paragraph describing the direct conversion of sunlight to electricity. I must have read that paragraph over ten times, thinking to myself, This cannot mean what I think it means, because if it *were* true everyone would be out dancing in the streets all over the world. My focus shifted from electronics to photovoltaics that day."

From the outset Firor realized that PV had an enormous advantage over other energy technologies. Namely, that solar was ubiquitous. Except at the poles during the winter, everywhere on the surface of the planet receives at least some energy from sunlight every day of the year. As PG&E gained experience through feedback from plants like Carrisa Plains, it became obvious that there was also a big difference between PV and conventional forms of generation in terms of economics. Normally, the more capacity you built, the cheaper your costs: a bigger boiler doesn't require twice as much material. With PV, however, each increment in capacity cost the same. The only way to reduce the cost, as the government had noted, was to increase the volume of manufacturing.

By the late eighties PV was still four times too expensive to compete with ordinary coal-fired plants, and triple the cost of gas, the most expensive conventional fuel used to generate electricity. But generation represented only one part of the overall picture. Having generated electricity, you then had to transmit and distribute it. Transmission was done via high-voltage cables slung between steel pylons. Distribution was the final link in the chain, the substations at which the electricity was stepped down in voltage by transformers for delivery, via the familiar power lines strung between wooden poles, to customer premises. It was in the distribution end of the system, Firor saw, that PV really made economic sense. Solar would be good for the grid.

Weinberg listened carefully to her idea. Then from the distribution planning department he summoned a young engineer named Dan Shugar. "Find out what is the value of placing photovoltaics within the distribution system," he told him. Though an electrical engineer by training, Shugar didn't really know what photovoltaics were. He read up on the subject. "Immediately it seemed like a great solution to a problem we had," he said. Namely, PG&E had all these substations for distributing electricity that would work perfectly well for 90 percent of the year. But at times of peak consumer demand, during hot summer afternoons, when commercial air conditioning and agricultural water pumping needs were at their greatest, the substations would overload. With the sun beating down on them the transformers would overheat, reducing their working life, and the quality of power delivered to consumers would drop. You could of course replace the transformers, but that was expensive. Or, you could put in a new line, but that was even more expensive.

Now, Shugar understood, here was an attractive new alternative. You could put some PV right at the substation, where it would lower the demand, especially at times of peak load, when the sun was at its hottest. By taking just 5 percent of the load off the transformers you could stave off the need to replace them for up to ten years. That would save the utility a great deal of money. "Grid support" was one term Shugar and his colleagues coined to describe this new concept; "distributed generation" was another. To test their hypothesis, under the aegis of the PVUSA program, PG&E built a 500-kilowatt plant at its Kerman substation near Fresno in California's hot, dry Central Valley. At an installed cost of $6.50 per watt the PV system was not cheap, but it was cheaper than the alternatives, repeated replacement of components or a total overhaul.

The test at Kerman proved beyond question that PV could be of substantial value to utilities, bolstering the weakest links in the chain that linked generator to consumer. It pointed the way to a future in which power would no longer be a commodity delivered, as Weinberg put it, "like water in an irrigation system, from a large reservoir to many customer sites." Future distribution systems would work more like computer networks, "with many sources, many consumers [and] continuous reevaluation of delivery priorities."

Meantime, Shugar devised a new planning methodology that would create incentives for developers to install renewables in high-cost areas. This, however, "was *way* too sophisticated a concept for the utility." Besides, by the time the Kerman installation began operating in April 1993, senior management at PG&E had become distracted. California was proposing a radical deregulation and restructuring of its electricity industry. Under the proposal, utilities would divest themselves of generation. This would prove a disastrous move that would leave them exposed to exploitation from predators like Enron, which in turn would lead to the California electricity crisis of 2001.

With no further need to investigate new technologies, PG&E disbanded its R&D group. Carl Weinberg retired. Many of the young researchers from his crew would go on to become the next generation of leaders in the PV industry. They included Dan Shugar, who would join PowerLight, which by 2004 would be the world's largest designer and installer of commercial solar systems.

Government and utilities had assumed that if PV were to be useful, then it would have to find a place within the existing paradigm of electricity generation. Locked into this mind-set, it was hard for utilities to follow the idea that PURPA had unleashed to its logical conclusion, namely, that if anybody could be a generator, then why not the homeowner? But as we shall see in parts 2 and 3, by the late eighties in Europe and Japan, the idea of installing solar panels on your roof then selling electricity to the power company was being taken very seriously indeed.

6

END OF "SOLAR SOCIALISM"

*T*he seventies had given the nascent PV industry a dream start. In 1972 the budget for research on all forms of solar energy at the US National Science Foundation was a measly $2 million. By 1978 the Carter administration was requesting Congress to allocate $105 million for research on photovoltaics alone.

In 1971 solar power cost an exorbitant $100 a watt (assuming you could find someone to sell you the cells). By the end of the decade, you could walk into a store and buy panels made by the likes of ARCO Solar for $10 a watt. This order-of-magnitude drop in price had been achieved in part through large-scale government procurement programs. The Department of Energy's aggressive goal was to make PV competitive with the cost of generating electricity by conventional fuels like coal and gas. The department was shooting for 50 cents a watt, and an annual production of 500 megawatts, by 1986. Audacious, perhaps, but no more so than putting men on the moon had been the previous decade. "A general consensus appears to be developing among the participants that the goals are reachable and may even be far too modest," wrote one cockeyed optimist in 1977. By 1981 the cost of a solar-generated kilowatt-hour of electricity had reached around $1. A huge advance, but still up to twenty times more than the cost of generation using fossil fuels.

In 1979 Jimmy Carter had committed the US to deriving 20 percent of its energy from renewables by the year 2000. Some boosters were even more bullish. They reckoned that, come the millennium, solar *on its own* could produce a quarter of the nation's total energy requirements. In new US homes, solar panels would be just as common as that other recent domestic electronic marvel, the microwave oven. Also in 1979 the American Physical Society produced a more sober prediction. Its panel of experts estimated that solar cells were unlikely to generate more than 1 percent of US electricity by the turn of the century. It turned out even that figure was over-optimistic. By 2006 solar was supplying *less than one hundredth of a percent* of US electricity.

In 1980 Ronald Reagan was elected in a landslide. The prevailing ideological winds shifted 180 degrees. American voters had had enough of Carter's interventionist style of big government. For President Reagan, as he made clear in his inaugural address the following January: "Government is not the solution to our problems, government *is* the problem." The Reagan administration wasted no time pulling the plug on Carter's solar programs. Federal government support for solar energy came screeching to a halt. Among its first targets was the Solar Energy Research Institute in Colorado, which Carter had visited with so much fanfare in 1978. SERI's budget was eviscerated, from $124 million in fiscal 1980 to just $59 million in fiscal 1982. Scientists who had left tenured university jobs to work at SERI were turfed out with half a day's notice and no severance pay. Funding was dropped for outside researchers, including two who went on to win Nobel Prizes in other fields (Ahmed Zewail, 1999, and Alan Heeger, 2000). "It was a scorched-earth policy," solar architect Steven Strong told author John Berger. "It was not just a lack of enthusiasm for renewables. It was [a] deliberate, vindictive, and orchestrated campaign to snuff [research] out in as forceful and as vehement a manner as they could."

All in all, 370 staff were dismissed. Also axed was Denis Hayes, the thirty-six-year-old anti-war activist Carter had handpicked to head the institute. A former student leader, Hayes had been responsible for organizing the inaugural Earth Day, April 22, 1970, in which some 20 million Americans participated. This initiative had thrust environmental concerns into the political arena for the first time. His dismissal, Hayes told a reporter, "was the bleakest day of my professional life; they descended on the [institute], fired about half our staff and all of our contractors." During the administration of Reagan's successor, George H. W. Bush, the institute would lose even its nominal connection with solar. Renamed the National Renewable Energy Laboratory, its budget would continue to be cut.

The nation's newly assembled solar infrastructure was likewise swiftly dismantled. Don Aitken was director of the Portland, Oregon–based Western Regional Solar Energy Center, which was responsible for implementing Department of Energy solar programs. "We had two years to set [the center] up and get it going," Aitken recalled. "Then Ronald Reagan came in and within six days all our contracts were cancelled and we were all fired." In his first energy budget, for fiscal 1982, President Reagan trimmed funding for solar energy projects to $303 million, down from the $707 million Carter had requested for 1981. At the same time, he increased funding for nuclear energy to $1.6 billion, a $300 million hike. The biggest cut in solar spending was for demonstration projects, the lifeblood of many start-ups.

As of 1979, one analyst estimated, private sources had invested upwards of $100 million in the fledgling solar industry, while the federal government had spent a total of around $400 million (compared with $50 million during the sixties). "Mr. Reagan's supporters argue that his reduction of Government funding for solar development will put an end to the creeping 'solar socialism' of the Carter era," wrote Thomas Friedman in the *New York Times* in August 1981. "Through its largesse, they say, the previous Administration attracted a great number of people who were enamored with the solar idea, but produced equipment that only the Department of Energy could afford to buy."

The cuts in spending should not have come as a surprise. In announcing his candidacy in November 1979, Reagan had explicitly detailed his energy policy. Out went the despised energy efficiency. "First," the candidate declared, "we must decide that 'less' is not enough." Since early in the Cold War, the very idea of saving electricity had been considered un-American by some right-wingers. When Russian projections for electricity generation were higher than those of the US, why, then it was the patriotic duty of every citizen to consume more. Progress could best be measured in terms of increases in energy usage. America had not conserved its way to greatness. Conservation was thus seen, almost by definition, as socialist.

Reagan's solution to the energy crisis was "more domestic production of oil and gas." Or, as a subsequent, more succinct Republican slogan would put it: Drill, baby, drill. "We must also have wider use of nuclear power within strict safety rules, of course," the candidate continued. "In years to come solar energy may provide much of the answer but for the next two or three decades we must do such things as master the chemistry of coal." Sure enough, for nearly thirty years, federal funding for research on renewables would run a distant third to the spending lavished on fossil fuels and nuclear. Though discredited by accidents, leaks, and siting controversies, the well-heeled, well-connected nuclear mafia remained a force to be reckoned with in Washington. Their mantra was that solar was not yet a practical technology. Many years later this view would be mocked in a prayer voiced by the animated buffoon Homer Simpson: "And Lord, we are especially thankful for nuclear power, the cleanest, safest energy source there is. Except for solar, which is just a pipe dream."

Strident pro-nuclear voices on the right included Dixie Lee Ray. The former chairwoman of the Atomic Energy Commission (which had been responsible for regulating the nuclear industry) and subsequent governor of the state of Washington was an unapologetic champion of nuclear over solar. "Solar could be thought of," she said condescendingly, "as a flea on the back of the nuclear elephant."

Then there was Edward Teller, the Hungarian-born "father" of the hydrogen bomb, who would be the most fervent apologist for Ronald Reagan's Star Wars, his sci-fi fantasy of a space-based missile shield. While solar might one day be feasible, Teller argued in a 1979 interview, mass production of cells would be necessary in order to make the technology economic. Such production would create, he asserted, wrongly, "incomparably more pollution and more danger than nuclear reaction."

Reagan was famously lucky, nowhere more so than with fossil fuels. By the time he took office, the price of oil had already begun to fall from the historic highs it had reached in late 1979. By 1986, the price had plummeted to around $20 a barrel, where it would remain for most of the next twenty years. The price of natural gas had similarly plunged. As a result, interest in alternative forms of energy dropped off the radar. "Natural gas was not only much cheaper than coal," wrote Mitchell Landsberg in the *Los Angeles Times*, "it was practically cheaper than air. Why mess with alternatives?" During the dream-time of the Reagan, Bush, and Clinton administrations, energy consumption was simply not an issue. Almost thirty years would elapse before the US elected another president who would endorse solar energy. Meantime, America went to sleep.

In the US, as in no other part of the world (or at least not to the same extent), solar energy has long been anathema to conservatives. To understand the reasons for this curious polarization, and the visceral dislike of the American right for "solar socialism," it is instructive to look at the membership of what by 1978 had come to be informally known as "the solar lobby." For the most part, the lobby was what might be called a citizens movement. "Solar appealed to the counterculture, to the disaffected, to society's rebels, and to those simply wanting independence from the system."

The lobby included environmentalists and anti-nuclear activists. Most prominent among them was Barry Commoner. Born in 1917 to Russian immigrant parents in Brooklyn, Commoner was a college professor, a microbiologist who had been radicalized by the effects on the environment of radioactive fallout from the atmospheric testing of nuclear bombs. In Washington, DC, in 1979 Commoner addressed some 70,000 protestors—including California's then (and future) governor, Jerry Brown—attending the largest anti-nuclear rally ever held in the nation's capital. "Remember May 6, 1979," he urged the throng, optimistically, "the day the solar age was born!" The same year Commoner published a polemic, *The Politics of Energy*. In it, he railed against President Carter's national energy plan as "dangerously dependent on nuclear energy" and "repressing the true potential of solar energy."

"For Commoner, a shift toward solar energy would take power—political and electrical—out of the hands of the wealthy and distribute it more equitably. As Commoner was also keenly aware, nuclear power required centralization and hierarchy, while solar energy could be easily distributed and controlled more locally and democratically. Commoner's final chapter in *The Poverty of Power* [1976] turned into a blistering critique of private enterprise and the capitalist system." In 1980 Commoner ran for president against Carter and Reagan, on the Citizens Party ticket. He finished fifth, with 233,052 votes (0.27% of the total).

Other members of the solar lobby had even higher profiles. Notably Tom Hayden, the student leader and anti–Vietnam War radical from the sixties. In 1982 he began his career in the California state legislature by pushing, among other things, for solar power. Hayden's then wife, the conservatives' *bête noire*, Jane Fonda, was an equally outspoken solar proponent. Fonda starred in *The China Syndrome*, a thriller in which cover-ups compromise the safety of a US nuclear power plant. By an eerie coincidence the movie opened on March 16, 1979, just twelve days before the real-life accident at Three Mile Island.

While promoting the film in media interviews Fonda took advantage of the widespread panic to criticize nuclear power. Springing to the defense of nuclear's reputation for safety and reliability, Edward Teller overstrained himself and suffered a heart attack. "I was the only one whose health was affected by that reactor near Harrisburg," Teller wrote in a *Wall Street Journal* op-ed piece. "No, that would be wrong," he added bitterly. "It was not the reactor, reactors are not dangerous. It was Jane Fonda."

By the mideighties, solar energy as a social movement was moribund. The American PV industry was in bad shape. Domestic shipments of solar panels by US manufacturers were declining. State and federal tax credits for renewables had been allowed to expire. The continuity that was so vital to sustain a growing market had been broken. In 1985 the last Department of Energy funded solar projects were completed. In 1986 there would be no new starts. This was not so serious for ARCO Solar and Solarex, whose customers were mostly commercial enterprises. Indeed there were some, Bill Yerkes for one, who believed that government involvement with all its delays and red tape had actually hurt the solar business.

For American PV producers the only bright light on the horizon was a rapid growth in exports. By 2000 overseas sales would account for 85 percent of US production, mostly to off-grid applications in developing countries. Meanwhile Japanese manufacturers, spurred by sales of solar-driven products like calculators and watches, were coming on strong. In 1986 the consumer electronics maker Sanyo overtook ARCO Solar to become the world's largest overall producer of individual solar cells. But the American firm remained the largest manufacturer of solar panels.

By 1989 annual sales of PV had reached around $125 million worldwide, with ARCO Solar accounting for around a quarter of the total. By then, however, the parent company was rapidly "losing patience" with solar power, according to an interview given by Bob Anderson's successor as ARCO chairman, Lodwrick Cook. The market was not growing fast enough for his liking. That August, ARCO announced it was selling its solar subsidiary to Siemens of West Germany. Though Siemens did not disclose the actual purchase price, industry insiders put it at between $30 million and $50 million. The giant power equipment manufacturer believed that, as a result of its acquisition, it would be well placed to fully develop what it saw, in stark contrast to ARCO, as "a fast-growing market." For Ken Zweibel, a manager at the Solar Energy Research Institute in Colorado, the sale of the 350-person company to a foreign firm was short-sighted. "It's selling American leadership to the highest bidder in a critical new technology," Zweibel complained to a reporter.

As it happens, 1989 was also the year when the West German government began the first of several large-scale solar subsidy programs. The same year, for the first time, Bonn and Tokyo outspent Washington on solar R&D. Their emphasis was not on water pumps, or telecom relay stations, or remote villages in developing countries, or central generating stations, but on putting PV panels on the roofs of the homes of ordinary people. How this radical departure came about is the subject of parts 2 and 3.

Part 2: Hundreds, Thousands, . . . Millions?

EARLY ATTEMPTS TO PUT SOLAR ON ROOFS IN
THE UNITED STATES, SWITZERLAND, AND JAPAN

7

LIKE A SNOWBALL DOWN A MOUNTAINSIDE

*W*ell before he built Solar One, the world's first house to mount photovoltaic panels on its roof, Karl Böer's life had been, as he himself put it, with considerable understatement, "a little bit unusual."

When he enlisted in the Luftwaffe as a patriotic seventeen-year-old, Böer had wanted to be a fighter pilot. Instead he was selected for his skill in flying to join a squadron whose desperate mission was to bring down Allied bombers over Germany by smashing into their rudders in mid-air. The pilots trained on very high-speed gliders with shortened wings. The gliders could dive at speeds of up to 190 miles per hour. To fly these frightening little machines, you lie flat on your belly, chin up, hands in front clutching the stick, feet on the pedals. It was, Böer recalled, "a most exciting kind of acrobatic flight training that we went through."

Luckily for him the mission was abandoned when it was learned, the hard way, that the bombers' rear gunners would shoot down the attackers before they got close enough to do any damage. In 1944 Böer was released from military duty to become a student. By 1961, at just thirty-five, he was already a distinguished professor and director of a department of physics that had been specially created for him at Berlin's prestigious Humboldt University. On August 13 that year the Berlin Wall went up, leaving his office on the East German side and his home on the West.

Two weeks later Böer left to attend a conference at Cornell University. In the US he received job offers from four different groups. He accepted a one-year research professorship at New York University. Back in Berlin Böer asked his superiors to grant him a leave of absence. They refused and he resigned. Abandoning the snug harbor of tenure was an unheard-of thing for a German professor to do. But Böer did not want to have to move to East

Berlin, where under the communist regime he would lose his freedom. It was characteristic of his restless intellect to seek new challenges.

After his year at NYU, Böer moved to the University of Delaware. He figured that at a small school he would have more elbow room than at a larger, well-established university like Stanford or MIT. American professors are encouraged to consult for industry. This arrangement suited the practically oriented Böer, who reckoned that between 1962 and 1970 he consulted for eighteen different companies. His clients included NASA's Jet Propulsion Laboratory. JPL was having problems with light detectors succumbing to damage caused by radiation on long, outer planetary missions. A possible substitute was degradation-resistant devices made from a compound called cadmium sulphide. Cadmium sulphide, as it happened, was Böer's specialty. His early work on the compound had been applied to the light-detecting sensors used to turn streetlights on and off. Now at NASA he realized that his pet material had much larger potential, applied to the conversion of energy from light to electricity.

In 1970 Böer took a six-month sabbatical to ponder the implications of solar energy. "We are running out of oil," he reasoned. "It just makes no sense to burn the damn stuff while we have sunshine coming to the Earth at a rate which is more than sufficient to provide all the electricity for all the industry and households in the United States. So why the heck not use it?"

On his return, like a good academic, he published a paper, "Future Large Scale Terrestrial Uses of Solar Energy." But Böer was also keen to put his ideas into practice. He proposed to test them by setting up a laboratory, which would do experiments, and a commercial company, which would develop and commercialize cadmium sulphide solar cells. The company would also serve to provide jobs for his students. It was not enough for professors, Böer maintained, merely to clone themselves. In his view there were already far too many graduates with PhDs in theoretical physics working as bartenders. His immediate superiors at the university warned that the kind of unprecedented academic-industrial collaboration he had in mind would never fly. Once in motion, however, Böer is not an easy man to stop. He composed a hundred-page, rigorously argued research proposal-cum-business plan and took it straight to the university president. "Needless to say I got all the money from the university, which I wasn't supposed to get, maybe ten times as much as any of my colleagues, and a large contract from the National Science Foundation." It was worth $1 million, an almost incredible sum for the time.

Within a year Böer had established the Institute of Energy Conversion at Delaware with himself as director and chief scientist. Though the name was deliberately chosen to indicate broader interests, the institute was effectively the world's first laboratory dedicated to the research and development of PV. He also tapped the oil company Shell for $3 million to fund his new start-up, Solar Energy Systems, with himself as chairman and CEO. "This is a guy who was always ahead of his time," solar cell researcher Larry Kazmerski commented admiringly. "He recognized that we scientists clutch onto this stuff until we're dying, but he realized that it had to be used to show it was useful."

Böer built Solar One on a suburban street in Newark, the University of Delaware's hometown, five minutes' walk from the institute. Seen from the front it resembled a standard A-frame house. In truth, however, it was only half a house. A contemporary visitor described Solar One as looking "as if it had been sliced neatly along the ridge of its roof with the northern half thrown away." The huge south-facing roof held twenty-four large (4-by-8 foot) solar panels. It was pitched an angle of 45 degrees, to catch the maximum sun in winter. The panels were not only designed to convert sunlight into electricity, but also into heat. "If you convert solar energy into electricity, the overall efficiency comes out to be at best 20 percent and, more realistically, 10 percent," Böer told the visitor. "But we can have 50 percent of the energy converted into heat." It was a total system approach. "When this converter sits directly on your roof, you can process the heat and do something useful with it. In the wintertime, obviously, you can heat your house. In the summertime, hopefully, you can air-condition your house."

Solar One was inaugurated, by the governor of Delaware, in July 1973. That first month thirteen hundred visitors showed up to marvel at the futuristic residence. Pictures of it appeared in newspapers and on televisions all over the world. Böer's timing was propitious. Three months later, Arab petroleum producers slapped their embargo on the US. "People said, Well, you had your ear on the ground—how did you have a connection to those oil sheiks?" he recalled, laughing. "I said, I didn't have any of that, I started for very egotistical reasons." Böer wanted to be more than just a professor, he wanted to show the world how to harvest the limitless energy from the sun.

To assist him in capturing solar energy, Böer hired over a hundred scientists and engineers, many of them experts in the field. Most renowned among them was Maria Telkes, the Sun Queen as she was sometimes called, who had been working on applications of solar energy for more than three decades. Telkes was born, to aristocratic parents, in Budapest—then still part of the Austro-Hungarian Empire—in 1900. In 1925, having completed a PhD in physical chemistry, she immigrated to the United States. In 1939 Telkes moved to the Massachusetts Institute of Technology to join the world's first research center on the use of solar energy. During World War II she designed a solar evaporator for converting seawater into potable water for use by shipwrecked sailors.

Between 1939 and 1962 four solar-powered houses were built at MIT, more than anywhere else. Telkes was involved with two of them. In addition, she designed an experimental solar heating system for the home of a friend in Dover, Massachusetts. All were based on "passive" technology. That is, storing solar energy in water or chemical salts, which could then be drawn upon to heat the house. Ultimately, however, MIT concluded that solar heat collectors were not economical by comparison with fossil fuel–powered heating, and the research was discontinued.

In 1978 Telkes retired from UD, though as we shall see, she remained active as a consultant in the solar field. By then, Böer had fallen foul of a new dean at the university. The latter, claiming that it was not possible for the former to serve two masters—industry and academe—had ousted him from his own institute. Solar panels made from cadmium sulphide/copper sulphide proved unsuitable for outdoor applications. Exposed to sunlight, the material degraded. Within three years, the panels had lost 60 percent of their capacity; after five years, they were as good as dead.

Solar Energy Systems changed the focus of research to developing thin-film cells made of silicon, forming a joint venture with Motorola. Silicon was not Böer's field. He felt that a partnership between an oil company—his backers, Shell—and an electronics company was doomed to failure. Shell bought him out. Three years later the venture went bankrupt. Despite these setbacks, Böer was undaunted. He had always known that the transition to solar energy would not come about overnight. "It's like a snowball down a mountainside," he told me. "You don't know what it'll be like by the time it gets to the valley."

In 2009, aged eighty-three, though long since retired from the university, Böer was still active in the field, doing research and publishing papers on cadmium sulphide solar cells. He derived considerable satisfaction from the fact that it was a layer of his pet material that, he claimed, boosted the efficiency of the cells made by a company called First Solar, which had recently become the world's largest manufacturer of solar cells.

Solar One was built as a laboratory, not as a home to live in. The same was also true of the experimental MIT houses. The first actual homes to be powered by PV were "wilderness houses." These were typically weekend homes and hunting lodges located in areas beyond the reach of the electric grid, where bottled gas provided the heating. Credit for designing the first homes that were powered by photovoltaics and connected to the grid goes to Steven Strong.

Strong was born in 1951 to an artist mother and an electrician father. By the age of seven he had built himself an electrified tree house. When the first oil crisis stuck in 1973, Strong was fresh out of engineering school, working on the Trans-Alaska pipeline. How wasteful, he thought, to invest all this money in infrastructure for transporting oil from the ends of the earth. Especially since oil was a resource that would last, at best, a few decades. There had to be better ways to apply his budding skills.

Strong discovered photovoltaics in one of the technology's first terrestrial applications, powering microwave telecom relay stations across the African desert in Cameroon. "I was blown away," Strong told *Time*, "Nothing was wasted. The energy was entirely renewable." Strong returned to his native Massachusetts, took some architecture courses, bought himself a drafting table, and set up shop as Solar Design Associates in an unused bedroom of his mother's house. The young idealist's goal was to create energy-conscious buildings powered by renewables. He began by designing solar-heated homes. Then, in 1978, with the first photovoltaic panels from the likes of Solarex becoming commercially available, he got his first big commission. Granite Place was a 270-unit midrise apartment complex in Quincy, Massachusetts. In addition to a large solar thermal array, Strong persuaded the owners to let him install a 5-kilowatt system on its roof. It was the world's first significant residential solar system. But it was an add-on, simply a bunch of solar panels lined up in rows on a flat roof.

The following year, his one-man firm won a contract from the Department of Energy to design a house in which the PV system would be integrated into the building. The Carlisle House as it was called featured passive solar heating—designed with the help of Maria Telkes—plus 126 solar electric panels capable of generating a whopping 7.3 kilowatts mounted on its southern-facing roof. More accurately, the PV panels *were* the roof. Strong would later confess that "the house looks like a large solar array behind which a living space has been organized." But it was a harbinger for future solar roofs in which, because it is completely integrated, the PV is no longer seen as such. In effect, it is invisible.

The Carlisle House was designed to draw utility power from the grid when necessary. Conversely, when the solar cells were turning out more power than the house could use, the excess power would be fed back to the utility. A small meter mounted on the wall of the dining room told the story in kilowatts. When utility power was drawn it ran forward. But when the PV was pumping out excess power, it ran backward. "The financial advantages of this house can best be explained at the end of the month or a whole year, when the energy bills and credits are tallied," Strong explained in his 1987 book, *The Solar Electric House*. "For the people who live in the Carlisle House, a reasonable program of energy conversion could easily bring the tally to an even draw—and maybe even put them into the black." In a grid-connected system, the utility and the homeowner would become partners of sorts.

Strong had obviously been thinking hard about the relationship between the two. With Granite Place he had been so preoccupied with getting the solar system installed that he had forgotten to inform Boston Edison, the local utility, about his plan to feed excess wattage into its distribution network. Strong mentioned his concern to the building's co-owner, a developer of Irish descent named Peter O'Connell. The latter did not hesitate. He asked Strong whether the solar system was ready to turn on. On being informed that it was, O'Connell simply threw the switch. Nothing went bang, everything worked as planned. In this world, O'Connell advised his young architect, it is always easier to obtain forgiveness than permission. But with the PV panels eminently visible on the building's roof, Strong felt obliged to ask the utility for guidance on interconnection to their grid. The power company responded with a thick binder that dealt with megawatt-scale installations. Intimidated, Strong asked O'Connell what he should do. "Ignore it," the latter responded blithely.

In June 1979 Strong had been at the White House attending the dedication by Jimmy Carter of the solar water heater on the roof of the West Wing. In return O'Connell invited Carter to attend the grand opening of Granite Place that September. Once the president had accepted, the developer also invited various local dignities including the governor, the state energy secretary, and senior executives from Boston Edison. But Carter had to cancel at the last minute, sending Denis Hayes as a replacement. In his speech, the director of the Solar Energy Research Institute conferred his blessing on the utility for allowing power from the building's PV panels to be fed into its grid. The state energy secretary said essentially the same thing. When the utility executives' turn to speak came, they had little choice but to praise the project, too. Interconnection was, for the moment at least, no longer an issue.

The happy idea of partnership, of having a single meter that ran backward as well as forward, and of being able to receive checks instead of electricity bills at the end of the month, was not one that utilities would long endorse. Meantime, power companies were delighted to bask in the positive publicity that flowed from being seen supportive of renewable energy. This was especially welcome at a time when so much bad publicity was associated with the shutting down of malfunctioning nuclear plants like Boston Edison's Pilgrim power station on Cape Cod Bay. In 1983, the utility commissioned Strong to build a solar-powered energy-efficient house in Brookline, Massachusetts. Impact 2000, as the house was dubbed, subsequently become the subject of a series on public television, a wonderful PR coup for the power company.

For the most part, however, utilities were not so accommodating. They thought they were doing solar homeowners a favor by allowing them to hook up to their precious grid, because intermittent, hence unreliable, electricity from renewable sources like solar and wind was to them low quality. "Junk power" was the disparaging term some utilities used. "According to the scheme usually advanced," wrote Tracy Kidder in the June 1980 issue of *Atlantic Monthly*, home owners would "sell their excess electricity at half price to the power company during sunny days (*to sweeten the deal for the utilities*), and buy electricity from the power company at full price when the sun wasn't shining" [emphasis added].

Buy low, sell high—that was much more to the utilities' liking than even-steven, one-for-one, a scheme initially known as "parity pricing." This would become clear from the world's first solar subdivision, construction of which began in Glendale, Arizona, in 1985.

John Long had originally wanted to build an entire hundred-dwelling solar-powered town in the sun-drenched desert west of Phoenix. But the housing market collapsed and, with federal tax credits for solar about to expire, a hurriedly cobbled-together twenty-two-home subdivision—named (what else?) Solar One—was the best the builder could manage.

Long was known as an innovator, a rarity in the highly conservative building industry. He ran a fully equipped laboratory to try out new ideas, like mass-production techniques, that would lead to improving the quality and affordability of the houses he built. His company, John F. Long Homes, was said to have been the first builder in the US to switch from ceramic to plastic piping. In particular Long's passion was frugal use of energy and water.

Born in 1920 in the "Valley of the Sun," as Phoenix is sometimes known, the first son of German immigrant parents, Long spent his youth milking cows and doing chores on the family farm. Such hard work, he believed, had conditioned him for the real world. In the early fifties, when Long started his business, Phoenix had less than 100,000 inhabitants. By the mideighties the city's population had ballooned to almost a million. In the postwar boom Long built small homes that veterans could afford. His firm would ultimately be responsible for constructing tens of thousands of homes.

Long's involvement with PV began in 1980, when he built an experimental solar model home as part of a Department of Energy demonstration program. The roof of the tract house was shingled with a 6.6-kilowatt system designed by ARCO Solar. The house produced about twice as much electricity as it consumed. For the Solar One subdivision, Long's company spent a whopping $1.6 million on 2,600 solar panels that were capable of delivering almost 200 kilowatts. Unfortunately, judging that homebuyers would not appreciate the look of roof-mounted solar systems, the builder elected to install the PV on the ground on two vacant lots at the entrance to the development. Not only was this a waste of land, it also left the panels vulnerable to repeated vandalization by rock-throwing kids.

Long had to fight Salt River Project, one of Arizona's largest utilities, for permission to connect the PV array to its grid. He promised buyers of Solar One homes free electricity. In fact, what transpired was better than that. Some residents actually received payments from the power company for the excess electricity they generated. In 1990, Long sold the PV array to the Solar One homeowners association for a nominal hundred dollars. With the residents' politically powerful protector no longer in the picture, the utility pounced. The standard meter ran forward when the residents were pulling power from the grid and backward when the PV produced excess power. Salt River Project replaced it with two meters, one for measuring power purchased, the other power sold. The rates could be adjusted according to the time of day, for peak and off-peak periods. In theory, this was fair enough. In practice, the problem was that the utility chose to rig the rates in its favor.

In 1991, during sweltering hot Phoenix summer afternoons, when in theory the utility should have been grateful for every extra ounce of juice, Salt River Project paid owners just 2.4 cents a kilowatt-hour. It charged them 9.5 cents, a differential of almost four to one. There were also disputes between homeowners and the utility as to how much electricity their PV was actually producing. Though the inverters had teething troubles, the solar array itself operated reliably. "The panels work fine," resident Rosemarie Williams complained, "the problem is getting the Salt River Project to stop screwing us."

"The problems of the Solar One homeowners association show the lengths to which a utility will go to sabotage a project it doesn't understand," concluded authors Daniel Berman and John O'Connor in their 1996 book, *Who Owns the Sun?* But perhaps the power company understood the potential threat to its business all too well.

As federal funding for solar research and technology dried up during the Reagan and Bush administrations, so too did interest at the state level. Even in Arizona, which enjoys more sunny days than other any state, solar was by the early nineties seemingly in terminal decline. In 1985, the last year of the federal tax credits, there had been some 700 solar-related businesses active in Arizona. Together they employed around 4,000 workers. By 1992 the Grand Canyon State's solar industry had withered to just 150 full-time employees. The state's Solar Energy Commission had been created in 1975 by its forward-looking governor Raul Castro. In its heyday, the commission had underwritten millions of dollars for solar R&D. In 1993 it laid off its last employee.

"In an old school a block from the State Capitol sit 62 electric solar panels donated to the state Department of Commerce's Energy Office [in 1987] by Phoenix developer and solar-energy advocate John F. Long," wrote John Dougherty in the June 30, 1993, edition of the weekly *Phoenix New Times*. Though some of the panels were cracked and shattered, many only needed to be dusted off. "That's one of the beauties of solar electric panels," Dougherty continued, "they need very little maintenance and can last just about forever."

"For several years Energy Office employee John Beimfohr tried to convince his superiors to take the panels outside and put them under the sun to demonstrate that solar energy is not a distant technology." But the decision makers showed no interest in turning the donated solar hardware into a high-profile educational tool. Beimfohr's job as solar communities planner for the state was eliminated in 1991. Funding was also terminated at the University of Arizona's Solar Energy Research Facility, "where more than 100 solar panels still sit on a [university] building rooftop, unused." The article noted that the former director of the facility, Don Osborn, had "moved on to the Sacramento Municipality Utility District to oversee the development of community-wide solar-powered homes, a program that has become a model."

We shall meet Don Osborn in chapter 9. First, however, we must travel to Switzerland to learn about Project Megawatt, the initiative that most influenced Sacramento's model program.

8

ALPHA MALE

*T*he Quiet Achiever was not the most gainly conveyance to behold as it trundled across the Australian outback. To be brutally frank, it resembled a mobile bathtub into which four bicycle wheels had somehow been inserted and on top of which a ping-pong table had been stuck. But though it looked peculiar, the Quiet Achiever had a serious purpose. Namely, to prove that a purely sun-powered vehicle could provide an alternative to gasoline-fueled cars. The ping-pong table was actually a brace of solar panels.

In just twenty days, over Christmas and New Year's 1982–1983, the Quiet Achiever crossed the Australian continent, traveling the 2,538 miles from Perth to Sydney at an average speed of 19 mph. The idea for the car had come from its co-driver, Hans Tholstrup, a thirty-six-year-old Danish adventurer. There was, Tholstrup admitted, no obvious benefit from his epic journey. But then, he added, there had been no obvious immediate benefit from the Wright brothers' first manned flight at Kitty Hawk, either. "We all now know what that led to," he said pointedly. If his venture motivated just one more idea in the development of solar power, then all his efforts would have been well worthwhile.

The early eighties were an era of solar feats, a return to the derring-do that marked the pioneering days of motorized travel in the last decade of the nineteenth century and the first decade of the twentieth. In July 1981 the Solar Challenger, an experimental, entirely solar-powered aircraft had flown 163 miles from Paris to Manston, an airport southeast of London. "People ask what the practical value is of flying a plane on solar power," the Solar Challenger's designer, Paul McCready, a fifty-five-year-old Californian physicist, told the *New York Times*. "I can't think of anything practical about it," he conceded. Nonetheless, like Tholstrup, McCready believed that his aircraft's successful cross-channel flight might somehow serve to advance the cause of solar power as a potential replacement for fossil fuels.

He was right. Among those deeply impressed by the flight of the Solar Challenger was Urs Muntwyler, a twenty-six-year-old Swiss electronic engineer. Muntwyler had just returned from Africa, where he had installed a PV system for a dispensary in Rwanda. "There I had the idea that, to become cheaper, solar energy must go into the mass market in industrialized countries," he recalled. Looking for a way to promote the use of solar energy in Switzerland, Muntwyler came up with the idea of staging a rally. A road race for solar-powered cars like the Quiet Achiever would pass through the centers of towns and villages, connecting with the public along the way.

The Tour de Sol, the world's first solar car race, was held over five days in late June 1985. "We told the Swiss police authorities that it was a kind of bicycle race," Muntwyler confessed. "In practice, it was more like Formula 1 on public streets, which is forbidden by the constitution in Switzerland. But by then, we couldn't be stopped." Fifty-eight vehicles registered for the 229-mile event. Most were little more than souped-up bicycles with solar panels attached. But the line-up also included some fancy-looking streamlined machines. They indicated that auto makers and local engineering schools were serious about solar racing.

The Tour de Sol was a huge success. Thousands of enthusiastic onlookers lined the back roads to cheer on the competitors as they swished quietly by. The race was won, easily, by the Solar Silver Arrow, a sleek fiberglass-over-steel racer. Emblazoned on its side was a legend that combined two names, one familiar, the other not: "Mercedes Benz Solar Powered by Alpha Real." For the German car company, the victory represented much wonderful publicity for very little outlay. For Alpha Real, a start-up founded by a daring Swiss entrepreneur, Markus Real, coverage in the local media gave the previously unknown company the profile it needed to go public. Four years later Real would generate even more headlines with his company's next adventure. Project Megawatt would put photovoltaic systems on the roofs of hundreds of houses and chalets all over Switzerland, by far the largest such application in the world thus far.

Cecile Warner is an American researcher who first met Markus Real at PV conferences during the mid-1980s, a time when solar R&D in the US was flagging. "Markus would pitch his ideas and talk about the projects he was working on, and there was just this sense of *awe*," she laughed, "like, Wow, he's so far ahead of us!" Real was and is far from your stereotypical Swiss conformist: he dances to his own drum. "You think of the Swiss as keeping the trains running on time, but Markus is usually four hours late for any-thing," Warner warned me. Then Real would show up flashing his friendly grin and, unable to resist his easy-going charm, you would instantly forgive him for his tardiness. "Markus is just the classic entrepreneur," Warner con-tinued. "He loves to start things, he has lots of great ideas, and the charisma that it takes to convince other people that he knows what he's up to." It was not just having the out-of-the-box ideas. Real also had the ability to seduce more linear thinkers into going along with him for the ride. Warner herself was among those who fell under Real's spell, taking a sabbatical year from the Solar Energy Research Institute in Colorado to go and work at Alpha Real in Zurich.

Markus Real was born in 1949, into a family of engineers. His father worked for the Swiss electrical engineering firm Brown Boveri, building steam turbines for coal-fired power stations around the world. Real spent his first four years in Australia, returning home to Switzerland unable to speak his native language. At university in Switzerland he too studied electrical engineering. But Real spent most of his student years goofing off, indulging his passion for sailing. In particular he loved racing eighteen-footers, a cate-gory of agile, extremely fast skiffs, thin shells rigged with huge sails. Taking months off from his studies was, Real recalled, "easy at that time, because it was 1968," the peak year for student riots on campuses across Europe. "Pro-fessors were glad if you didn't make too much noise." Somehow he managed to graduate, much to the astonishment of everybody including himself, with a good degree. Real tried working at Brown Boveri, his father's company, quickly concluding that a desk job was not for him. He formed an agency chartering sailboats, which enabled him to spend the summer sailing around the Greek islands.

The oil crisis of 1973 got Real thinking about alternative forms of energy, solar in particular. Fascinated, he read whatever he could about it. He build some primitive solar collectors that could produce hot water. In 1979, returning to Switzerland from a spell in South America, Real happened to hear that the Swiss Institute for Nuclear Research was looking for someone to work on solar thermal technology, using mirrors to concentrate light and boil water to create steam. At the institute Real came across photovoltaics for the first time. By comparison with solar thermal, which merely drove con-ventional turbines, "PV was something new, something special," Real said. "I was convinced that what we had here was a revolution in power generation." In those days, the nuclear institute was "swimming in money." To make sure that this lavish level of government funding was maintained, it was necessary that all money allocated be spent. At the end of each financial year, the procurement people would rush round asking researchers if they needed to order any new equipment. Real responded that, yes, he'd like a kilowatt's worth of photovoltaic, please. The panels were installed at the institute in 1980. His initial intention was to operate them in stand-alone mode. Then Real heard about Steven Strong's work with grid-integrated buildings in Massachusetts. He decided that hooking up the system to mains power was the way to go. "We hired an engineer to make an inverter and secretly built up this 1-kilowatt array on the roof, put in the inverter, didn't ask anybody [for permission], and started feeding electricity into the grid." For Real, as for others before him, observing the panels in operation was like falling in love. "I saw the beauty of photovoltaic—it just sits there and turns the meter. Every time I went to lunch I looked at the meter, and it kept turning, even at low insolation," that is, when the sun was behind the clouds.

In 1984, after five years in his cushy job at the institute, Real quit to set up Alpha Real. He took the name from his car, an Alfa Romeo, altering the spelling to reflect his love of Greece. That September Urs Muntwyler and two colleagues announced the inaugural Tour de Sol. Real leapt at the chance to participate. His first impulse was to partner with Alfa Romeo. On second thought, he decided that that would be confusing. In November Real wrote a proposal and sent it to Mercedes Benz. Eight short months later, the team had designed and built the Solar Silver Arrow and won the inaugural Tour de Sol.

Front-page newspaper coverage gave Real the profile to take his company public. In conservative Zürich, Switzerland's financial capital, home of the gnomes, this was unheard-of. "Normally a company has to have a fifty-year track record and an annual turnover of 50 million," Real said. "Then along comes Mickey Mouse with no track record and no revenues." Somehow, Real was able to convince not only investors but also customers like utilities, who gave him development contracts. Soon the fledgling firm had a staff of twenty-seven engineers. "It was easy," Real recalled of the early days. "You didn't have to be very good, you just had to do it, because nobody else was doing it."

In 1986 Real visited the US, where he saw the large, central station–type multi-megawatt PV arrays at Rancho Seco near Sacramento. He became convinced that this was not the way to go, that the best place for photovoltaics was on people's roofs. "It makes absolute sense," Real told writer John Perlin. "The roof is there. The roof is free. The electrical connections are there." In addition to which, in densely populated Switzerland, it would be hard to find enough land to build central power stations. The following year, 1987, Real initiated a research program to develop "a standardized, roof-mounted 3-kilowatt PV system that would stimulate high-volume demand throughout Switzerland."

"High volume" meant 333 identical systems, amounting to 1 megawatt of solar power in total. He called his plan Project Megawatt (333 × 3,000W = 1,000,000W). It was an opportune moment for such a proposal. The nuclear disaster at Chernobyl in April 1986 had shocked the Swiss, who depended on nuclear power for 40 percent of their electricity. In response to the threat of radioactive fallout from the crippled Ukrainian reactor, Swiss authorities promptly banned fishing on Lake Lugano. They recommended that pregnant women, breast-feeding mothers, and small children should avoid fresh milk and vegetables. Panic-buying of evaporated milk, bottled water, and frozen vegetables ensued. In a 1984 referendum, Swiss citizens had voted 55 percent against a nuclear-free future. Now, in the wake of Chernobyl, popular sentiment turned against atomic power. In a second referendum held in 1990, the Swiss voted 55 percent in favor of a ten-year moratorium on the construction of new nuclear power plants.

In entering and winning the Tour de Sol, Real had demonstrated a genius for PR. With Project Megawatt, he excelled. He had a leaflet printed and mailed out to potential customers. "Alpha Real Seeks 333 Power Station Owners," the leaflet said. "All you need is a roof exposed to the sun." On June 15, 1989, he held a press conference to announce the project. "All the newspapers came, even the most conservative ones." He remembered one cheeky journalist telling him, "Mr. Real, you are the only PR company that has an engineering department!" Blanket coverage did the trick: next day, Alpha Real's phone was ringing off the hook. Within a few weeks the company had taken orders for its 333 systems. Now came the hard part— making good on the promise. There were of course teething troubles: inverters would die, fuses would blow, circuit breakers would trip. But by the end of 1990, the systems were all installed; one by one the problems—especially the infant mortality of the inverters—were fixed.

Alpha Real's initial customers were mostly doctors, dentists, and lawyers—professionals who could afford to pay for the pricey equipment.

People like Martin Vosseler, for example, founder of the Swiss chapter of Physicians for the Prevention of Nuclear War. It was relatively easy to sell systems to such enlightened, well-heeled people, "the tip of the iceberg," Real called them. "They wanted to be pioneers, to have this thing." But Alpha Real also attracted many not-so-well-off customers, environmentally conscious souls who felt that they should install a solar system simply because it was the right thing to do.

Swiss law had long allowed small producers of hydroelectricity to export their excess output to the grid. There was nothing utilities could do to stop producers of solar electricity from doing the same. The question was, How much to pay for this new, unasked-for source of power? Most utilities grudgingly agreed to pay two or three cents per kilowatt-hour, much less than the wholesale rate. "Their attitude was, we don't know when the power is coming, why should we pay good money for that?" Real said. Meanwhile, the companies continued to charge their customers twelve or fifteen cents per kilowatt-hour for conventionally generated power. For many solar pioneers the issue of being paid never arose because they consumed all the power they generated themselves. Some, however, including three of Real's friends, a doctor and two lawyers from Lucerne, grumbled that the differential was unfair. These were not environmental activists, they were men of influence.

Switzerland has over a thousand, mostly small municipal utilities. Their board members are typically drawn from the communities they serve. In Lucerne, as it happened, Real's friends were acquainted with the director of the local power company. At a Rotary Club meeting or some such function, they buttonholed him and made their case. "The director thought, these are my friends, why should I quarrel with them over such a small amount?" He made the logical decision: Let the meter go forward and backward, let the buying and selling price be the same. The breakthrough had been made; a precedent set for parity pricing of solar energy.

The systems installed under Project Megawatt were, necessarily, retrofitted to existing roofs. But like Steven Strong before him, Real understood that if PV was to grow and multiply then, for economic as well as esthetic reasons, it would be necessary to develop solar systems that could routinely be integrated into roofs during construction. "Markus recognized early on that you really needed to design the building from the ground up, to involve architects in the design process," Cecile Warner said. "If you did that, you would have a better outcome." Alpha Real developed a prototype roof tile that incorporated solar cells. It could be interconnected to other identical tiles to form a photovoltaic roof. The company also began working to build PV into glass facades.

In order to manufacture these purpose-built components, and to scale up from 300 to 3,000 and 30,000 systems, it was necessary to bring in an experienced, well-capitalized industrial partner. In 1991, Alpha Real formed a joint venture with Glas Trösch, Switzerland's biggest glass manufacturer. The idea was that Real and his team of engineers would design the products while their much larger partner would produce, distribute, and sell them through its existing logistics and sales networks. Sounded good in theory, but in practice it turned out that the glassmaker's sales force found selling solar systems was much more complex, and took far longer, than selling glass. "So they lost steam in selling our product because it wasn't ready, it was too early."

Like many pioneers, Real was ahead of his time. Also, as with many early phase entrepreneurs, the idea of duplicating what he had already achieved, albeit on a larger scale, did not excite him. He had demonstrated that his idea was practicable. Now it was time to let others take over. "For the first two years, we had the lead, in Switzerland, in Europe," Real said. "But then my thoughts were drifting to new fields, I wanted to do new things, different stuff, and leave the field to Glas Trösch. Around this point, others also came along."

In 1994 Real quit the company he had founded and moved on to tackle new challenges. The initiative he had set in motion would fail to take hold on a national level. Partly this was because it was premature, but partly also because of Switzerland's radically decentralized political structure. In such a fragmented system it was relatively easy for individual utilities or municipalities or even cantons to introduce innovative energy incentives and policies, but very difficult to propagate them nationwide. In Switzerland such changes require not only the assent of parliament, but frequently also a public referendum. This would not be the case in other countries, notably Japan and Germany. Emulating Project Megawatt's example, their governments would, as we shall see, establish much larger solar rooftop programs of their own.

Media coverage inevitably brought Project Megawatt interest from overseas. The Japanese were particularly curious. They sent groups consisting of government bureaucrats, and representatives of industry associations and individual companies. At first the tiny Swiss outfit found the attention flattering. "We showed them everywhere, gave them coffee, everything," Real said. Soon, however, the visitors became a distraction. The engineers began to complain that they had work to do. "So we decided to charge them; every time we charged a little bit more, and still they came." Real lost count of how many Japanese groups had visited his company.

But Real's ideas about how PV should best be deployed also spread via less formal channels. At a mountain chalet owned by a close friend, he happened to meet an unusual-looking American who was vacationing in Switzerland after attending an international conference on photovoltaics in Italy. It seemed that this person was general manager of the municipal utility that served Sacramento, California's state capital. As it happened, Real had visited the 2-megawatt PV arrays this utility had recently erected. Now, the manager told him of his plan to greatly expand the PV plant, advancing the conventional argument that economies of scale would make the solar panels cheaper. Real told him he was wrong, that concentrating the panels in one location was not the way to go. Better to distribute them on the roofs of homeowners, as Project Megawatt had successfully demonstrated. Real was used to convincing people with his arguments. But though the manager listened carefully to everything the Swiss said, he seemed unmoved, saying very little. Real left feeling miffed: he had seldom encountered such an unsympathetic audience.

How wrong he was. The quiet American was David Freeman, President Carter's former energy advisor. As soon as he got back to Sacramento, Freeman summoned his staff. "If the Swiss can do a megawatt a year, so can we," he told them. A megawatt a year—that sounded like a lot of solar. Shortly thereafter, Real was bemused to get a call from his friend and fellow-sailor Dave Collier, the engineer who had been responsible for installing the photovoltaic arrays at Sacramento and who had showed him around. "What did you say to Dave Freeman?" Collier wanted to know.

9

GLORIOUS SMUD

*D*avid Freeman's nickname was "the utility repair man," because he had made a career out of fixing utilities that were broke. Seldom had there ever been a power company as broke—literally and figuratively—as the Sacramento Municipal Utility District, SMUD as it is prosaically known locally, when Freeman rode into town in June 1990 to fix it.

The nation's sixth-largest muni, SMUD was relatively big, serving over half a million customers. But SMUD was just a tenth the size of the Golden State's giant investor-owned utilities, Pacific Gas & Electric and Southern California Edison. Historically, SMUD had derived most of its power from hydro. But in 1963, with dams tapped out, to meet the growing demand for electricity the utility commissioned a nuclear plant. Three years later construction began at Rancho Seco—Spanish for "dry ranch"—a sparsely settled area of rolling hills twenty-five miles southeast of the city. It was completed in 1974, becoming the fiftieth nuclear plant built in the United States. Nuclear plants were supposed to be expensive to build but, once running, expected to produce electricity that was cheap. Rancho Seco had certainly been expensive to build, having cost $350 million, double the original estimate. But because accidents and management blunders kept knocking the plant out of commission, it was also ruinously costly to run.

Rancho Seco was a twin of Three Mile Island, a fact that made local residents nervous when in March 1979 the Pennsylvania plant suffered a partial core meltdown. Then radioactive iodine was discovered in milk supplied to a Sacramento dairy from cows that grazed in pastures next to the plant. SMUD management blamed fallout from Chinese nuclear bomb tests. In fact, Rancho Seco was leaking contaminated water like a sieve. Concerned local citizens formed an activist group, Sacramentans for Safe Energy (SAFE). The leakage was fixed and things settled down. In the early eighties Rancho Seco became famous as the site of the largest central-station PV plant in the world. A megawatt's worth of solar panels were installed in 1984, with a further megawatt added two years later. An iconic photograph showing the solar arrays in the foreground with the nuke's giant cooling towers in the background served to publicize SMUD's supposedly forward-looking energy policies.

Then in late December 1985 an accident that nearly caused its reactor to melt down forced Rancho Seco out of action. Management attempted to cover up the mess but it came to light after the Chernobyl disaster the following April. Though a public utility governed by an elected board of directors, SMUD had hitherto been run more like a private club. That changed in November 1986 with the election to the board of Ed Smeloff, a community organizer who had worked with Tom Hayden's Campaign for Economic Democracy. Smeloff had run on a platform of reassessing the economics of nuclear and investing in the new technology of renewables. It took the new director just nine months to conclude that closing down Rancho Seco for good made sense. Meanwhile, the hapless plant continued to burn money, operating at just 40 percent of capacity. Every day Rancho Seco was idle, the utility had to purchase $262,000 worth of power from outside sources.

Local media led by the *Sacramento Bee* were scathing in their coverage. "The never-ending series of mishaps are beginning to look like a very high-budget Marx Brothers film, with Harpo in charge of warning the city should there be an emergency," sniped a local TV news channel. Between 1985 and 1988 electricity rates almost doubled. In 1989 citizens were outraged to discover that SMUD management had ordered $220,000 worth of "distinctive attire"—blazers, neckties, and so on—to boost its staff's sagging morale. A further $130,000 had been budgeted for dry cleaning. Irate callers jammed talkback radio switchboards.

In 1988 SAFE activists gathered enough signatures to put the closure of Rancho Seco on a ballot. The nuclear industry responded vigorously, pouring money into a massive campaign—"Give the Ranch a Chance"—to save the plant. The initiative was narrowly defeated. But as the utility lurched from one lengthy outage to the next, a second ballot was held. In June 1989 Sacramento became the first community to close a nuclear reactor by public vote. With SMUD more or less in free-fall, on the brink of financial collapse, one option was to ask PG&E to take over. But rather than going cap-in-hand to San Francisco, Smeloff seized the opportunity to rebuild the organization. To effect the transition, to rescue the utility's finances and its image, a dynamic leader was required.

After a nationwide search the board finally settled on David Freeman to be SMUD's new general manager. On New Year's Day 1990 Smeloff called to offer him the job. When Freeman arrived in the state capital six months later, he soon found out how poisonous relations between the utility's executives and its customers had become. "A bunch of crooks," was how one local resident described his new employer to him. A prospective neighbor hissed at Freeman when he told her where he was going to be working.

With his trademark cowboy hat, Tony Lama boots, glow-in-the-dark white sideburns, and Southern drawl, David Freeman looked and sounded like a throwback from the Wild West. In fact Freeman had grown up poor in Chattanooga, Tennessee. His father, an immigrant from Lithuania, had been an umbrella repairman. Like that other high-profile Jewish cowboy, Kinky Friedman, Freeman had honed out of adversity a razor-sharp wit. His accessibility and mastery of the sound bite made Freeman an immediate hit with the Sacramento media. He soon won the approval of the citizens, among other things, by auctioning off the utility's "distinctive attire" on local television.

Born in 1926, a civil engineer turned lawyer, in the early sixties Freeman had gone to Washington intent on prosecuting civil rights cases for the Kennedy administration, but his accent disqualified him—Southern juries will think you're a traitor, he was told. Energy was his second choice. Way back in 1967 under President Johnson, Freeman had been the first person in the US government to be given responsibility for energy policy. As we saw in chapter 1, *A Time to Choose*, the Ford Foundation report he authored, became the basis for Jimmy Carter's energy policy. Freeman had hoped to be appointed secretary of the newly formed Department of Energy, but he had been blackballed, he believed, by the oil lobby. As a consolation prize Carter named him head of the Tennessee Valley Authority, the largest utility in the US. For the rest of his long career, Freeman would work in the public sector, an indefatigable servant of the people.

The TVA had already built three large nuclear plants, and had fourteen more under construction. Freeman immediately acted to stop construction of eight of these units, arguing that their skyrocketing costs would trigger rate increases for customers and bankrupt the utility. His fundamental insight was that adding capacity willy-nilly was too expensive. Far better to reduce the demand for electricity through efficiency measures. At SMUD his first steps were to freeze rates and to set a goal of zero growth in the electricity supply. To this end he instituted a series of innovative programs. One was paying people to trade in their old power-hog refrigerators for frugal new ones. Another was a plan to plant half a million shade trees near people's homes. These would provide relief from Sacramento's scorching summer sun and thereby reduce air conditioner usage. By late 1992, 37,000 refrigerators had been exchanged and 55,000 trees planted. Soon, instead of hissing at him, people would stop Freeman on the street and thank him for cutting their electricity rates. In fact, the utility had not cut the rates, just reduced people's usage so that their bills became lower. "You do the right thing," Freeman drawled, "and the Lord, she'll reward you."

"He's an amazing guy," marveled Dave Collier, an engineer who had worked for SMUD since graduation. "Before I met him I really had my doubts as to exactly how much of a difference one guy could make. But Dave Freeman came along, this little guy, kind of soft-spoken, he managed to turn the utility around almost overnight. . . . He was absolutely worth his weight in gold."

Energy efficiency was one part of Freeman's vision. The other part was reducing reliance on fossil fuels. He set an ambitious goal of sourcing three-quarters of electricity from renewables (including hydro) by 2000. Under his leadership SMUD would become, by far, the nation's largest solar photovoltaic utility. Though SMUD had been one of the first to implement large-scale central-station solar power, it had been a long time since the utility had actually done anything with PV. To lead the push into renewables, Freeman brought in an expert in the field. When he was contacted by SMUD in 1991 Don Osborn was heading up the solar energy research facility at the University of Arizona. "My first reaction was, Why would I want to do renewables through a utility?" Osborn recalled. "It didn't seem like a favorable fit, because utilities were not at all known for being serious about renewables." Then the SMUD folks told him about the energy efficiency programs that David Freeman had instituted and how the general manager was intent on doing the same with renewables. They flew Osborn over for an interview. He was hired with a free hand to establish SMUD's solar program.

Osborn had been concerned that the utility would not take renewables seriously. Now he found that his biggest problem was developing a program that was ambitious enough to meet Freeman's expectations. "That was a problem which was wonderful to have—it was like you were trying to push open a door that had been stuck, then all of a sudden someone pulls from the other side and it swings wide open," Osborn told me. "Dave used to yell at me, More! Sooner! Faster!" The strategy they adopted was to gain technical and commercial experience with distributed generation by deploying PV across a range of applications. SMUD would verify what Carl Weinberg's group had recently demonstrated at PG&E, namely, that distributed generation was actually good for the grid. The baton had been passed on.

By far the biggest application was scattering small-scale grid-connected solar systems across homeowners' rooftops. Unlike Project Megawatt, where individuals would pay for the panels, under SMUD's PV Pioneer program the utility would own the solar systems and the electricity they produced. The role of the homeowners was simply to lend the power company their roofs, and to pay SMUD a small monthly surcharge—a four-dollar "green fee"—for the privilege. Despite this impost and the lack of any financial incentive, the program was soon oversubscribed—more customers volunteered to help SMUD do the right thing by the environment than there were panels to go round. Homeowners were proud of their rooftop systems, seeing them as a highly visible expression of their environmental commitment. Between 1993 and 1999 the program installed a hundred systems a year, a total of six hundred.

In 1999 the program entered a second phase in which customers were allowed to purchase their own systems and sell their excess power to the utility. A key feature of this program was that it reduced what Osborn called the "hassle factor," the paperwork and bureaucratic hoops through which utilities would force customers who wanted to install PV to jump. "We made it very easy for a customer to go solar," he said. "It was not something they had to study or learn very much about, they didn't have to get deeply involved with the implementation. Once they made the decision to do this, it was basically taken care of for them."

In addition to the residential applications, Osborn, Collier, and their group at SMUD also blazed new trails in the implementation of PV. "We have serial number one of everything," Collier laughed. One ingenious application was particularly noteworthy. On steel poles above a section of the huge parking lot at Cal Expo, where each year the state fair is held, they installed the world's first large-scale (500 kilowatt) car port. This had the dual advantage of providing shade for parked cars as well as power for the utility. It was noticeable during events held at Cal Expo that this section of the parking lot filled up first.

By 2001 Osborn's group at SMUD had installed over a thousand PV systems, totaling nearly 10 megawatts' worth of solar electricity. This was a serious amount, representing over half the grid-connected PV in all of the United States at that time. "For many years, we were the name of the game," said Collier. The orders that SMUD placed were some of the largest the solar industry had seen. Innovative companies like Berkeley-based PowerLight received their first contracts from the utility. The scale of the demand from SMUD changed the nature of the industry, Osborn thought, taking it a step closer to the long-cherished goal of mass production.

David Freeman would remain at SMUD for three and a half years. "During that time his energy level never flagged," Ed Smeloff wrote. "He was always looking for the next new thing." Freeman left SMUD in 1994, moving on to another utility that needed repairing. Though Freeman had worked wonders during his time at Sacramento, he had not managed to reach into the guts of the organization to institutionalize his changes. Once he had gone, the old bureaucratic utility mind-set reasserted itself. Energy efficiency and solar power came to be seen, at best, as icing on the cake—just good PR—and, at worst, as a distraction from the main business. In particular, utility economists could not get their heads around the notion that distributed solar power was intrinsically more valuable than central station power.

They did not acknowledge the obvious, that PV was a clean, renewable resource that did not pollute the environment. Nor that its ability to be installed anywhere—on a rooftop, a carport, or a substation—in any quantity—from watts to megawatts—made PV much more flexible in responding to requirements for added capacity. Nor that, by being located at or near the point of use, solar electricity did not incur losses during transmission from a remote central station. Nor that PV produced power at precisely the time when it was most valuable, during the spike that occurred during the afternoon in the hottest days of the summer. (In Sacramento, the difference in demand for electricity between a 95 degree day and a 105 degree day could be as much as 300 megawatts, or about 15 percent of SMUD's peak load.)

These were just some of the multiple advantages—"stacked benefits," in the jargon of solar electricity. But getting utility economists to acknowledge stacked benefits was well-nigh impossible. Their analysis was still based on old Soviet-style command-economy planning. To them a kilowatt-hour was a kilowatt-hour, regardless of whether it was generated by a central station or a rooftop system. Freeman used to scoff that, if you asked utility economists to evaluate the world on a net present value basis—a standard measure used to appraise long-term projects—they would conclude it was not worth saving.

During the late nineties SMUD underwent the kind of wrenching management shakeup that large organizations periodically inflict on themselves. Osborn and Collier had to fight constant battles with skeptical midlevel managers to keep the solar program going. Eventually, in 2002, worn down by the in-house wrangling, they left to form their own company, Spectrum Energy, installing large-scale PV systems mainly in California.

SMUD's pioneering solar program had demonstrated what a utility could do. For nearly a decade Osborn had been the pied piper of photovoltaics. Among other things he helped put together a consortium of utilities that was active during the early nineties spreading the good word about PV around the country. For a while it seemed that utilities in other territories might attempt to replicate SMUD's achievements. By the late nineties, however, "all the momentum we created at SMUD was gone," said Smeloff, who himself resigned as a director in 1997. "The air just came out of the ball." Electricity generators in California were distracted as they struggled to deal with the turmoil that deregulation brought. Money was no longer available for experimental programs. Anything that added cost was ruled out by utilities fearful of losing customers to a third party. It was time to retrench, to go back to basics.

But in pockets at SMUD the solar spark still burned bright. One of Don Osborn's last initiatives at the utility had been to set up Solar Advantage, a demonstration project whose aim was to push the incorporation of PV into new homes built in the district. This was based on a recognition that fully half the growth in demand for electricity in California was driven by new homes. Also, by an awareness that, from an esthetic point of view, it was much better to integrate PV during construction than to retrofit a solar system as an obtrusive add-on to an existing roof. In 2001 SMUD began working with seven local builders, sourcing the PV, coordinating its delivery, installation, and marketing in around forty new and model homes.

Solar Advantage was a success. It showed builders that they could use solar to differentiate their houses in the marketplace and customers that, in an energy-efficient PV-powered home, they could save up to 60 percent on their electricity bills. But builders are a conservative bunch: they do not like to be told what to do. It was thus a masterstroke to hire one of their own to transition the new house initiative from demonstration to commercialization, from model homes to whole subdivisions.

Before joining SMUD Wade Hughes had owned his own construction business. He knew from personal experience the nuts and bolts of putting houses together. Hughes could talk to builders in their own language. He persuaded them that incorporating solar was not so much about meeting a state government mandate. "'Mandate' for our builders is kind of what you would call a four-letter word," Hughes said. Rather, it was the right way to build homes for twenty-first-century owners. SMUD's Solar Smart homes initiative began in January 2007. Builder partners were paid around $7,000 for each solar system they installed. But the utility's involvement went beyond handing out incentives. SMUD helped market Solar Smart homes via broadcast and Web-based advertising. It sent out flyers to its 560,000 residential customers, inviting them to tour a Solar Smart home.

The utility also published comparisons. "For example, the average Solar Smart bill in April and May of 2008 was $24 for both those months, whereas the average non–Solar Smart alternative was around $85," Hughes said. "Those are real numbers that people can wrap their heads around and understand the benefit of living in Solar Smart homes."

Homebuyers have the opportunity to pay via their mortgage the $10,000 or $12,000 it costs the builder to integrate the solar system. That increases mortgage payments by $50 or $60, but because of the savings in utility bills, buyers actually end up paying less. "So by living in this house you'll not only save $10 a month, you'll also have a home that is more comfortable and environmentally friendly," Hughes said. "It really is as close to a no-brainer as you can get."

By the end of 2008 contracts had been signed with around a dozen builders, including local firms and big national outfits like Miami-based Lennar, for the construction of over 4,100 Solar Smart homes. Representing over 30 percent of all new homes in the utility's service area, this exceeded initial expectations. Solar Smart homes were spending half as long on the market as a standard home, a boon to builders in a soft housing market. "We can really envision a time where it would be the exception for builders not to build a Solar Smart home," Hughes said. "The case is that compelling."

Out on building sites every day working to get solar onto homeowners' roofs had turned Hughes into "one of the zealots," as a colleague affectionately described him. Hughes was able to persuade builders, he reckoned, because he himself believed passionately in what he was doing. "It's a funny thing that happens to a person when they feel like they're doing something larger than themselves, and that they have a purpose that goes beyond themselves," he said. "I suspect that's a big part of what's happened to me: at some point along the way I became convinced that this is in fact a good thing to do, not just for me, or for SMUD, or for any builder, but for the world."

For California in 2009, four thousand plus solar-equipped houses was still a lot, even for a state that mustered over half of all PV installations in the United States. But though it was still leading the way SMUD was only one, relatively small utility in a single state. Each year, the US replaces around 2 percent of its housing stock, about 2 million homes give or take. There was still a very long way to go.

Meanwhile, other countries were leaving the US in their dust. During the same period, 1991 to 2009, large-scale government programs in Japan and Germany were deploying solar systems on a massive scale, involving tens and even hundreds of thousands of roofs. Before investigating these programs and their very different origins, first let us look at early attempts in the United States to develop what has turned out to be one of the most important elements in propagating solar. Namely, the policy mechanism used in the US to incentivize and reimburse customers for purchasing rooftop PV systems.

10

BEWARE THE CAMEL'S NOSE

*A*s a student researching the causes of the "wind-rush," the sudden surge of investment in wind energy in California during the early eighties, Tom Starrs had what he modestly described as "a minor epiphany." Namely, that the main driver for investment in renewable energies had virtually nothing to do with any recent advances in the technology. Rather, it was energy *policy* that played the most important role. Investment in wind had been rooted in the tax breaks that state and federal law had made available to developers. "That was a really eye-opening thing for me," Starrs said. "It made me realize that we can use policy, legislative and regulatory policy, as a vehicle for encouraging the development and use of different energy technologies."

Armed with this insight, in December 1992 Starrs invited himself to a meeting of the Photovoltaics for Utilities group (PV4U) in Stuart, a picturesque little harbor town, the self-styled "sailfish capital of the world," on Florida's Treasure Coast. The group had been formed in Tucson, Arizona, the previous year at the urging of among others SMUD's Don Osborn. Its goal was to develop a strategy for stimulating greater near-term use of photovoltaics in the utility market, hence drive down the cost. In attendance were around 120 representatives from electricity generators, state regulatory commissions, the Department of Energy, PV industry, and consumer advocacy groups. That first meeting had been unique—magical some said—because, for once, the stakeholders had met not as adversaries, but as potential collaborators. At a workgroup consisting of maybe twenty people at the Stuart meeting, Starrs stood up and introduced himself. He explained that he was a graduate student at the University of California at Berkeley looking for a meaty topic into which to sink his teeth. "I sat down, and this guy literally in front of me, who I didn't know, had never seen before, leaned back in his chair, and sort of whispered out of the corner of his mouth—'net metering!'"

Starrs had no idea what the stranger meant. Nor that this was Steven Strong, the architect who as we saw in chapter 7, had the previous decade designed and built the world's first grid-connected solar electric house and who by now probably had more experience with PV-powered houses than anyone else in the US. When the session was over Starrs got together with Strong. The latter explained what he meant by the term "net metering." The basic idea was simplicity itself. It exploits the fact that the rotating aluminum disk on the garden-variety electric meter used to track the number of kilowatts a household consumes in a given period—usually a month—has the ability to spin backward as well as forward. This ability meant that net metering of solar electricity could be introduced for residential customers with no change to the existing equipment. For once, as one wag put it, "Murphy was on our side."

Net metering is essentially an accounting mechanism based on parity pricing: any excess electricity generated by photovoltaics (or other form of generation) flows out to the grid. It is automatically credited to the customer at the same—that is, retail—rate as electricity flowing in from the grid. The meter spins backward, effectively erasing a portion of the total charged. "Net" simply means the final figure read out at the end of the billing period. Starrs was entranced by the concept. It seemed to him that net metering was the obvious way to simplify the often-byzantine process of connecting small systems to the grid. Also, to provide an answer to a complex question: What is the value of electricity generated and delivered within the distribution system? As Starrs knew from his work in the policy arena, it pays to keep things simple.

The son of a career diplomat in the US Foreign Service, Tom Starrs grew up chasing his dad around the world, to postings all over Europe and Latin America. He became interested in energy issues as an undergraduate at UC Santa Barbara. Starrs learned that over 90 percent of our energy comes from fossil fuels, which are nonrenewable sources. Sooner or later, they would run out. "So when I started learning about these renewable technologies, they just really captured my interest." By temperament and training Starrs had always seen himself as a strategic thinker and a problem solver. "I'd rather say yes to something than no, that's fundamental to my nature," he told me. Instead of railing against conventional technologies, Starrs preferred to be proactive in promoting the use of alternative approaches, to help the transition from coal and oil and gas to a sustainable energy future.

Success in the field would be measured by the number of megawatt hours of electricity renewables produced. Up until the early nineties, that meant mostly supporting incentives that promoted large-scale utility projects. "But there was originally for me, and remains, a certain appeal to the idea of people effectively generating their own electricity and meeting their own energy needs with an indigenous resource that falls on the roof of the building they own."

Starrs wrote the first-ever paper on net metering. In June 1994, he presented the concept at an American Solar Energy Society conference in San Jose, California. The paper caused quite a stir. "During the presentation, people were getting the gist of what I was saying, leaving the room, going out into the hallway, grabbing colleagues and dragging them back into the room. Although I started with an audience of 30 or 40 people, by the time I finished twenty minutes later, there were probably 80 or 100 people there. Afterward I was just barraged with questions and business cards. That's when it first hit me that, for whatever reason, this issue really resonated with people. In particular, the people at this conference who almost by definition were interested in creating better and more opportunities for developing and using solar power."

Fast-forward six months to another working group, this time at the First World Conference on Photovoltaic Energy Conversion at Waikoloa, on the Big Island of Hawaii, which 900 people attended. The purpose of this group was to propose net metering as a policy option. Specifically, to write a bill and see if they could get it submitted to the California state legislature. The question arose, Who should draft the bill? In retrospect Starrs would realize that as an under-employed graduate student with a law degree, he had been set up. "Because the heads of the other six or seven people around the table all immediately swiveled to look at me. It was clear that they'd either loaded the deck in advance or, to be more charitable, they knew that I had a legal background and therefore was most likely to be able to actually draft a piece of legislation."

With the assistance of two colleagues, Les Nelson, then the executive director of the California Solar Energy Industry Association (whose fifty-odd members were almost all installers of solar thermal water heaters, there being in the midnineties no firms in the state specializing exclusively in grid-connected PV systems) and Howard Wenger, then a consultant, formerly a member of Carl Weinberg's group at PG&E, Starrs set about writing a bill.

There was already, as we have seen, federal legislation that required utilities to buy electricity from other companies. But Jimmy Carter's Public Utilities Regulatory Policy Act was primarily concerned with *wholesale* power produced by large commercial facilities rather than *retail* power produced by small residential ones. PURPA's key concept was "avoided cost." In other words, buying bulk power from outside facilities was beneficial for utilities because it saved them from having to build expensive additional generating capacity. PURPA was just a framework. How the law was implemented was left up to individual states. Since 1978, when Congress passed the act, some fourteen states had attempted to come to grips with the rate-setting issues residential generation raised. The result, inevitably, was a mishmash of ad-hoc approaches. These tended to leave connection and remuneration up to the discretion of individual utilities. It was an invitation to have more-or-less arbitrary decisions made.

What was needed, Starrs believed, was "not to rely on the largesse of a utility that decides to be charitable and offer net metering voluntarily. Rather, it was legislation that would compel utilities to make this option available." At the same time he realized that strategically it was important to differentiate owners of 2-kilowatt systems whose main purpose was to meet their own electricity needs and had some leftover power to sell, from PURPA-compliant companies that had built megawatt plants specifically to sell power to the utilities.

Allowing the connection of "qualified facilities," as the power plants of nonutility generators were known, had been a bitter pill for utilities to swallow. The idea of having another intrusion on their traditional core competence, Starrs thought, would be anathema for them. Faced with the prospect of having to pay retail rates for electricity fed into the grid by owners of residential systems, the power companies would undoubtedly baulk. It smacked of losing share in a business where they were used to owning 100 percent of the market. Net metering thus represented a potential competitive threat. To preempt this problem, Starrs came up with a clever concept. Net metering would not constitute a *sale* of energy to a utility; rather, it would be an *exchange*. Utilities were used to swapping power, they did it with other utilities all the time. Under this scheme, however, there would be no cutting of checks for customers at the end of the month.

In addition to compelling utilities to accept net metering there was also another, equally important, set of issues to codify in the legislation.

Namely, so long as utilities exercised discretionary control over who could interconnect systems to their precious grid and under what circumstances, there would be little point in having net metering. If the owner of, say, a 50-megawatt wind farm wanted to interconnect with the grid, then he could afford to hire a specialist engineer to negotiate terms and conditions with the utility. But for an ordinary householder who wanted to hook up his 2-kilowatt rooftop PV system, a recalcitrant utility would be a deal-breaker. "In order for this to work," Starrs said, "we had to develop a set of rules that would simplify the process of interconnecting these systems in a way that more or less eliminated the utilities' project-specific discretion over interconnection."

The proposal was that, so long as the PV system's inverter—the device that converts continuous direct current output by the panels to alternating current in sync with the grid—met certain technical specifications, then the utility would be obliged to accept that inverter as the interface. The power company would not retain the ability to impose any additional requirements regarding interconnection. There was legal precedent for this argument. For decades AT&T had battled in the courts to maintain its monopoly on what equipment customers could plug into their wall socket. The phone company argued that interconnection of telephones made by anybody other than its manufacturing arm, Western Electric, would compromise the stability and reliability of its network. Starrs had studied the epic anti-trust telecoms lawsuit in grad school. He knew how eventually the regulator had ruled that any manufacturer willing to meet certain standard specifications could make and sell devices to the consumer. "I wanted to come as close to that model as we could on the electricity side," he told me.

Having drafted legislation, in order to get it passed by the California state legislature in Sacramento, it is advisable to recruit a powerful sponsor as the bill's "author." Few lawmakers were more powerful than senator Al Alquist. The legendary eighty-six-year-old New Deal Democrat was a long-time supporter of innovative energy initiatives. In 1974 Alquist had coauthored the law that created the California Energy Commission. Now, he shepherded the net metering bill through Senate Energy Committee hearings. The other key person was Catherine Lynch, an experienced lobbyist who bird-dogged the legislation from proposal to passage.

During the hearings it was fortunate, as Starr's colleague Les Nelson commented wryly, that "nobody had a clue what net metering was." But everyone could understand it because net metering was such a simple idea. As such it would be easy for utilities to administer. To further sweeten the pill, the advocates included limits in their legislation. Net metering would not have to be for all of their customers, just some of them: it would be limited to PV systems of 10 kilowatts or less. But the size of any individual system was not that important to a utility. What really mattered was the aggregate capacity of all systems, because that had economic implications. To allay this fear, the bill would cap the total output of net metered systems at just 0.1 percent of the peak demand of each utility.

"This was an incredibly effective approach because it undermined all the utilities' arguments," Starrs said. They had blustered that PV systems were unsafe. In the light of the vast amount of data accumulated by PG&E and latterly by SMUD, whose headquarters were little more than a stone's throw from the capitol building where the bill was debated, this was plain silly. But as Starrs well understood, right from the start, even when utilities raised objections on technical or safety grounds, the fundamental issues were always economic. "The risk was the long-term risk of having a significant portion of their customer base start generating its own electricity, and the implications that would have on the utilities' sale of energy and ultimately their ability to capture revenues from those customers."

The 0.1 percent figure sounded minuscule. It meant that California utilities together would have to accept up to 50 megawatts of PV. For the utilities, 50 megawatts out of 50 gigawatts was not going to be a big deal. But for the nascent market for grid-connected PV, which at that point was probably less than a couple of megawatts, it was potentially huge. "So my argument was, Come on—let's take this and run," Starrs said. "If we can get 50 megawatts' worth of this, by the time we get to 50 megawatts the entire industry will be different enough that, at the very least, we'll be in a better position to go back and reopen this. Then we can fight to increase 0.1 percent to 0.5 percent, for example. Which is, by the way, exactly what we did." (As of late 2009, the figure in PG&E territory was 3.5 percent.)

PG&E's long-time lobbyist in Sacramento was virulently against new technology, especially nonutility technology. At the hearings he opposed the net metering bill, arguing strenuously that the utility's customers had no business generating their own power. The lobbyist was accustomed to having the legislature vote the way the utilities wanted. On this occasion, however, he lost. The bill passed unanimously in an otherwise strongly bifurcated legislature.

Over the next eight years Starrs took his model on the road, fighting tire-lessly to get net metering laws onto the books of other states. Ultimately more than forty states would sign up to the idea. "The reason this work was so successful is that none of the utilities really took the issue that seriously," he explained. "It was never going to amount to anything, not going to cost them anything, so why not create a little bit of goodwill in the legislature by supporting it?" In time, however, utilities across the country would come to see net metering as the proverbial camel's nose under the tent. "They recognized that, if the technologies evolved over the years, this could end up really coming back to bite them from an economic perspective."

In his dealings with various state legislatures, Starrs marveled at how often solar initiatives won bipartisan support. Democrats, having environmentalists as part of their constituency, had historically been pro-renewables. But despite the longstanding hostility toward anything to do with renewable energy issues in the nation's capital, outside the Beltway, "we tended to gather quite a bit of Republican support. Net metering fit quite well with some of the Republican policies based around energy independence; those themes really resonated with those folks." In addition to which, people in general tended to be favorably disposed toward PV. "My personal experience has been that virtually everybody loves solar energy," Starrs said. "People's eyes light up at the mention of solar power, they just seem to like the idea of it, and look for reasons to support it."

Pat Redgate has been in the solar business longer than most. His Long Beach–based firm, Ameco Solar, began installing solar thermal heating systems back in 1974. However, until the passage of the net metering bill, Redgate said, "there was no such thing as PV for homeowners, unless you were some kind of really crazy person with a lot of money living in the desert. We used to say that 'PV' stood for 'potentially viable'—and that was us in the industry saying that." There was no chance photovoltaics would ever take off, Redgate thought, "unless there was some way you can sell the energy back, simply, to the utility.

"Of course the only reason we got net metering wasn't because of the herculean efforts of Tom Starrs and Les Nelson and all those other people—

although I will give them credit for getting it passed—the real reason was that the utilities in California were in a mood to deregulate. The environmentalists told them, Well, give us net metering (and a few other things) and we won't object to deregulation. And the utilities were so eager for that vision of unfettered income from deregulation that they said, OK, we'll give you net metering as long as it's restricted to 0.1 percent.

"So then that happened, it became law, then nobody paid any attention to it, nobody cared where the electricity came from, it was just traded and bartered and sold like a commodity." By the turn of the century, only about five hundred Californian homeowners had applied to take advantage of net metering. Then in mid-2000, as a result of predatory market manipulations by energy traders, Californians started to suffer from rolling blackouts. The following summer the power crisis reached its nadir. Consumers were being instructed to turn out their lights and do their laundry at off-peak times. Within a few weeks of the first blackout, Redgate started receiving inquiries from people wanting to buy grid-connected photovoltaic systems. At first he told them that PV didn't make economic sense, there were cheaper ways to ride out blackouts. But these individuals were insistent. "They wanted protection against blackouts and shortages, they didn't want to suffer. And solar was sexy." Finally, Redgate concluded that "it would just be better if I ignored what I considered to be logic and started selling people what they wanted. When you get right down to it, net metering is absolutely crucial," he said, "but it wasn't the thing that made the solar industry take off, it was higher prices and shortages."

In fact what *really* made the solar industry take off was large-scale government support programs. These, as we shall see in the next few chapters, were the motive force first in Japan, then in Germany. The Japanese relied on a model similar to the one that had driven initial investments in renewables in the US. This would ultimately prove fatally flawed. The Germans, by contrast, would create their own unique model. It would become by far the most successful way of promoting the spread of solar the world had ever seen.

11

LAND OF THE RISING SOLAR

*I*n Japan, the oil crises of the seventies had a different name. The Japanese called them "oil shocks" (*oiru shyokku*). With good reason because, as a country with almost no indigenous energy resources, Japan was uniquely vulnerable to embargoes slapped on fuel supplies. In 1973 the Japanese depended on imports for 99 percent of their oil. Imported oil provided 77 percent of their total energy. A new policy was clearly required.

In Japan formulation of policy has historically been the province not of the government, but of the elite central bureaucracy in Tokyo. Ever on the lookout for opportunities to extend their turf, the mandarins of the Ministry of International Trade & Industry (MITI) quickly donned their thinking caps. By early 1974 they had drawn up a plan dubbed the Sunshine Project. The program was designed to foster the development of alternative sources of energy. Its goal was to provide substantial amounts of electricity from renewables by the turn of the century. As its name suggests, the Sunshine Project emphasized solar energy. However, its priority for the first few years was not solar photovoltaic, which at the time was still in its infancy for terrestrial applications, but on the older technology of solar thermal.

Solar thermal works using mirrors to focus sunlight onto pipes containing a liquid. The heat thus generated produces steam, which drives a turbine just like in a conventional power station. Solar thermal, it turned out, was not an appropriate solution for Japan. The technology runs best in desert regions where the sun shines more or less continuously. But Japan's skies are often blanketed by clouds, especially during the rainy season and subsequent humid summer.

By 1981 it was clear that solar thermal was not going to work. In the interim PV had progressed from being a space-only curiosity to an industry that, while still tiny, was growing fast. That year, concerned that its budget for solar research might be cut, the ministry swiftly shifted its bets from solar thermal to solar electric. It also set up an agency, the New Energy Development Organization (NEDO), to coordinate research. Thus, at precisely the time when the Reagan administration was slashing US federal funding for solar research, the Japanese government was investing heavily in the development of PV. Solar would continue to enjoy large and stable R&D budgets throughout the eighties and nineties. Japan would be the only industrialized country to sustain commitment to renewable energy research even after oil became cheap again.

In addition to public research institutes, the Sunshine Project involved participation by private Japanese firms. The corporations were attracted not only by lavish government research grants but also by the national commitment to introduce a substantial amount of solar power by 2000. With Tokyo signaling that PV was going to be a pivotal technology, firms could look forward with confidence to future large markets.

During the eighties NEDO demonstration projects provided almost the only market for Japanese manufacturers of solar panels. Such projects included the 1-megawatt central station PV plant built in 1985 at Saijo, a small city on the backwater island of Shikoku. This remote site was chosen because it was one of the few places in over-crowded Japan that had enough space available to accommodate such a large-scale array.

Another improvised location was Rokko Island, an artificial island created from reclaimed land off the port city of Kobe. There, in 1986, Kansai Electric Power built a high-density subdivision consisting of 180 dummy homes, each furnished with its own 2-kilowatt solar system. Some of the "homes" were also equipped with electrical appliances—televisions, refrigerators, and air conditioners—which could be switched on and off by computer. They provided, as one bemused observer put it, "a lavish lifestyle for their phantom occupants." The project's goal was to examine the effect on the grid with multiple residential PV systems connected. The Rokko Island experiment allowed the technical issues of distributed generation to be explored in a systematic way under controlled conditions.

Like its counterparts in the US and Europe, the Japanese electric power industry had long contended, without much evidence, that connecting power sources not managed by the utilities themselves could cause the grid to become unstable. Since the provision of stable and reliable power was their core mission, utilities were reluctant to open their grids to small generators. The results of the field tests conducted at Rokko Island by the power industry's Central Research Institute conclusively demonstrated that the reliability and safety of the grid could be maintained even when multiple PV systems were connected.

Though large in scale, the Saijo and Rokko projects were one-offs. The domestic Japanese PV market remained a minuscule niche. As such it was not big enough to sustain the interest of heavy electrical equipment manufacturers like Hitachi and Toshiba, which, by the end of the decade, had withdrawn from the solar field. These conglomerates are based in Tokyo, where proximity to the central government enables them to enjoy its grace and favor.

During the eighties it seemed to Western critics that Japanese government ministries and private corporations were just two tentacles of the same nefarious entity, the so-called "Japan Inc." In this view the Japanese economic miracle had come about largely as a result of industrial policy formulated by MITI's farsighted bureaucrats. Firms colluded with each other to carry out the government's vision. This stereotype was a grotesque oversimplification. In particular, it ignored a vibrant sector made up of smaller, entrepreneurial firms. It was to this sector that the three largest Japanese manufacturers of PV belonged. All three are based in the Kansai region of western Japan, three hours by bullet train from the nation's capital. From there, they have long looked outward, to overseas markets. Two of them, Kyocera and Sanyo, had been since the midseventies committed to the development of solar technology, funding their research independent of government largesse. The third firm, Sharp, had an even longer history of photovoltaic R&D. Research on solar cells began at Sharp in 1959, just four years after their invention at Bell Laboratories. In 1961 the company announced a PV-powered transistor radio. Two years later Sharp began mass production of solar cells for use in powering navigation buoys and subsequently, on a larger scale, lighthouses.

The main focus of Japanese consumer electronics companies like Sharp was portable, handheld products. The transistor radio was the first of these, the electronic calculator the next. In 1976 Sharp introduced the world's first solar-powered calculator. In 1979, at its R&D center near the ancient capital of Nara, the company built a hybrid house that like Solar One was powered and heated by solar panels. Throughout this time, outlay on R&D of photovoltaics far exceeded any income. The reason Sharp was willing to continue losing money over several decades was the commitment to solar technology of its top management. In particular, of the company's visionary founder, Tokuji Hayakawa. "I believe the biggest issue for the future is the accumulation and storage of solar heat and light," Hayakawa wrote in his 1970 autobiography. "While all living things enjoy the blessings of the sun, we have to rely on electricity from power stations. With magnificent heat and light streaming down on us, we must think of ways of using those blessings. This is where solar cells come in. . . . For example, if we can install solar systems on roofs, homes could be self-sufficient in power." By 1970 Hayakawa was almost seventy-eight years old and had long since retired from the day-to-day management of his company. Though his vision of a solar-powered future was still influential on his successors, by the midnineties Sharp had fallen far behind its domestic rivals in PV production.

Although much younger than Hayakawa, Kazuo Inamori (b. 1932) was also an outsider who had built up his company, Kyocera, from scratch. Strong-willed and with philosophical leanings, Inamori was proud of the fact that he had never had to ask Japanese officialdom for help. Indeed, he was highly critical of interference from government bureaucracy. In articles and books he argued that, so far from serving the people, ministries were more interested in protecting their turf and safeguarding the jobs of their own. In consensus-loving Japan, these were provocative views. The mandarins would have their revenge.

Kyocera began development of solar cells in 1976. Product shipments commenced three years later. The company's first big order was for panels to power a microwave telecom relay station located in the Peruvian Andes. Kyocera's initial success had been based on manufacturing ceramic packages for silicon chips made by the likes of Intel. In the late eighties the company diversified aggressively into Japan's newly liberalized telecoms market. Interviewing Inamori in mid-1993, I asked him where he expected the company's growth to come from over the next ten years. Chip packaging, telecoms, and solar cells, he told me. This was prescient, given that the Japanese government subsidies program that would drive the growth of the solar market would not be announced until the following year. But Kyocera had been the supplier of solar panels to Markus Real's Project Megawatt. Inamori could see which way the solar winds were blowing.

The third major Japanese solar manufacturer was Sanyo. Like Sharp, Sanyo had applied solar cells to power calculators. Unlike Sharp, these cells were made not from crystalline silicon, but from a thin film of noncrystalline "amorphous" silicon that could be sprayed onto plastic or glass. (We shall take a closer look at thin films in part 5.) Though amorphous silicon used less than one-hundredth the amount of material, it was not nearly as efficient at converting sunlight into electricity. Happily, amorphous silicon thrived under fluorescent light. This made it ideal for use indoors, in offices and classrooms. Sanyo launched the first amorphous silicon–powered calculators and watches in 1980. The cells were developed by a team led by an unusually determined researcher named Yukinori Kuwano. During the late eighties I interviewed Kuwano several times. On each occasion he was bubbling with enthusiasm for some new application of Sanyo's amorphous silicon. Most memorable among them was a blanket made from flexible solar cells. It was undergoing trials to power lights at a field hospital run by a team of eye surgeons who were performing operations in remote parts of Nepal.

But the solar-powered calculator remained by far the dominant application, racking up annual sales in the hundreds of millions. In 1986 those sales catapulted Sanyo into first place among solar cell producers. That year consumer products like calculators and watches accounted for two-thirds of the Japanese PV market. During this period Sanyo also produced a traditional, wave-shaped Japanese roof tile made from transparent glass into which strips of amorphous silicon had been integrated. Before roofs made from such tiles could be connected to the grid, however, it would first be necessary to eliminate some bureaucratic red tape.

MITI regulations stipulated that any power system capable of generating more than a puny 30 watts had to be installed by a qualified senior electrical engineer. This was just one of many barriers supported by the politically powerful electric power industry that stood in the way of the large-scale deployment of residential rooftop PV systems.

Two factors, one internal, the other external, helped bring about change. Within the ministry itself the Office of Alternative Energy Policy lobbied hard to get regulatory barriers removed. The chief of the office made shrewd use of mass media like television and newspapers to drum up support for reform. He argued that renewable energy technologies like PV would never achieve their potential unless the procedures for grid connection were simplified.

Industrial leaders like Sanyo's Kuwano chimed in, complaining that archaic government rules were holding companies back. In June 1990, to head off rising criticism, MITI moved swiftly to simplify the regulations and issue new installation guidelines. At the same time the Japanese Federation of Electric Power Companies announced that, in line with the government's targets for the introduction of renewables, its members would voluntarily introduce a net metering program. Starting in 1992 they would pay the same rate for electricity they purchased as they charged their customers. The power companies took this step not because they were eager to promote the spread of renewables, but because they wanted to avoid being obliged to introduce net metering by government mandate. Being voluntary the arrangement could be terminated later, at a time of their choosing.

Kuwano would himself be the first to take advantage of the permissive new regulatory environment. In July 1992 he installed a 2-kilowatt net-metered PV system on the roof of his own home in Osaka. (It was, however, a conventional retrofitted crystalline system, not one made from roof tiles containing his precious amorphous silicon.) At international conferences the irrepressible Kuwano began proposing a grandiose scheme for a future global solar network. It would interconnect the grids of different countries with low-loss superconducting transmission cables. Such a network would make solar power available everywhere anytime, he explained, even after the sun went down.

The external factor prompting a change in energy policy was an awareness of recent European initiatives to put solar on roofs. We have already seen the intense Japanese interest in Markus Real's Project Megawatt in Switzerland. Though noteworthy, this was a private-sector-driven initiative. What really caught the attention of the bureaucrats at MITI was a solar proposal made by their counterparts at the German Ministry of Research & Technology. The 1,000 Roofs Program was announced in September 1990 (see chapter 13). In the Diet, members of parliament demanded that Japan should adopt a similar subsidy program. This was embarrassing for the ministry. When it came to formulating industrial policy, MITI had written the book. Now the Germans were beating the Japanese at their own game. In 1992 MITI dispatched a fact-finding mission to Bonn. Camera-toting officials visited German companies and installations, photographing and videotaping everything they saw.

In 1993 Walter Sandtner, the official responsible for setting up the German PV program, was invited to Tokyo to address a high-level meeting. It was attended by some eighty representatives of government, research institutes, and industry. "I spoke and they were *highly* interested," Sandtner said. "They had innumerable questions." Meantime, Kyocera and Sanyo sent their own representatives to Germany looking for potential sales. "They observed in a very very detailed manner what we were doing," he said. Kyocera's boss Inamori particularly impressed him. "He was a highly original man," Sandtner recalled.

In 1994 MITI launched the New Sunshine Project, a subsidy program that, Sandtner believed, was closely modeled on its German predecessor. Bonn had provided subsidies of up to 70 percent of the cost of residential PV systems. Tokyo would offer smaller but still generous subsidies of 50 percent. The ministry's plan originally called for equipping a modest 700 roofs. This was more than twice as many as Project Megawatt, but only a third of the final German total of 2,250. Even this number was highly controversial, however. There was no precedent for the government providing subsidies to owners of individual properties. Japan's powerful Ministry of Finance was reluctant to fund such a radical departure.

MITI persisted with its proposal. The ministry argued that, unlike the German initiative, which had merely been a technology demonstration program, their program would create a large *market* for PV. It would give solar cell manufacturers an incentive to invest in production facilities. Once established, mass production would rapidly bring down the price of PV systems, leading to more orders. At this time the ministry was under pressure to bring renewable technologies to market. By 1993 the Sunshine Project had been going for twenty years. It had consumed around $5 billion of Japanese taxpayers' money. There was precious little to show for all this investment in terms of commercial products.

In the event, as MITI had anticipated, Japan's three major manufacturers of PV rose to the challenge. Assured of government support they quickly invested in equipping lines for the mass production of solar panels. Firms began marketing 3-kilowatt residential systems for around $50,000. The question now was, Would there be enough well-heeled Japanese customers willing to pay $25,000 even though there was little chance of getting much return on their investment? Indeed there would. In its first year the program attracted over 1,000 applicants, far more than expected. In 1995 the program was expanded to 1,000 roofs: this time, well over 5,000 homeowners applied. In 1997, in the light of such a strong market response, MITI shifted gears. The ministry expanded its program to 10,000 roofs while reducing the subsidy to 33.3 percent.

Who were these consumers who were prepared to pay above the odds for solar electricity? As you would expect, and as surveys showed, they were mostly affluent individuals who wanted to contribute to protecting the environment. Some liked not having to pay electricity bills. The fact that the government was supporting the program was also important to them.

By the turn of the century the Japanese had installed over 50,000 rooftop systems, up from just one—on Kuwano's house—in 1992. Japan had shot to number one in PV, with a 40 percent share of a rapidly growing market. Kyocera had overtaken its rivals to become the world's leading manufacturer of solar panels. Of course there had been hiccups along the way. In early 2000 it became known that about 20 percent of the PV systems shipped by Sanyo between 1996 and 1998 were defective. (It wasn't that the panels didn't work, just that they generated slightly less electricity than advertised.) As punishment for perpetrating this fraud on unsuspecting Japanese consumers MITI suspended the company from participation in its program for three years. Blaming the mislabeling on a US subsidiary, Sanyo's president resigned. His successor was none other than Yukinori Kuwano. Under Kuwano's dynamic leadership, Sanyo would go on to develop some of the world's most efficient crystalline silicon solar cells.

Also in 2000 MITI revealed that, prior to participating in the rooftop program, Kyocera had misappropriated government research funds designated for the development of solar-powered cars. A red-faced company was forced to refund the money and issue a groveling apology. It too was banned from participation in the program for three years, clearing the way for Sharp to take over the top spot.

In its peak year, 2004, the MITI program installed systems on the roofs of 60,000 Japanese homes. By then the program's annual budget had climbed to $200 million from $17 million in the first year, while the level of subsidy had dropped to just 3 percent. In Japan residential systems represented well over 90 percent of solar installations, whereas in the US, for example, household roofs accounted for only about one-third of solar supplied to the grid. Thus far, it had a been a textbook case of market creation. System costs had dropped by about 30 percent while sales had soared. Japan was seemingly on course to achieve its goal of a self-sustaining, subsidy-free solar market. Then things started to go wrong for the ministry (which, since a reorganization in 2001, had changed its name to the Ministry for Economy, Trade & Industry, or METI).

In 2005, despite strong opposition from PV makers, METI phased out subsidies. After more than thirty years of continuous support, the Japanese government had finally pulled the plug. The time had come for the solar market to go it alone. The only remaining incentive was net metering. But as Japanese electric power companies contemplated changes to their rate structures in anticipation of market deregulation, it was not clear whether a voluntary concession like net metering would remain in effect. With subsidies and by implication government support eliminated, the price of PV systems rose. Domestic sales immediately began falling; by 2007, they were down by almost half. Demand dried up.

Though a total of 400,000 solar systems had been installed, that represented less than 1 percent of all Japanese households. Clearly, there was still immense potential for growth. But according to Takashi Tomita, the visionary engineer who over three decades had led Sharp to preeminence in the PV industry, the Japanese market for residential solar systems was close to saturation. Japan was rapidly running out of affluent, environmentally conscious customers who could afford to lose money by investing in solar. Meanwhile, Sharp was grappling with problems of its own. Anticipating continued rapid growth, the company had greatly expanded its production capacity. Unfortunately it had neglected to negotiate long-term delivery contracts for the crucial raw material, silicon. With the solar industry growing at between 30 and 40 percent a year, an acute shortage of silicon feedstock had developed. Those without signed contracts in place missed out. Sharp was unable to run its production lines at full capacity. In 2007, as punishment for his failure to procure sufficient silicon, Tomita was forced to step down as head of Sharp's solar division.

The following year the company was knocked off its top spot as the world's number one PV maker by Q-Cells, a German start-up. Nipping at Sharp's heels in third place was another aggressive newcomer, the Chinese firm Suntech Power. Meanwhile, Kyocera was also going backward, slipping to a distant fourth. Japan's overall share of the global solar market dropped, from 45 percent in 2005 to just 25 percent in 2007, barely ahead of China's 22 percent. Sharp's individual share plummeted from 22 percent in 2006 to 12 percent a year later. The reason for this rapid reversal of fortune is clear. In 2004 the German parliament passed a law that guaranteed installers of solar systems a profit on their investment. As a result the German PV market took off like a rocket. Between 2003 and 2007, installed capacity in Japan more than doubled. During the same period, however, capacity increased in German by a stunning *nine times*. By 2007 Germany had installed twice as much solar as Japan.

Other European countries, notably Spain, followed the German example. By 2007 Europe accounted for 70 percent of demand for solar, versus just 9 percent for Japan. Japanese companies were still responsible for almost 30 percent of solar panel shipments, but these were mostly intended for export. Humbled by this blow to national pride, METI hurriedly cobbled together a revised subsidy program. Introduced in early 2009, the new incentives offered customers a rebate of around 10 percent of system costs for systems up to 10 kilowatts. The aim was to attract 35,000 applications in the first three months. In fact, fewer than 22,000 homeowners applied for the subsidies. With the Japanese economy struggling, consumers were leery of making big investments. Especially since according to the government's own forecasts, the cost of PV systems was expected to drop by half in the next three to five years. The Japanese government set an ambitious target of increasing solar generation to 28 gigawatts by 2020, up from just over 2 gigawatts today.

For their part Japanese manufacturers like Sharp announced ambitious plans to scale up their production capacity. They would construct giant gigawatt-scale plants with the capacity to crank out a billion watts of PV a year. But their competitors around the world (especially in China) had equally grandiose plans. It seemed to analysts unlikely that Japan would ever regain its once-dominant manufacturing position in PV.

To ginger up the market METI talked of compelling power companies to buy excess power generated by residential PV systems at twice the retail rate. This would be paid for by placing a small surcharge on the rates paid by every consumer. After all, that was how the Germans paid for their solar electricity. Amid all this discussion of revitalization, there was initially no mention of Japan attempting to replicate Germany's solar "feed-in tariff." That was odd, considering that the feed-in tariff was widely recognized as the crucial policy mechanism responsible for Germany's astonishing success in deploying far more photovoltaic than any other country. How the feed-in tariff was conceived, developed, propagated, passed into law, and implemented is the subject of part 3.

Part 3: Hier Kommt Die Sonne

HOW CLOUDY GERMANY CAME FROM NOWHERE TO LEAD THE SOLAR RACE

12

LEAVING THE ARMY TO FIGHT

*L*unchtime on December 21, 1989, in Aachen, a small city (population 250,000) in Germany's far west, just across the border from Holland and Belgium. On a pint-sized plaza across from an ultra-modern, glass-fronted, four-story bookstore, just down the hill from the Kaiserdom, the oldest cathedral in northern Europe and last resting place of Charlemagne the great Frankish king-emperor, a band of dogged enthusiasts have set up a demonstration.

It consists of twelve rectangular solar panels oriented roughly in the direction of the sun (hard to tell where the sun actually *is* on this overcast, drizzly day). A cable snakes between the panels and various electric power tools are laid out on a trestle table. They include a drill and a handheld jigsaw, with which one of the enthusiasts is fashioning curved shapes from plywood boards.

The point is to show Christmas shoppers and passersby that even under the most inauspicious conditions imaginable on this, the shortest day of the year, when the daylight is poor and the weather miserable, solar power still works. "But people were not interested," the group's leader, Wolf von Fabeck, recalled. "No one stopped to watch." He buttonholed a smart-looking youngster and asked him whether he thought the demonstration interesting. "I'm not stupid," the youth retorted. "You've got a battery hidden in there somewhere." Others were similarly skeptical. "They didn't want to believe in solar energy," von Fabeck said, "even though they saw it with their own eyes."

Back in the late eighties, solar power—*photovoltaic* the Germans call it, laying the stress on the fourth syllable—fo-to-vol-TAH-ik—was still very much an exotic technology, virtually unknown except to a handful of tree-hugging romantics, long-haired engineers, and idealistic third-world development specialists. In all of Germany hardly half a dozen houses had solar panels on their roofs.

Fast-forward twenty years and *Wunderbar!*—a seemingly miraculous change has taken place. In the federal republic today more than half a million homes generate clean energy from rooftop solar systems, most of them purchased within the last few years. In 2009 alone Germany installed almost 2.5 giga (= billion) watts of PV, the equivalent of several large-scale power stations.

In less than a decade, from out of nowhere, a vibrant new German solar industry has sprung up. Consisting almost entirely of entrepreneurial start-ups, the industry has located most of its factories in the former East Germany, where jobs are hard otherwise to come by. Solar start-ups employ over 50,000 people—far more than the German coal industry—and they continue to grow, even during a global recession.

Somehow, Germany has managed to leapfrog rivals to become the world's leading supplier and user of solar technology, eclipsing Japan (the former front-runner) and leaving the US (with a market share of less than 10 percent) in its shadow. Indeed, the federal republic is home to more than half of *all* solar installations anywhere.

Installations come in all shapes and sizes. They range from small, kilowatt-size systems on top of individual houses to large, megawatt-size generators on the roofs of Lutheran churches ("protecting God's creation"), and of Bavarian pig barns (selling clean power is more profitable—and more predictable—than making bacon). There are even bigger installations, too. For example, solar farms consisting of serried rows of giant trackers in what were formerly fields. Also, "great walls" of photovoltaic running alongside the autobahn, where they double as noise-prevention barriers.

Canny entrepreneurs use satellite photographs to spot unoccupied south-facing roofs. They contact the owners, offering to rent the spaces so that they can install revenue-generating solar panels on them. Enthusiastic Germans will stick photovoltaics on anything that does not move. They joke that if a dog goes to sleep in the garden for half an hour, it will wake up with a solar panel on its back.

What is the origin of this unprecedented high-tech success story in what Donald Rumsfeld contemptuously referred to as "old" Europe? How has a country that is blanketed by clouds two-thirds of the year, that gets about as much sunshine as Anchorage, Alaska, shot ahead of the pack in switching to solar? The short answer is because of a simple, highly effective policy mechanism known as a "feed-in tariff."

In essence, the German-style feed-in tariff takes the uncertainty out of investing in solar. Homeowners, farmers, and small businesses become entrepreneurs, recouping their costs plus earning a profit by selling *all* the electricity that their panels generate, at a premium rate, guaranteed for twenty years. The feed-in tariff is the engine that has turbocharged the growth of the world's largest solar industry.

Though the solar feed-in tariff was first enacted in 2000 by the federal German government, the impetus and hands-on experience on which it was based came from grassroots activists working at the local level. Like any successful innovation, the feed-in tariff is claimed by many fathers. It is generally agreed, however, that its birthplace was Aachen; the person with the best claim to paternity, Wolf von Fabeck.

Wolf von Fabeck is not an entrepreneur nor does he fit—by any stretch of the imagination—the stereotype of the sandal-wearing, weird-beard greenie. He is in fact a soldier by profession from an army family that goes back many generations. His father, grandfather, and two of his brothers were all military men, as was his wife's father. Born in Potsdam, just south of Berlin, on May 9, 1935, von Fabeck is of the generation that came of age when Germany was recovering from the ravages of war. In 1956, aged twenty-three, von Fabeck joined the Bundeswehr. This was a very different organization from Hitler's Wehrmacht. The latter has been castigated for unquestioningly following orders. The guiding principal of the "citizens' army," as the Bundeswehr is sometimes known, is by contrast inner leadership (*Innere Fuhrung*). The modern German soldier owes his primary allegiance to justice and legality rather than the dictates of his superiors.

In fulfilling his post-military mission, von Fabeck has embodied a combination of traditional and modern soldierly virtues. A tenacious warrior challenging the complacent orthodoxy of Germany's monopolistic electric power industry, he has shown both self-control and inner leadership. Like any great general von Fabeck has also displayed the moral fortitude that inspires others to follow him into battle.

"He is a very disciplined man, very well organized," said Harry Lehmann, an environmental scientist who has known von Fabeck since the early eighties. "This gave him the power to organize with a small amount of money, and just a few people supporting him."

"He is not a man for compromise, once he has made his mind up," said Sven Teske, a Greenpeace activist who first encountered von Fabeck during a pro-solar campaign in the early nineties. "He's very strong, which is great for pushing stuff uphill."

Staffan Jacobsson is a Swedish academic who met von Fabeck in Aachen in 2001, the year after the enactment by the German government of the solar feed-in tariff. "What was really fascinating," Jacobsson recalled, "was how an ascetic, low-key person with a small office in a basement could have such a vast impact on society. He should perhaps have a Nobel Prize."

If you bumped into von Fabeck in the street you would never pick him as a revolutionary. He is lean and rangy. His long, handsome face is topped by a high forehead that is accentuated by a receding hairline. He wears a casual wide-check shirt and slightly scruffy-looking black jeans. Around his neck half-moon glasses hang on a cord, from his belt dangle various items, including a pouch which contains his diary. In it, a calendar lists his forthcoming speaking engagements, of which there are many.

I meet Wolf von Fabeck in early November 2008 in his somewhat gloomy office in what is actually not a proper basement, being only slightly below street level. It is located in a five-story apartment building, one of many, of various vintages, squeezed together on Herzog Strasse, a quiet Aachen side-street about ten minutes' walk from the medieval city center. Given his long-running battle against the utilities and fallings-out with former colleagues, I had half-expected someone who would be prickly and brusque to the point of rudeness. Instead what I find is a humble, unassuming man who is hospitable and accommodating. After several enjoyable hours answering my questions and discussing his achievements, he invites me for lunch at one of his favorite restaurants just outside Aachen. A former farmhouse, the eatery is set in a clearing by a duck pond. The golden leaves on the trees may be past the peak of their autumnal glory by the time we get there, but they are still very beautiful. This idyllic location serves as a reminder of why Germans are so fond of their forests.

Unusually for a soldier, von Fabeck was also an academic. For most of his career he lectured in mechanical engineering at the Army University of Applied Sciences in Darmstadt. Gyroscopes—spinning devices that are particularly useful in navigation systems—were his speciality. In 1980 he published a monograph on the subject. The army transferred him to Aachen, where there was a laboratory for testing mechanical equipment delivered to the military from industry. From this experience he derived a lesson: "Making money is important for industry, nothing else matters."

Like many Germans, von Fabeck is a nature lover. In particular he is a keen walker: coastal rambles are what he enjoys most. In summer 1984, von Fabeck underwent what he described as "an ugly experience" while holidaying on the North Sea island of Baltrum. This is a resort whose unspoiled beaches and national park attract hikers and bird-watchers, especially from the heavily industrialized Ruhr region nearby. By the early 1980s the damage inflicted by acid rain—which in turn was mostly caused by sulphuric acid–laden smoke from coal-fired power stations—was hard not to notice. A 1983 survey showed that fully a third of West Germany's trees had been affected. *Waldsterben* it was called, meaning "forest death."

Taking his morning constitutional on Baltrum, von Fabeck was horrified to discover that during the course of a single stormy night, the vegetation had suffered significant damage. In just a few hours all the leaves on the northwest side of the trees and bushes had fallen off, leaving the foliage on the other side of the plants still healthy. "I tried to clarify how this had happened, and finally was forced to conclude that it was the result of air pollution," he wrote. "A major source of air pollution was—and is—the energy supply. Even the high-chimney policy then in force could not prevent pollutants falling back to earth. [Until the late 1970s building ever higher chimney stacks was the primary way of dealing with air pollution in Germany and elsewhere.] Hence for me arose the question of finding a replacement for the fossil energy supply."

Von Fabeck first came across the concept of solar energy in the foreword by Carl Friedrich von Weiszacker to a book on the limits of the nuclear industry. Von Weiszacker (1912–2007) was the last-surviving member of the team that had tried to build an atomic bomb for the Nazis. In 1957 he was one of eighteen scientists who signed a declaration urging West Germany not to develop nuclear weapons. In his foreword, von Weiszacker wrote that he would not have been in favor of atomic energy, either, had he known how carelessly the technology would be handled. He explicitly mentioned solar energy as a possible alternative. Intrigued, von Fabeck decided to investigate this photovoltaic stuff for himself. Having purchased a solar panel, he tested it in the kitchen of his house, hooking it up to run his wife's food processor. The blades of the machine rotated, albeit only very slowly. But, von Fabeck noted, "at least it worked." As an engineer, it was natural for him to do some back-of-the-envelope calculations. He estimated that to generate enough juice to run the food processor at full power, he would need twelve solar panels. How to obtain the other eleven?

In the late eighties a solar panel cost around $1,000, which meant a dozen would cost $12,000—way beyond the von Fabeck family budget. We can't afford it, his wife told him. The solution came from his pastor: Why not start a club to pool the funds you need? In November 1986, together with seven others, von Fabeck founded the Aachener Solarenergie-Förderverein (SFV), the solar energy supporters association of Aachen, with himself as its director. The group's original members were all men. They included a professor, a military friend, his Lutheran pastor, and a Catholic priest ("so we had God's blessings from both sides!"). Next month following his ugly experience with forest death, von Fabeck took early retirement from the Bundeswehr, to devote himself full-time to the solar cause. In effect, he left the army in order to fight.

The ravages of acid rain had been the original stimulus that prompted von Fabeck to turn renewable energy activist. Now, however, an epochal event occurred that marks the starting point of Germany's long journey to preeminence in renewables in general, solar in particular. On April 26, 1986, six months prior to the formation of the SFV, a reactor at the Chernobyl nuclear power plant in Ukraine spun out of control and melted down, spewing a cloud of radioactive debris that drifted across much of Western Europe.

Although other countries like Finland and Sweden were more severely affected by the toxic fallout from Chernobyl, the response was most extreme in Germany. "Anxious West German citizens were glued to their TV sets and radios, hungry for any scrap of information they could get about the disaster and its implications for them," author and political analyst Paul Hockenos wrote. "Frantic, [they] wanted more facts and demanded to know what precautions to take. But the government seemed as bewildered as its citizens. Bonn played down the incident. 'We're nine hundred miles away from the site of this accident,' said the federal interior minister, Friedrich Zimmerman. 'It's entirely out of the question that the German people are in any danger.' Scientists and other experts contradicted him, one after another on news and call-in shows. . . . Children should stay indoors, everyone should take iodine pills, vegetable gardens should be covered with plastic tarps."

There was a reason for this overreaction. Germans had had a long history of anti-nuclear activism. It dated back to 1974, when citizens launched the first full-fledged grassroots action against a planned nuclear reactor in the tiny town of Wyhl (population 3,612) located near the Swiss border in the far south of the country. Against all odds, the activists had succeeded in blocking its construction. During the seventies many similar local citizens' initiatives sprouted. They linked up, forming a loose network of protest groups, coordinating their campaigns with one another. The symbol of the movement was a laughing yellow sun; its polite slogan, *Atomkraft? Nein Danke* ("Nuclear power? No thanks"). Then in March 1979 at the Three Mile Island plant near Harrisburg, Pennsylvania, the worst-ever civilian nuclear accident occurred. Many Germans extracted a lesson from Three Mile Island: nuclear power was not safe.

They resented the fact that in the wake of the first oil crisis, all three of West Germany's main political parties had hastily signed up to nuclear power, without properly assessing its risks, and without asking them, the voters, for their consent. But there was also another, existential component to the citizen unrest. In December 1979 NATO announced its decision to deploy new US nuclear missiles on West German soil. "Suddenly, it was apparent to Germans on both sides of the front line that, in a nuclear war, their countries would be the first targeted and destroyed. Germany would be the battlefield for the superpowers' nuclear holocaust." During the "Hot Autumn" of 1983, literally millions of Germans took to the streets to try and stave off deployment of the new weapons, by then known as Euromissiles.

On April 1, 1986, just a few weeks before Chernobyl, at the site of a planned nuclear fuel reprocessing plant in Wackersdorf, Bavaria, more than a hundred thousand protesters clashed with police armed with water cannons and tear gas. "They were no longer mainly long-haired *muslis* (tree huggers), punk rockers, and guitar-strumming Protestants. Senior citizens, mothers, housewives, teachers, and children were the ones telling the government that they were not prepared to accept the risks of nuclear power."

For many of the protesters Chernobyl was the last straw. Between 1976 and 1985 public opinion had split down the middle on the question of nuclear power. Following the Ukrainian catastrophe this changed dramatically: within two years opposition to nuclear power topped 70 percent. On the political front the Social Democratic Party committed itself to phasing out nuclear power. The Greens—the little ecology party that had been the voice of the grassroots campaigns and that in 1983 had for the first time won seats in federal parliament—called for the immediate shutdown of Germany's twenty nuclear reactors. *Aussteig!* ("Withdrawal!") was their rallying cry. But nuclear power was responsible for providing 20 percent of Germany's electricity. What could replace it?

By the end of the eighties most Germans agreed that profound changes were needed in the way they used energy. At the same time in cities, towns, and villages around the country citizens had lost faith in their government's ability to bring about these changes. Some decided to take matters into their own hands.

The double whammy of forest death and nuclear catastrophe (reinforced by early rumblings of climate change) led to pressure on a reluctant federal government in Bonn to change direction on energy policy. "Until the end of the 1980s and in fact beyond, renewable energy faced a political-economic electricity supply that was largely hostile. The electricity supply system was dominated by very large utilities relying on coal and nuclear generation. These utilities were opposed to all small and decentralized forms of generation, which they deemed uneconomic and foreign to the system." Now, however, the conservative Christian Democrat government in Bonn faced opposition in the form of a succession of private members' bills, some of them drafted by its own deputies, reflecting the high level of public concern. In the face of this onslaught, fearful of a backbench revolt, the government capitulated.

Until that point the calculation of the tariffs that electric utilities were allowed by law to charge their customers rested on two pillars. One was security of supply: in other words, the utilities were permitted to invest in whatever it took to maintain sufficient generation capacity—a combination of "base-load" power stations providing constant power plus "rolling reserve" plants to handle peaks—and thereby ensure that the provision of electricity was adequate to meet demand. The other pillar was that the infrastructure should be cost-effective. But in the wake of the acid rain crisis (and the Greens' election results), Bonn hurriedly enacted strict environmental protection laws. As a result, by 1988, almost all coal-fired generating plants had been retrofitted with scrubbers to eliminate sulphuric acid. The problem was that this filtering equipment was very expensive. It was thus not economically efficient, nor did it provide a more secure supply.

In 1989, under pressure from backbench members of parliament, the federal Ministry of Economic Affairs modified the framework regulating the calculation of electricity tariffs to permit utilities to take into account any environment-friendly investment they had made so that they could recover their costs, even though that increased the prices they charged their customers. The new decree also allowed utilities to (voluntarily) enter into contracts with private suppliers of renewable energy, even if that electricity cost more than power generated using conventional means by the utility itself. It was on these new legal provisions that von Fabeck and the SFV would subsequently base their petition to oblige the local utility in Aachen to pay for citizen-generated solar power. Since the process began in Aachen, it became known as the Aachen model.

Solar was—and is—the most expensive of the alternative technologies. Harry Lehmann dubbed PV the "prima donna" of renewables. Lehmann is a Peruvian-born particle physicist turned energy scenarist and peace movement activist who now heads the Division of Environmental Planning & Sustainability Studies at the Federal Environmental Agency in Dessau. In the eighties he lived upstairs from the basement out of which the SFV operated.

"The prima donna is the most expensive singer in the opera," Lehmann explained, "she's the soloist, the one that everyone looks at and talks about. What happened in the early days was that, although there already were a lot of alternative technologies—like small hydro or wind or biomass—they were not attractive enough, they didn't have the charisma." Hydroelectric plants dated back to the nineteenth century, windmills likewise had a long history in Germany, while biomass generators merely changed the fuel, substituting woodchips and the like for coal. Solar, by contrast, was the ultimate space-age renewable. PV had no spinning turbine blades, no moving parts at all in fact.

"Photovoltaic was the symbol, the icon of a new technology era in energy," Lehmann said. "In addition, it was clearly a decentralized technology. At the time we were thinking that we would switch from a centralized energy system to a decentralized one, and photovoltaic was also a symbol of decentralization." A decentralized system represented a means of producing power that was owned and controlled by the individual: it would be *people's power*, an attractive idea for a generation that had come of age in the late sixties.

Lehmann did not have access to a roof of his own. He and some teacher friends clubbed together to purchase their first photovoltaic system. They installed it on the roof of the local Waldorf (Steiner) school in Aachen. Though capable of generating only half a kilowatt, the panels were very expensive. "If I think about the price we paid back then in the eighties, I cry today," Lehmann laughed. "But as part of a movement, you have to do that— some people bought a car, we bought some photovoltaic and put it on a school."

Other early purchasers of PV included concerned citizens like Dietrich Lohrmann, a professor of history at the technical high school in Aachen. He was inspired by von Fabeck's idea of individuals owning their own power stations. The state of North Rhine-Westphalia was offering a 50 percent subsidy on PV systems. In late 1991 Lohrmann decided that, rather than buying the second car that he and his family of four needed, he would spend 39,000DM ($23,400) of a small inheritance from his grandfather on a system for the roof of their home. In its first year of operation, the system produced nearly 1,500 kilowatt-hours of electricity, 650 of which were fed into the grid. The local utility paid him just 100 deutschmarks ($60) for this power. But Lohrmann had no thought of making any return on his investment. The system would be, he said, "our contribution against the looming climate catastrophe." His "second car" was, and would continue to be, a bicycle.

In addition to its iconic appeal, there was also another, more pragmatic reason for choosing solar. In the south of Germany, snowmelt cascading down mountainsides can be channeled through hydroelectric plants. In the north, especially near the North Sea, coastal gales propel wind turbines. But in the flat, landlocked middle of the country, there is not so much water or wind, especially during the summer months. For Aachen, that meant solar was more or less the only viable renewable option.

From the outset, von Fabeck and his little band of enthusiasts realized they would have to pay more for solar than for other technologies (as it turned out, almost six times as much as for wind energy, for example). But they were willing to bite the bullet, to make the conceptual leap and say, We want solar, and we're prepared to pay whatever it takes to get it. But *how* would they pay for it? That was the question von Fabeck and his colleagues set out to answer.

In his kitchen, then subsequently via public demonstrations, von Fabeck had shown that photovoltaic was practicable. But the notion that you could stick solar panels on the roofs of houses and have them function reliably year-in, year-out was far from being widely accepted, even among specialists. Thus, in addition to being expensive, photovoltaic was also exotic. But the main problem was, how to reduce the cost? Like others before him, von Fabeck understood that people would not pay for PV so long as the panels were too expensive, and if no one was buying, manufacturers would not invest in new factories, ensuring that prices stayed high. It was what Germans call a "devil's circle." On the other hand, if enough demand for solar panels could be drummed up, then the price would drop as the manufacturing volume rose. But producers would not make the necessary investment in building production capacity unless their order books were full. They needed to be certain that there would be sustained demand for their products.

How then to persuade large numbers of citizens to buy solar systems? Though an idealist, von Fabeck is also by training an engineer, a practical man, a problem solver. He understood that for most people the decision to invest would be made on financial considerations, not moral ones. The likes of Dietrich Lohrmann or Harry Lehmann and his utopian pals might be prepared to dig deep into their own pockets to pay for solar panels with no thought of getting their money back. But such dedicated environmentalists were very much the exception to the rule.

"We were idealists, but we were not many," von Fabeck told me. "Most people are not idealists, they act in their own economic interest." If the solar movement was ever to go mainstream, then financial incentives would be necessary. "We had to change the market environment so that nonidealists would be rewarded for doing the right thing." Persuading people to put money in solar systems would be much easier if they knew they would be reimbursed for their investment.

Every Wednesday, SVF members gathered in the semi-basement at Herzog Strasse to discuss ideas for how the market environment could be changed. Their thinking evolved slowly, taking several years to reach maturity. "It was all small steps," von Fabeck said, "one after another." His original idea was that purchasers of solar systems should get "fair-cost compensation" (*Kostengerechte Vergütung*)—a 100 percent rebate—for the equipment they bought. Though some members favored this idea, others were skeptical that such a radical proposal would be taken seriously. They were right.

In September 1989 the group sent a letter to the Federal Ministry for Economic Affairs in Bonn outlining their idea. They did not even receive an acknowledgment from the bureaucrats, let alone a reply. But this was just a preliminary skirmish. The real battle, against the local utility in Aachen for the right to receive fair compensation for citizen-generated solar power, would not be joined for another three years. Meantime, in 1989, as we will see in the next chapter, solar energy was about to get an unexpected boost, from the federal government in nearby Bonn.

13

A PERSUASIVE DEMONSTRATION

*T*he assignment came as a complete surprise to Walter Sandtner. A lawyer by training, Sandtner had previously been responsible at the federal government's Ministry of Research & Technology in Bonn for maintaining links with international scientific organizations. Now his minister, Heinz Riesenhuber, asked the forty-six-year-old bureaucrat to take over as director of the ministry's division of renewable energies. Sandtner's first reaction was to protest that this had always been a job for a technical specialist. But the minister was adamant: he didn't want another geek, he wanted someone who could talk to the public in layman's language, someone who moreover had the political nous to soothe fractious parliamentarians.

By 1989 Riesenhuber had been a minister in Helmut Kohl's conservative coalition government for seven years. Something of a dandy, he was instantly recognizable by his trademark bow tie. Indeed, Riesenhuber had once campaigned for reelection using a poster depicting only a bow tie in the German national colors of black, red, and gold. Lately, however, the minister had come under fire for his aggressively pro-nuclear energy policy. In particular, behind schedule and over budget, the fast-breeder reactor project he championed had become an embarrassment to the government.

In the Bundestag, Germany's parliament, the Christian Democratic Party was being criticized for neglecting renewable energy. Some deputies were making a fuss because the research ministry had no technology demonstration programs under way. Also, because the budget allocated by parliament for the development of renewable energy technologies was not all being spent. Lately, these deputies had begun demanding that the minister take action. But the technocrat in charge of the division, though an able man, was critical of renewables, thus reluctant to waste taxpayers' deutschmarks on their development. The minister himself had dismissed the likes of wind and solar power as too expensive and not worthwhile.

Riesenhuber told Sandtner that, in order to allay criticism from parliament and address public concern, he had to do two things. One was to set up a high-profile program, the other to make sure that all the money earmarked by parliament really was spent. Sandtner went away scratching his head. What should I do? he asked himself. Then he had an idea. Sandtner summoned some renewable energy specialists. How many houses equipped with photovoltaic systems are there in Germany? he asked them. Just six, they replied. To Sandtner, that seemed like a good starting point. "Let's make a thousand!" he said.

A thousand was a nice round number and also easy to remember. The goal of his 1,000 Roofs Program would be to demonstrate that solar technology worked. Sandtner convened a meeting at which he and the specialists discussed details. Afterward, the experts sent him a three-page memo listing the reasons why such a program would not be feasible. The technology was immature; in particular, no suitable inverters were available, nor were industrial standards for photovoltaic systems in place. Sandtner reconvened the panel. Perhaps he had not made himself clear the first time. His minister would not take no for an answer. "I don't want to hear why it's not possible," Sandtner chided the experts, "I only want to hear what we have to do to make it possible."

Minister Riesenhuber announced the 1,000 Roofs Program at a press conference in mid-1990. The response from the public was immediate and overwhelming: much to Sandtner's astonishment, 60,000 requests for information poured into his office. "I got a special secretary who had nothing to do except type the addresses and put the information materials in the envelopes," he said. Meanwhile, his phone rang off the hook. One call was from a group of engineers in the northern city of Münster who told him that they had scheduled a meeting that very evening for seventy of their members specifically to discuss his program. "The reaction for me was a great surprise," Sandtner confessed. "I saw the commitment and the enthusiasm of the people."

Those keen to participate included teachers, environmentalists, and technically minded people curious to know how PV worked. Eventually, out of around 4,000 applicants, 2,250 were accepted. The expanded total came about because of a deal between federal government and states. Each of the eight larger states would get 150 systems, the three city-states (Berlin, Bremen, and Hamburg) 100. Since reunification had just taken place, the five states from the former East Germany were also included. The installations were all heavily subsidized, much to the chagrin of the hardline neoliberals at the Ministry of Economic Affairs, whose catch-cry was "no subsidization of technologies unfit for the market." The federal government paid 50 percent of the costs, with the states chipping in 20 percent (except for the small southwestern state of Saarland, which refused). The remaining 30 percent, paid by the participants themselves, was tax deductible.

Research in Germany on solar cells had begun in earnest in 1981. Its originator was Adolf Goetzberger, a semiconductor physicist who transistor co-inventor Bill Shockley had recruited to work for him in what was not yet known as Silicon Valley. He established with eighteen researchers the Fraunhofer Institute for Solar Energy Systems in the southwestern city of Freiburg. People thought they were crazy. "In the first three or four years, we had very bad times," Armin Räuber, one of Goetzberger's original colleagues, told me. "Nobody wanted this strong engagement with solar energy, especially not the people from the Ministry of Research & Technology." Also in 1981, Werner Kleinkauf and his students at the University of Kassel began investigating ways of integrating PV into the electric grid. They too were met with derision. "In the early days, people laughed at us," said Günther Cramer, then one of Kleinkauf's students, now CEO of SMA Solar Technology, the world's largest maker of inverters.

During the eighties the federal government initiated two solar demonstration projects, the largest of their kind in Europe. More than seventy individual installations were ultimately completed. Yet by 1990 the total amount of photovoltaic deployed around the country was a measly 1.5 megawatts. At that time, the German PV community mustered fewer than a hundred people.

"It was very very difficult to convince people to use solar energy," recalled Udo Moehrstedt, a physicist turned businessman. Moehrstedt had dabbled in PV for some years selling a charge regulator he had invented to the off-grid market, mostly for use in hunting huts, caravans, and boats. (A charge regulator is a device that controls the amount of voltage coming from a solar panel to a battery. It prevents damage caused by overcharging.) "People didn't know what solar was," Moehrstedt told me, "they had completely no idea of what could be done." Some way of spreading the good word about photovoltaic was clearly required. Moehrstedt came up with the idea of holding a solar symposium. It would cause the community to snowball: "You invite people," he explained, "you inform them, they inform others, and so it goes on."

In April 1986 Moehrstedt organized the first PV symposium at his home base of Bad Staffelstein, a small spa town (population 10,600) in the picturesque Upper Franconia region of northern Bavaria. The venue was Kloster Banz, a former Benedictine monastery that dated back to the eleventh century. Its salient feature is a church, a baroque masterpiece, complete with twin onion-domed towers and magnificent ceiling frescoes. That first gathering attracted seventy-five attendees. Thirty of them were speakers, the other forty-five came to listen. Participants were mostly company people from the car industry, the utilities, and the highway authority. A few researchers from the Fraunhofer were also invited.

Four days later, the meltdown at Chernobyl occurred. After that, Moehrstedt recalled, he had no problem whatsoever attracting people. As he had hoped, attendances snowballed. The second year 140 people came; by the third year, the number had swelled to over 800. Harry Lehmann, the Aachen-based particle physicist turned renewable energy scenarist we met in the previous chapter, was one of those who took advantage of the opportunity to learn on offer there. "I remember going to Staffelstein in the beginning and having my first lesson on photovoltaics, about how these bloody things work," he said. "The program was filled with basic lectures, so that people who knew more or less nothing had the possibility to get their first glimpse about all that."

The cloisters were also a great place for networking. "I met a lot of people that we from North Rhine–Westphalia [state] hadn't met before," Lehmann said. "Staffelstein was important, because it was a German meeting. It brought people together, all these regional and local organizations had a point where they could meet." Like at any good conference, what happened after hours was more important than the formal program. "Staffelstein is a very small town," Moehrstedt explained, "there are no attractions, no bars, nothing. So in the evenings you sit in the monastery and you talk and you drink until two or three in the morning. You meet everybody, and outside business you can talk relaxed. That is the spirit of Kloster Banz—very informal."

As the title he chose for his program—1,000 Roofs—indicated, Sandtner had from the outset insisted that the proper place for photovoltaics was on top of people's houses. That way, he explained, "we could show to Germans and the whole world that PV doesn't need any extra territory, we only use the space which is already there." This in turn led him naturally to the idea that the photovoltaic roofs in the program should all be connected to the grid. The subsidies would be paid to owners, Sandtner insisted, not for the purchase and installation of their PV systems, but for the production of clean electricity they fed into the grid. The reason for choosing this approach was partly that Germany, like California a few years earlier, had had an unhappy experience subsidizing the construction of wind turbines. In particular, one large-scale prototype had cost the government 100 million marks. It ran for a few days, then broke down, kaput. Having pocketed their rebate, the developers had no incentive to keep the machine running. Sandtner was determined not to repeat this mistake.

To ensure that utilities were obliged to pay for the electricity thus generated, however, the law would need to be changed. By a happy coincidence, a private members' bill governing payments for electricity fed into the grid by owners of wind turbines and small hydro systems had recently been drafted by two government backbenchers. Sandtner was able to tack a clause covering solar onto their one-page bill.

The minister for economic affairs and the leaders of the ruling Christian Democrats initially opposed the bill. But the initiative turned out to be remarkably popular with some of their constituents. Owners of small hydro plants were typically conservatives. So too were farmers who saw the potential for gaining additional income through locating wind turbines on their fields at a time when EU subsidies for cereals were shrinking. In the Bundestag strict party discipline prevails: anything that threatens it is a serious matter. Concerned that the challenge to its leadership from its own backbench deputies was getting out of hand, the government capitulated.

In December 1990 the German parliament passed the Electricity Feed-In Law (*Stromeinspeisungsgesetz*, mercifully abbreviated StrEG) unanimously, with support from all parties. The law obliged utilities to allow grid connection and guaranteed producers of renewable energies up to 90 percent of the retail price for electricity. This was a good deal for wind and hydro owners, but nowhere near enough to recompense purchasers of solar systems. Still, the principles of compulsory grid interconnection and payment of elevated rates had been established. It was a major leap forward, one of the stepping stones to the solar-specific feed-in law that would be enacted fourteen years later.

For the moment, Germany's big utilities made no objection to feed-in tariffs. Who cared about one thousand, two thousand roofs? The numbers were too small to bother about. Besides, at this time the power companies were preoccupied with the challenge of taking over the electricity sector in the former East Germany.

In 1992 the 1,000 Roofs Program ended, having largely achieved its goal. It had demonstrated (yet again) that PV was a practicable technology. Thanks to the program the reliability of solar systems had greatly improved, especially in the area of inverters, which because prone to burn out had long been PV's Achilles' heel. SMA Solar Technology, a spin-off from Professor Kleinkauf's group at the University of Kessel, got its start in business building inverters for the program.

"The 1,000 Roofs Program was so important because it was the first time we saw that there was a market coming up," SMA's Cramer told me. Previously, inverters had been hand-soldered in batches of ones and twos. The money the company earned from sales to the government program funded the R&D to develop an inverter that could be mass produced. SMA Solar Technology would go on to dominate the global market for PV inverters.

The subsidies program had also been good news for system installers. The welcome injection of cash enabled small firms that had previously been living hand-to-mouth to build up some muscle. Prior to the program, PV had been a cottage industry catering to what Olaf Fleck, CEO of Sunset Solar, one of Germany's first installers, called "stupid customers." Stupid meaning not that customers were unintelligent, just that they were prepared to pay ridiculous amounts to realize an idealistic vision. The solar market was minuscule. "At this time I knew every individual solar module I was sending out," Fleck recalled. "Not just the module, I knew the name of the customer who bought it, his first name, family name, birthday, account number, family status, and so on."

After the program, it became routinely possible to order a PV system for a roof anywhere in Germany and have it installed within a few days. But the Ministry of Research & Technology, unlike its Japanese counterpart, was not in the business of market creation. Two thousand-plus roofs was not enough to justify investments in new production facilities by Germany's indigenous solar cell industry, which in those days consisted of just three firms. Unlike their US counterparts, none was a solar-only specialist. The largest was the giant power equipment maker, Siemens. In 1989, as we saw in part 1, Siemens had acquired ARCO Solar. The company proceeded to shut down domestic solar manufacturing in Munich, leaving only a small R&D facility. Its focus was on markets in the developing world, such as Nepal and Vietnam. The other makers were divisions of Nukem, a uranium services company, and DASA, an aerospace conglomerate. In 1994, the two divisions merged to form Applied Solar Energy (ASE).

The solar manufacturers had been hoping that there would be a follow-up to the 1,000 Roofs Program, but in the event nothing substantial was forthcoming. In 1993 Hermann Scheer, a Social Democrat deputy who had been the most vocal of the critics Sandtner had been tasked with appeasing, proposed a 100,000 Roofs Program. This failed to win government support. Sandtner argued that such a large-scale subsidy program would be enormously costly. Domestic firms would not be able to meet the demand, which would mean opening up procurement to American and Japanese PV suppliers. Why should German taxpayers have to pay billions of marks to foreign firms? By 1995, lacking new orders, it looked as if ASE, Germany's last-remaining domestic photovoltaic producer, might also abandon its homeland and move production offshore.

In 1996 Walter Sandtner was transferred, ironically enough, to work in the nuclear field. His successor as head of the ministry's renewable energies division was, like his predecessor, a skeptic. For all his unflagging efforts in promoting solar energy Sandtner received no official recognition, no promotion within the bureaucracy. There would however be one last surprise in store for him: in Vienna in 1998 he was awarded the Edmond Becquerel Prize for outstanding contributions to the development of solar photovoltaic energy, an accolade normally given only to distinguished academics in the field.

For PV in Germany the midnineties would be, as Greenpeace's Sven Teske put it, "the dark ages." But the flames that Sandtner had fanned were not entirely extinguished. The focus now shifted from government subsidies back to grassroots activities, to Wolf von Fabeck and his fight against the municipal utility in Aachen for the right of solar producers to earn fair compensation for the clean electricity they generated. Joining forces with him in this battle was Eurosolar, a pro-PV ginger group founded by Hermann Scheer and represented by Harry Lehmann.

In solar's long struggle for acceptance as a viable replacement for the generation of electricity by fossil fuels and nuclear we have thus far encountered mostly activists and entrepreneurs. What solar advocacy had conspicuously lacked, what it now for the first time found, in the doughty efforts Hermann Scheer, was a political champion. Before returning to Aachen, let us first look at the career and achievements of the world's most eminent pro-solar politician and polemicist.

14

SOLAR PROLIFERATOR

*I*n January 1986, just three months before Chernobyl, Hermann Scheer published a book. Entitled *The Liberation from the Bomb*, its subject was nuclear disarmament. Scheer was then spokesman on disarmament and arms control for the opposition Social Democratic Party in the Bundestag, the German parliament.

The book called for an end to dependence on nuclear energy. But its author also flagged an open question hanging over the anti-nuclear movement. Namely, that the movement was pinning its hopes on energy efficiency and conservation; it had not identified alternative sources of energy. Accordingly, Scheer included in his book a digression in the form of a chapter on the need to develop renewable technologies. It was a topic on which he was not formally qualified to pontificate. Scheer was merely, as he would later—with characteristic immodesty—confess, "a curious reader of scientific publications with the ability to identify key elements in them." His thoughts on renewables were not rehashed from those of other people: they had all sprung from his fertile imagination. "I am a person who always has his own ideas," Scheer said with a chuckle.

When the book was published debate was raging over Ronald Reagan's controversial Strategic Defense Initiative (SDI). This proposed to deploy an impermeable, space-based anti-missile shield that would make nuclear weapons impotent, hence obsolete. Apologists argued that Star Wars would eventually lead to the development of useful new technologies with non-military applications. Even in Europe there were enthusiasts who backed participation in the US program because it promised early access to any breakthroughs made. Scheer rejected this logic as flawed. Rather than hold out hopes for nebulous future spin-offs, better to address future technological requirements directly. What was needed most was a non–fossil fuel alternative to nuclear power. Scheer proposed an SDI in which the first two letters of the acronym would stand, not for "strategic defense," but for "solar development." Since photovoltaic technology already existed, what remained to be done was to promote the industry and increase the volume of production.

Though overall reaction to Scheer's book was favorable, some reviewers took exception to his digression. The influential weekly magazine *Der Spiegel* dismissed it as nonsense. Responses from environmental organizations and Green politicians were likewise negative. The experts advised Scheer to stick to what he knew—foreign relations and nuclear nonproliferation. But Scheer also received some positive feedback from scientists in the renewable energy field. They wrote him letters expressing their gratitude— "at last, a politician who understands!"—and recommended further reading in the form of papers and books. Scheer was stimulated by the commotion he had caused. It seemed to him that he was onto something.

Scheer had always had a well-developed sense of his own destiny. No ordinary life for him, he was fated to play an important role. Now, to his surprise, Scheer discovered there were no other politicians in Germany—or, apparently, anywhere else for that matter—who had espoused the cause of renewable energy. In politics, as in science, ambitious practitioners seek unpopulated fields, because in such places there is room for them to realize their ambitions. In solar Scheer had found such a field. "I concluded that if the issue was so important, then I should do that," he recalled. For him, solar energy became "a question of personal responsibility."

Over the next twenty-five years, Scheer would devote himself to furthering the solar cause. He turned himself into an expert on the subject, founding promotional organizations, writing numerous books and innumerable articles, speaking at conferences, debating skeptical opponents in public forums, giving lectures and interviews to print and broadcast journalists all over the world, badgering reluctant ministers, and most importantly, using his political skills to shepherd enabling legislation through the German parliament.

"He's a very dynamic leader, he has an aggressive way of making things work," said Karl Böer, a kindred spirit who became friends with Scheer after they met at an international solar energy conference in 1988. "Without Hermann Scheer, solar energy in Germany would be nothing."

"Without him, we would never have done it," agreed Olaf Fleck, CEO of Sunset Solar. "He was our anchor in politics, especially because he was not a Green, he was Social Democratic Party and this was *crucial*. . . . If it had been the Greens only, people would have gone, Ay-ay, those crazy Green guys. But Scheer was from a real party, a party with the most widespread support in society, and it was crucial that he was in there, fighting for the idea."

"He plays in this field a very important role indeed," added Walter Sandtner, the bureaucrat whose seminal 1,000 Roofs Program was initiated largely as a result of pressure from Scheer. "He has done a great job."

In November 2007 I traveled to Berlin to interview Hermann Scheer. On my way to his office halfway down Unter den Linden, the spacious tree-lined boulevard that culminates at the Brandenburg Gate, I witnessed an extraordinary spectacle. The might of fossil fuel manifested itself in front of my eyes. As I walked through the gate, passing under the statue of the goddess of victory urging on her four-horse chariot, I wondered why there seemed to be policemen everywhere. The reason for this elevated presence soon became apparent. King Abdullah of Saudi Arabia was paying a state visit to the German capital.

Speeding up the boulevard toward me came an enormous motorcade. It began with a few individual police cars and outriders. They were followed by ten large motorcycles ridden, line abreast, by policemen clad in emerald livery. Behind them streamed a lengthy procession consisting of green and white police cars and gleaming black Benzes with temporary blue rotating lights stuck on their roofs. The centerpiece of this cavalcade was the king's stretch Mercedes limo, identified as such by the Saudi green scimitar-with-Arabic-text flag flying from its hood. In his wake came dozens of other blue-light-bearing Benzes, police vans, still more motorcycles and, bringing up the rear, a solitary ambulance. Big Oil had come to town.

Herman Scheer turns out to be a rumpled, heavy-set, ursine man. His brown eyes are baggy and hooded. His salt-and pepper thatch has long since lost any semblance of order because of his tendency to run his fingers through it. Scheer suffers from an excess of nervous energy. Unable to sit still, he keeps sliding his chair backward and forward on its casters. He fidgets with a cigarette and a match, no doubt wishing he could light up. His large office reeks of tobacco smoke. On its walls hang two photographs: one of Scheer as a young man with his mentor Willie Brandt, the former West German chancellor; the other, a recent snap of him with Arnold Schwarzenegger, on which the governor of California has scrawled, "Thank you for all your good work." Behind his desk is a brightly colored abstract painting by his wife, Irm Pontenagel. In a corner sits a television showing live pictures from the debating chamber in the Reichstag, the nearby parliament building. The sound is turned off, but Scheer's eyes constantly flit across to the screen, checking what is going on.

Though now in late middle age, Scheer has something boyish about him as he recounts for the umpteenth time his remarkable achievements. His manner is pugnacious. He speaks slowly, in a low growl, in flawed but excellent English. He has a habit of repeating the last few words of his sentences for emphasis, a rhetorical tic that marks him as someone used to public speaking. As he warms to his topic, telling funny stories about his adventures, Scheer emits a wheezy smoker's laugh and an occasional, unexpected giggle.

Hermann Scheer was born on April 29, 1944, in Wehrheim, a small town just north of Frankfurt. Though his parents moved around a lot during his childhood, he grew up mostly in Berlin. His father was a manager for a health insurance firm. An autodidact from an early age, Scheer loathed school. "I hated to have teachers," he told me, "to feel dependent on them, to have to read what they wanted and not what I wanted." Politics was not his first choice of career. On graduating from high school in 1964, having considered and rejected the idea of becoming a sports journalist or an actor, Scheer joined the army. Not because he was attracted to soldiering as such, but because he wanted to become a general, a leader of men. Scheer attended the Bundeswehr's Officers School, served a year as a lieutenant, then quit. By then he had realized that an extremely bureaucratic organization like the army was no place for an ambitious, self-driven individual like him.

In his student days, Scheer was a sportsman. He swam and played water polo on the German national team. Water polo has been called the roughest game in the world. It combines swimming, soccer, and basketball with—under the water, where anything goes—wrestling, boxing, and mugging: good preparation for the rough and tumble of politics. An idealistic youngster, Scheer espoused the cause of social justice. In 1965 he joined the Social Democratic Party. At the University of Heidelberg he studied economics, sociology, political science, and public law. Scheer enjoyed the academic life at Heidelberg, Germany's oldest university, because it gave him the freedom to set his own curriculum. "I studied, but in an autodidactic way," he recalled. "I think in my studies I did not go to more than ten hours of lectures by professors."

During the sixties Heidelburg was one of the centers of student protest in Germany. Scheer arrived there in time to participate in the 1968 upheaval that roiled campuses across Europe. He is thus a member of what Germans call "the '68 generation," whose "mishmash of unorthodox Marxism, visceral anti-authoritarianism, and participatory democracy would jar loose the stodgier attributes of German-ness." With their utopian visions the sixty-eighters would imbue the German PV industry, as we shall see in chapter 18, with a distinctive character. Scheer was elected president of the students' parliament, a representative body. From Heidelberg he moved on to the Free University of Berlin, where in 1972 he received a PhD in economics and social science. By 1975 Scheer had risen to become vice chairman of the young socialists, the youth wing of the SDP, an ideal springboard for a career in federal politics. Following stints as an assistant professor of economics and a systems analyst at the German Nuclear Research Center ("although at no time did I favor nuclear energy, *at no time*") in 1980 Scheer was elected as a deputy to the federal parliament in Bonn.

In November 1969 West Germans had voted the conservative Christian Democrats out of power for the first time in the postwar period. Willie Brandt, the leader of the SDP, became chancellor of a coalition government. Brandt was a reformer. "We want more democracy," he told parliament in his first speech as chancellor. A passionate politician who could joke with ordinary people, play the mandolin, and enjoy his drink, Brandt was extremely popular with young Germans. Including Scheer, who followed the idealistic Brandt line in preference to the more realistic line of Helmut Schmidt, who took over as chancellor from Brandt in 1974. When Scheer published his controversial book in 1986, Brandt told him that, in his view, the most interesting chapter was the one on renewable energy. Though his mentor knew nothing about the subject, he sensed that Scheer had hit on something and urged him to follow up. For his protégé, it was acknowledgment that his gut feeling had been right.

Scheer would pay a price for his allegiance to Brandt. Given his experience and obvious abilities he was seen by some in his party as a future foreign minister. But Schmidt's successor, Scheer's exact contemporary Gerhard Schroeder, was not interested in visionaries. When the "Red-Green" SDP-Green Party coalition came to power in 1998 with Schroeder as chancellor, Scheer would be passed over for high office. For an ambitious man like him this was naturally a disappointment. But it meant that he was now free to devote his prodigious energy to promoting solar.

Having identified that the question of finding an alternative to nuclear and fossil fuels was of crucial importance, and decided that he himself would assume the role of renewables champion, Scheer embarked on a two-track strategy. The first track was within parliament. He resigned as chairman of the parliamentary arms control and disarmament committee in order to push legislation that would pave the way for a alternative energy industry. Scheer became the leader of the deputies who believed that something needed to be done to nurture renewable technologies. In addition to SDP and Green MPs, this nonpartisan group also included quite a few deputies from the Christian Democratic Union and other right-wing parties. The first fruits of Scheer's activities on this track were the 1989 decree that permitted utilities to take environment-friendly investments into account when calculating electricity tariffs and, the following year, the Electricity Feed-In law. The big, game-changing legislative innovations would follow the election of the first Red-Green coalition in 1998.

Scheer was concerned to avoid being pigeon-holed on the Bundestag's energy and environmental committees as an expert. He realized that it was necessary to transcend the specialist domain and establish a broad con-stituency for solar in society as a whole. "It doesn't make sense to fight always in the field, in the circle only of energy experts," he explained. "It was necessary to over-run them, to stimulate the public debate. And that meant that my role in parliament was not enough." Perhaps officer training school had taught him something after all.

The second track of his strategy was thus to venture outside the Bun-destag, to outflank the normal political structures and politicize the issue in a nonpartisan way. The public had to be enlightened about solar and other renewables and their potential to replace nuclear and fossil fuels. As a vehicle for raising public awareness, in 1988 Scheer founded the European Solar Energy Union (Eurosolar), a nonprofit, nongovernmental organization, with him as honorary president and his wife, Irm Pontenagel, as managing director. Though almost all of Eurosolar's initial members were German, as the name suggests, from the outset Scheer had set his sights on broader pas-tures. Back then there were no pro-solar activist groups at the national or international level, only apolitical organizations like the International Solar Energy Society, whose members were predominantly scientists. In Scheer's view, the goals of such organizations were altogether too modest. The white-coats did not want to change the world, merely to get enough funding to con-tinue their research. He thought the scientists were naive. For one thing their attitude was deterministic: develop the technology, then everything else would naturally follow. The boffins were blissfully ignorant of political barriers and vested interests. They seemed not to realize that the utilities and the nuclear and fossil fuel lobbies would do everything in their power to thwart them.

Eurosolar began with around a hundred members, although only a few were actively involved. One such was Harry Lehmann, who became the first editor of *Eurosolar*, the eponymous quarterly journal that the group began publishing in September 1989 to focus on solar technology and energy policy. There were of course existing publications that covered energy and energy policy. The problem was they were dominated by people from the nuclear and fossil fuel industries. If you submitted a paper about renewables to such journals, their referees would simply return it with the comment that it was unscientific. On its signature bright orange cover, the first issue of *Eurosolar* featured an illustration of an industrial-scale solar engine for converting sunlight to steam. It had been invented by a Frenchman, Augustin Mouchot, and demonstrated at the Universal Exhibition in Paris in 1878. The point was clear: solar power had history.

Eurosolar's activities included organizing conferences. Some were on technical topics, like solar architecture and urban planning. Others aimed at building a nationwide pro-solar infrastructure. "We made a network of all the people working in the field and the small NGOs in different cities," Lehmann said. To coordinate the activities of these groups, Eurosolar organized a twice-yearly get-together at which the activists and true believers would assemble to swap ideas. As we shall see in the next chapter, Eurosolar also lent its support to other pro-solar groups, like Wolf von Fabeck's SFV.

When he began his pro-solar crusade Scheer was already a well-known political essayist with two books and many other publications under his belt. Taking up his pen to promote the solar cause was thus a natural thing for him to do. Ultimately, between 1987 and 2005, Scheer would write five books and, by his calculation, over a thousand articles on the need to transition from nuclear and fossil fuels to solar. The first two books, *Die Gespeicherte Sonne* (*The Stored Sun*, on hydrogen as a solution to the energy storage problem, 1987) and *Das Solarzeitalter* (*The Solar Age*, 1989) were not widely noticed. But with his third polemic, *Sonnen-Strategie* (published in English as *A Solar Manifesto*, 1993), Scheer broke through to reach a mass audience.

In *Sonnen-Strategie* he argued that the discussion on solar thus far had been conducted, even by many of its supporters, too defensively and faint-heartedly. "All too frequently, solar energy is promoted as an alternative for the provision of small, even microscopic, decentralized energy supply options, leading the general public to conclude that it cannot compete as a substitute for the major conventional energy carriers." Such pusillanimity had to stop. The book set out to prove the feasibility of the following, radical assertion:

> It is not enough to increase the share of renewable solar energy to 10, 20, 30, 40 or 50% of human energy consumption. It would not overcome dangers to our existence, but would merely postpone the collapse of human civilization. *The goal for the century ahead must be the complete substitution of conventional sources of energy by constantly available solar energy—in other words, a complete solar energy supply for mankind.*" [emphasis in the original]

This was an ambitious claim that even some solar proponents regarded as over the top. Not since Barry Commoner back in the late 1970s had anyone made such an outrageous proposal. It upset Scheer's colleagues in the SDP, which had traditionally been friendly to coal and coal miners. But Scheer's claim was backed by the first 100 percent renewable energy scenarios his friend and colleague Harry Lehmann produced the following year. They showed that a world powered entirely by renewables was indeed technically feasible.

The conversion of mankind's energy supply system to solar must be, the book asserted, "the very top strategic goal of politics and economics—*Priority Number One*" [emphasis in the original]. The introduction of a global solar energy system was more important for humanity, Scheer went on, than the Industrial Revolution and the French Revolution had been for the economic and political development of modern times. History also provided a model for his solar strategy, in the shape of the massive effort during the second half of the nineteenth century that went into constructing railways. In some years, Scheer pointed out, Germany had allocated up to 10 percent of its gross national product to this task. What humanity had done before, it could do again, Scheer would subsequently write. "[W]hat is required is no more complex or more expensive than the development and production of satellite, aerospace, communications, medicine, or weapons technology—and it is less complex by far than nuclear technology. The assertion that it is not possible to arrive at a comprehensive energy supply using renewable energy is an insult to the creativity of physicists, chemists, and engineers."

What was particularly impressive about Scheer's writing was its scope. Plenty of books had explained the technology of solar. But few if any painted the big picture, placing the technology in its socio-cultural-economic context and proposing policy recommendations at every level from local to global. *Sonnen-Strategie* concluded with a call to action. "The solar revolution needs new supporters in society, in politics, in economics, in science, and in the media."

For many readers the book was an eye-opener. "It was a best-seller in Germany, selling 80,000 copies, and it mobilized many people," Scheer told me. "It helped bring renewable energy to the attention of politicians, businesspeople, and the general public. Many people say that this book was the incentive to change their opinion." *Sonnen-Strategie* was translated into English and seven other languages. Its successor, *Solar Weltwirtschaft* (*The Solar Economy*, 1999), did even better, appearing in eleven foreign-language editions. Between them, the two volumes are probably the most widely read books on renewable energy in the world. *The Solar Economy*, wrote American author Neville Williams, was "the most important, complete, and thoughtfully reasoned argument for a global solar economy yet published." It was also, he noted grimly, "nearly impossible to buy in the United States."

In addition to his phenomenal written output, to get his message across, Scheer also hit the road. By his own reckoning he delivered roughly two hundred speeches a year, to an ever-expanding audience. By the late nineties this included attendees at international PV conferences. "Scheer is a marvellous speaker," said former PV industry analyst Bob Johnson. "He goes a bit over the edge, he's a bit over the top—quite a bit—but nonetheless he's a great crowd pleaser." Once Scheer got going, there was no stopping him. "If he was slated to speak for half an hour, you could assume that he'd be up there for an hour and a half," laughed Johnson's former colleague, Paula Mints. "But he's very impassioned and it was kind of fun to watch him because he used to pound the podium and shake his head."

By no means were all of Scheer's lectures preaching to the converted or enlightening the uninformed. Some would be delivered to specialist audiences who were actively hostile to his message. Scheer understood from the outset that he would have to argue "in a broad way, not only on technological matters, but also politically, socially, ecologically, economically." He would in other words have to be able to defend his ideas in all fields, "because the counter-arguments come from all sides. And therefore, if you had one weak point in your argument, then you would lose."

This willingness to take on all-comers demands "very strong self-confidence." Happily, self-confidence was something with which Scheer had been bountifully endowed. As could be seen from an invited talk he gave in the late nineties to the prestigious German Physical Society, an organization which dated back to 1845 and was, according to Scheer, very old-fashioned. The society's past presidents included Max Planck, founder of quantum theory and one of the twentieth century's most important physicists. Scheer was fond of quoting Planck's dictum that "A new scientific truth does not triumph by convincing its opponents and making them see the light, but rather because its opponents eventually die, and a new generation grows up that is familiar with it."

The audience at the society that day consisted of roughly 150 professors. "I think at least a hundred of them were pro-nuclear, they had the opinion that renewables cannot work." Scheer took the scientists by surprise by commencing, not with political remarks, but by citing the second law of thermodynamics, which deals with ever-increasing entropy. He used the law to argue that renewables were humanity's only realistic choice, the only source of energy large enough to keep entropy at bay.

At the conclusion of his talk he asked the learned scholars if they had any rebuttals. Silence. Again he asked: If they had detected any flaws in his arguments, could they please let him know what they were. Still silence. Finally, Scheer managed to elicit a criticism: Switching to renewables would cost too much. He pounced. "That is not a scientific argument," he admonished his audience. "We are speaking about basic physics, not costs, because costs will change as a result of industrial development."

Economics was not their specialty, he chided the physicists. "How can you forecast the cost in the year 2030 of a technology which does not exist today but will exist in the future?" Scheer demanded. "Or, of a technology which is in the early stages of development, before mass production, how can you forecast the cost of that? No economist can do that," he continued, "because you cannot know the speed of improvements in productivity and it is impossible to forecast how many devices will be produced for PV. If you do not have this data, you cannot forecast the cost, and therefore this is an argument no serious economist can accept, and no one here is an economist!" *QED*. There were no further questions.

Such was the force of Scheer's argument and the power of his personality that few were brave enough to take him on directly. After another speech he gave, this time in front of five hundred students at a university, he called for questions. A student took the floor. How can solar be a realistic option if the energy payback time for a PV cell is longer than the module's working life? the youngster asked. "Where did you get this foolish argument from?" Scheer shot back. Then he heard murmurs around the room. "Immediately I realized that it must have come from a professor and that he must be here. So I asked the student, Was it from one of your professors?" More murmurs. "Is he in the room?" Louder murmurs, then some of the students muttered, Yes! "I asked the professor to come to the podium and reason his argument, on an empirical basis, explaining where he got his assertion from. He had no choice, he had to come forward. Then I told him, if a student hears this nonsense then it is OK for him to ask whether it is true, I can give him the answer. But if a professor, who is paid to have the knowledge to teach, presents such nonsense, then there is no excuse." It was "an execution, a real execution," Scheer crowed gleefully, "but he deserved it."

By the turn of the century Hermann Scheer had made himself the best-known public intellectual in the solar field. For all his tireless proselytizing, however, he had not achieved national prominence on the political landscape. Many Germans had probably never heard of him. He was after all just a deputy, a mere member of parliament, not a minister. As a seasoned political operative working behind the scenes, however, Scheer was highly effective. The crowning achievement of his political career would be the passage of the Renewable Energies Law in 2000. Together with its 2004 revision, this would provide the legal framework that supported Germany's unprecedented solar boom. How he and his colleagues managed to get this legislation enacted against the wishes of the government is the subject of chapter 17. First, however, we must go back to Aachen to see how the foundations for this epoch-making event were laid by grassroots activists working at the local level.

On October 14, 2010, just as this book was going to press, Hermann Scheer died unexpectedly, aged sixty-six. Solar had lost its greatest champion.

15

ESCAPE FROM STAWAG

*T*he city of Aachen dates back to the Romans, who took advantage of the hot springs they discovered there to build a spa. In the ninth century Aachen—known to the French as Aix-la-Chapelle—hosted the court of Charlemagne. During the Middle Ages the kings of Germany were crowned in the city. Following a devastating fire in 1656, Aachen declined in importance. Today the city is best known to horsey folk. Specifically, as the location of *Concours Hippique International Officiel,* the world's biggest open-air equestrian tournament, which each June attracts hundreds of thousands of spectators.

Aachen is located in North Rhine-Westphalia, Germany's westernmost and largest state in terms of both population and economic output. NRW centers on the Ruhr Valley, once the heartland of coal mining, an area that was for generations notorious for its belching blast furnaces, smoking chimneys, and soot-darkened skies. Industrial decline left the state a legacy of enormous ecological problems. By the late eighties a massive clean-up was under way. The locals wanted their skies blue again.

Aachen boasts a decent university augmented in the nearby town of Jülich by a technical college. In the mideighties this college became one of the first in Germany to offer courses in renewable energy technology. Politically, the Aachen branch of the Green Party was active. Local representatives of the Social Democratic Party were sympathetic to environmental initiatives. In 1989 a "Red-Green" majority with an SDP mayor took over the city council from the conservative Christian Democrats.

The local community included a coterie of concerned eco-academics and alternative-technology-savvy engineers, plus a cadre of anti-nuclear activists and peaceniks. The mix was leavened by a pinch of "crazy people," as Harry Lehmann described them, quickly adding that they were "*good* crazy people, who invented their own cars or whatever." Survival for such mavericks in conformist German society was never easy. What they had in common,

Lehmann felt, was "a certain sturdiness." Once committed to a cause they would stick with it, all the way.

Aachen was thus fertile ground for the radical ideas of Wolf von Fabeck and the Solarenergie-Förderverein. The first of these ideas, as we have seen, was that purchasers of PV systems should be fully compensated for their investment. Von Fabeck's original thought had been that solar buyers should present their receipts to their electricity supplier for reimbursement. He soon realized that that such a scheme would subsidize the purchase of even the most expensive hardware, regardless of cost, a wasteful use of public funds. It would also cause far too much organizational overhead.

Subsidies had intrinsic drawbacks. In 1989 eight German states were offering rebates to underwrite the purchase of PV. But such government-funded schemes would only pay for part of the cost, typically between 35 (Baden-Württemburg) and 70 percent (Berlin). They were not enough to incentivize most people to invest. Paid for out-of-tax revenues and offered on a first-come, first-served basis, subsidy programs were also prone to bureaucratic delays. Applications would get tangled up in red tape, annual quotas would be filled halfway through the year, forcing potential buyers to postpone their purchases. Worse, where the dispensing of large chunks of taxpayer money is concerned, budget cuts are always a danger. Legislatures have to renew such programs annually. A subsidy program introduced by one government can be slashed at the stroke of a pen by its successor. Uncertainty is thus introduced into the market, anathema for manufacturers. They hesitate to invest in new production capacity, the very opposite of what von Fabeck wanted. Subsidies were thus obviously not the way to go. What other ways of paying for PV were there? Inspiration, and a partial solution to the problem, came from Switzerland.

Burgdorf is a small town (population 15,000) in the canton of Berne in north-west Switzerland. The region is best known for producing Emmental, a hard cheese with large holes that is used to make the classic Swiss dish, fondue. Like their neighbors the Germans, the Swiss had also been deeply affected by the catastrophe at Chernobyl, voting for a moratorium on nuclear power in a 1990 referendum.

Like many Swiss towns, Burgdorf owns and operates its local utility. Burgdorf also has an engineering school. There, in 1987, professor Heinrich Häberlin decided to investigate photovoltaic technology, to see if it could contribute to a non-nuclear future. The director of the local utility, Theo Blättler, approved of Häberlin's initiative, paying him to hire an additional assistant to do research on PV. At other Swiss engineering schools, the focus was on building solar cars. Häberlin decided to concentrate on a more practical theme: grid-connected solar systems. In 1988 the researchers hooked up an experimental 3-kilowatt PV system to the local grid. It worked well.

Thus encouraged, and with popular support from the citizens of Burgdorf, in January 1991 Blättler introduced an innovative scheme that was designed to encourage local homeowners to install photovoltaic systems. The utility would pay a premium price of one Swiss franc (around 70 US cents) for every kilowatt-hour generated by PV systems installed during the next five years. All electricity thus produced would be fed into the grid and recorded by a dedicated meter. Payments would start from the day of commissioning and continue for twelve years. The scheme would be paid for, not by taxpayer subsidy, but from a fund created by setting aside 1 percent of electricity rates. Hence the name, "rate-based incentives."

Though generous, the incentives were not sufficient to cover total PV system costs. A top-up from the national subsidy program was needed to make up the difference. But rate-based incentives worked: during the next five years, Burgdorfers installed an average of thirty-four solar systems a year. By the turn of the century, Switzerland was generating more solar electricity per capita than any other country, while Burgdorf was producing ten times more PV per capita than the Swiss average.

Burgdorf had taken a great leap forward. Switzerland's highly decentralized political system is good for taking initiatives at the local level, but less so at propagating those initiatives nationally. Rate-based incentives would be picked up, not in Switzerland, but in Germany. "Burgdorf was paying higher tariffs long before we had the idea," von Fabeck admitted. "It wasn't a cost-covering feed-in tariff—the Swiss just paid a little bit more, and that was a good idea. But they didn't make a broader concept or principle from it. However, the Burgdorf example was an inspiration for us, which we wanted to take further." And take it further they did—much further.

Thus far, the reason most people had installed PV systems was for their own use, typically because they wanted to do something good for the planet.

Anything leftover went to the grid. Under the system pioneered in Burgdorf, by contrast, *none* of the electricity generated was for personal use, it was *all* fed into the grid. The intention was to help meet society's need for clean energy, not the individual's. This new scheme fundamentally altered the role of the private generator.

"The main idea," von Fabeck explained, "was that private owners of solar systems should be treated exactly as if they were utilities. Like any power company, owners should be able to feed the electricity they generate into the public grid and receive in return a payback rate that allows them to finance and maintain their systems, plus a small profit. Since everybody consumes the clean energy that the owners generate, everybody should pay for it, via their electricity bill. This way, the utilities would be responsible for reimbursing private producers. There would be no need for state-funded subsidies, or new taxes, or any government involvement at all."

Ultimately, it all boiled down to a simple equation. Given that (a) the rated lifetime of a household PV system is twenty years, and that (b) the average annual "insolation"—that is, the amount of solar radiation per unit of horizontal surface—is known, then (c) the number of kilowatt-hours a system will generate over twenty years can be calculated. The upfront cost of a typical residential PV system was around $25,000. The question was thus, How much would a utility have to pay system owners per kilowatt-hour over twenty years to compensate them for their investment?

Von Fabeck worked out that the answer was two deutschmarks ($1.20) per kilowatt-hour, around twenty times the cost of conventionally generated electricity. To pay for this, as at Burgdorf, a 1 percent surcharge on electricity rates would suffice. The annual increase in cost to an average family would be just 30DM ($18). "The initial thinking was that [the surcharge] would be too heavy a burden," recalled Hans-Josef Fell, the Green deputy who would author the federal solar feed-in tariff bill (and who we shall meet in the next chapter). "But Wolf von Fabeck was the first who calculated this, and he opened our eyes—it is nearly nothing for consumers."

Germany's electricity market is highly fragmented. The market was, and post-liberalization in 1998, still is, dominated by giant firms, the equivalent of US investor-owned utilities. Today, the "four sisters"—E.ON, RWE, Vattenfall, and EnBW—generate between them more than 80 percent of Germany's electric power. The remainder comes from local electric utilities, about a thousand of them, typically corporations whose stock is owned by the municipalities they serve. One such was Stadtwerke Aachen AG (STAWAG), which since 1901 had been providing Aachen's citizens with gas, electricity, and water. (Stadtwerke means "municipal utility," AG stands for *Aktien Gesellschaft*, which means "stock corporation.")

Now, armed with their two basic ideas—full-system cost compensation paid for using rate-based incentives—von Fabeck and the SFV launched their pro-solar campaign in earnest. One tactic they employed was to gather signatures from citizens for a petition to the city council to force STAWAG to implement their ideas. Another was to lobby local councillors directly. Suitable representatives were chosen, von Fabeck explained, according to appearance. "We said, You've got a beard, you talk to someone from the Greens; you know how to knot a tie, you talk to someone from the Liberals!"

The solar activists explained their concept to the politicians. To their surprise, the response was positive. Almost all the councillors, even the conservatives, thought that it was good. At a council meeting in September 1992, it was proposed that STAWAG should implement full-cost compensation for PV systems. The utility would pay a new tariff of two deutschmarks per kilowatt-hour, guaranteed for twenty years. Installations would be capped at 1 megawatt. The proposal passed by a majority vote, members of all parties except the knee-jerk neo-liberal Free Democrats agreeing to the plan.

The problem now was STAWAG. The utility flatly refused to implement the council's proposal. STAWAG was not opposed to solar per se, so long as it owned the means of production. Indeed, in 1991 the utility had replaced the south-facing glass facade of its twenty-year-old administration building in Aachen with an innovative PV system. This was a chessboard-patterned combination of glass and 4 kilowatts of translucent blue solar panels, an early example of what would later be known as building-integrated PV.

What the utility and in particular its CEO, Wolfgang Petry, a bigwig in the national society of municipal utilities, objected to was paying private producers the premium price that the council proposed. Such an outcome would be illegal, Petry insisted. In addition, there was also the issue of keeping accounts of who had generated what, a responsibility the utility was not willing to accept. STAWAG kept coming up with reasons for not complying with the city's instructions. Unfortunately, the way German stock corporations are structured, once the CEO decides not to do something, it is very hard for shareholders to overrule his decision. From the time the solar proposal was submitted to council until final approval was granted would take almost three years.

The battle generated intense public interest. To cover the debate Aachen was fortunate in having two local newspapers. If one didn't report a new development, then the other one would. Hundreds of letters supporting the rate-based cost-covering proposal poured in. The papers also published some letters opposing the measure. It turned out that nearly all of them were written by employees of the utility. This too was made public. The blaze of publicity surrounding the SFV's fight against STAWAG helped build support in Aachen. It also stimulated public awareness of PV in other cities and towns, in the state of NRW and elsewhere in Germany, where there were also groups keen to encourage the production of electricity from renewable sources. Soon, other municipalities would be instructing their utilities that they wanted to introduce pro-solar incentives along the lines of the model that Aachen had developed.

To Bob Johnson, a veteran Silicon Valley–based solar industry analyst, what the folks in Aachen were proposing sounded beguiling. "I'd heard about the initial details, and I was fascinated. The production subsidies they were proposing were far far better than equipment subsidies that we now enjoy in the United States. To my way of thinking they particularly addressed the question of, Why are we putting this equipment online? That is, to obtain clean energy, not photovoltaic equipment. Plus the characteristic of the scheme was that it would be paid for, not by adding to the federal tax rolls but as it was utilized, by adding a very very small increment like the equivalent of a tenth of a penny per kilowatt-hour to the electricity bill. The cost is spread over the entire customer base, so more people are encouraged to get onboard, as opposed to just paying for their own utility bill. I just thought it was a wonderfully healthy way for the grid to grow."

In Europe for a conference in Holland in 1991, Johnson took the chance to pop across to Aachen for lunch with von Fabeck. The latter was, Johnson thought, very logical. "When we met I asked him, How much of a subsidy are you planning to pay back to the people who put up their own systems? He gave me the figure, 2DM. I had to pause a few seconds to convert it from deutschmarks, then I said, Oh my God—that's over a dollar! He looked at me as if I were one of the enemy. From the mild-mannered retired electrical engineer he became much like a German general—I consider him the General Rommel of the PV movement, I didn't know he had actually been a soldier—and he said, Well, are you with me, or are you against me? He was very combative."

Until the liberalization of the German electricity market in 1998, the price that utilities could charge for electricity to households, small businesses, and farmers was controlled by state governments. In Düsseldorf, the capital of North Rhine-Westphalia, the Ministry of Economic Affairs was responsible for setting prices for the hundred and twenty-odd utilities in the region. Seeking to force the utility to accept its proposal, the city of Aachen appealed to the state economic minister, Günther Einert. Both sides made their case; Einert ended up ruling in favor of STAWAG. In his judgment of January 1994 he indicated the reason for his refusal was that the proposal would cause the price of electricity to rise too high: in his view, this would not be in the economic interests of German consumers. Einert also let it be known that he had no intention of using the electricity price to support "a club of hobby engineers and solar tinkerers," an unmistakable swipe at von Fabeck and the SFV.

Though the lengthy battle was frustrating for the pro-solar camp, it did ultimately have an upside. "[STAWAG's CEO] always submitted long letters to the minister, written by his experts, explaining the legal arguments why it was not possible to do what the shareholders wanted," said Harry Lehmann, who in addition to representing Eurosolar also headed the SDP's energy and environmental group in Aachen. "He had a lot of lawyers, he was always looking for reasons why he should not follow the advice of his owners to do this and that. But this was good luck for us, because these letters showed up all the holes we had in our concept. So we were able to adjust our concept. He slowed the whole thing down, but ultimately he made it legally clear, so that later no one could find anything else wrong with it."

Particularly contentious was the issue of whether a private producer should be allowed to make a profit. Von Fabeck had become convinced that it was not enough merely to recover one's costs: people needed the incentive of a return in order to persuade them to invest. They should, he insisted, be treated exactly the same as if they were utilities.

That had been the unexpected lesson of StrEG, the original feed-in tariff law, back in 1991. The most important provision in creating a market for renewables was to make the rate high enough so that the investor gets a return on the investment. The economic minister was utterly opposed to this idea. In his view, individuals who generated electricity from photovoltaics should be happy to get *any* money back. If after twenty years they had made a loss, then that was their problem.

"We had calculated that if someone invests in making electricity with solar, then he should get a return on his investment," Lehmann said. "But we didn't clarify how much the profit would be. The CEO argued that he could not agree to a decision that allowed people to become rich through renewable energy technology. So we had to look for something in the law relating to that. We found an agreement between all the states in Germany which said that every utility can make something like 6.5 percent per year profit. Then we said, OK, if every nonprivate utility can do that, then private producers should also be able to make 6.5 percent per year profit on the capital they put in. So although he was able to stop the game for a few weeks or months by finding legal reasons not to do something, because we were able to adapt and make changes, we ended up winning."

Following the minister's ruling, a group of NGOs led by the SFV and Eurosolar went to court to challenge his interpretation. The essential question was whether the law allowed electricity rates to be increased for environmental reasons. The court decided that German law does allow an increase, of up to 5 percent in the price of electricity, for environmental reasons. With the legal battle won, the ball was now back in the court of the Ministry of Economic Affairs. Here, the solar activists were aided by an unexpected ally within the state bureaucracy.

Dieter Schulte-Janson was the official at the ministry with the responsibility for regulating electricity prices in North Rhine-Westphalia. His initial concern had been with Aachen's proposal to fund the scheme by putting a surcharge on the rates paid by the consumer. Was that allowed? The utilities said no. His minister agreed with them. But Schulte-Janson bravely decided to disregard his minister's decision. He commissioned a legal opinion from a leading law firm. They agreed with his view, that it was acceptable to impose a surcharge to promote the production of electricity from renewable sources.

Fifteen years later Lehmann and von Fabeck remained deeply grateful to Schulte-Janson for having stuck his neck out for them. He was, Lehmann thought, a representative of the best type of civil servant. Schulte-Janson would not stand for any nonsense, he would do things properly, always adhering to the rules. But if the law allowed him to assist the underdogs in their struggle against the power of the monopolists, then he would do it. "I wanted to fight with people who are more like Davids against the Goliaths,"

Schulte-Janson told me. "The Goliaths really didn't need a voice, but Lehmann and von Fabeck were Davids and they still did need help. So I gave them a little helping hand, not more." For this, Schulte-Janson would not be rewarded; indeed, he was subsequently "sent to Siberia, so to speak," transferred from regulating electricity to water.

The court ruling, backed by Schulte-Janson's legal opinion, left the minister no choice but to allow Aachen—and three other cities in NRW, which in the interim had also applied for permission—to go ahead with their proposal. First, however, the minister wanted a method for calculating the cost of solar electricity. To hammer one out, Schulte-Janson set up a commission consisting of representatives of the four stakeholder groups: consumers, prorenewable organizations (with Eurosolar's Lehmann as their representative), utilities, and the regulator. The format Schulte-Janson chose for the negotiations was a roundtable. "If you have a roundtable," he told me, "you can be sure that you get a result which everybody stands behind and that was what I wanted. I didn't want to have fights, one against the other, I just wanted consensus."

The commission sat for three months. It worked out a formula for calculating the price of a kilowatt-hour of solar electricity. This was highly specific, including system cost, ratio of investment between owner and bank lender (40:60), and the amount of interest to be paid on that investment. An additional 1.5 percent was added to cover insurance and system maintenance. Annual rental of the second, feed-in meter was also included. Operating lifetime was established at twenty years. The figures added up to a final price for feed-in compensation for PV of 2.01 deutschmarks, exactly in line with von Fabeck's original calculation.

The methodology was published on November 29, 1994. A precedent had been set. During the next six years the model would be widely adopted. By 1996 around twenty-five German towns and cities, including some big regional centers like Hamburg and Munich, had implemented solar feed-in tariffs based on the Aachen model. The snowball kept growing until it triggered an avalanche. By mid-1997 the total number of cities and towns had reached forty-two—half of them in Bavaria—with a combined population of over 7 million. In just one year, between 1995 and 1996, installations of PV in Germany leapt from 100 kilowatts to almost 2 megawatts. "With the support of residents, politicians, solar advocates, and the PV industry," Anne Kreutzmann, an Aachen-based activist, predicted, "a federal law could be written to make rate-based incentives available for all of Germany." She was right. In 2000 Wolf von Fabeck's Aachen model would become the basis for the Renewable Energies Law that would establish solar feed-in tariffs at the national level.

Ironically, Aachen was not the first to implement the model that bears its name. Other cities with more compliant utilities leapfrogged ahead. First to adopt rate-based incentives, as we shall see in the next chapter, was the small northern Bavarian town of Hammelburg, home of Hans-Josef Fell, who would subsequently author the federal Renewable Energies Law.

While Wolf von Fabeck and the SFV deserve much of the credit for the dramatic increase in PV installations, Lehmann felt that it was important to acknowledge that they had counterparts all over the country. "The reason why solar is a success in Germany is not because it was happening in one place, but because it was happening in ten, fifteen, twenty places," Lehmann said. "There was a movement, a grassroots movement in the real meaning of the word 'grassroots.' Suddenly local newspapers all over Germany were writing about these groups, then the movement spread. You know, grassroots movements are not controllable, you can't stop them—the people are very motivated. And when ideas are good, they propagate through society very fast."

16

ELECTRICITY REBELS

*A*ll his life people had been telling Hans-Josef Fell that what he wanted to do was impossible, and all his life he had been proving them wrong. Even within his own party, the anti-nuclear, pro-environment Greens, Fell was often seen as an extremist because he had been one of the first to advocate 100 percent renewables as the solution to Germany's future energy needs. But in addition to being a visionary, Fell is also an eminently practical person. He would repeatedly demonstrate to doubters that his radical ideas were doable—by putting them into practice himself. For example, in 1984, on a hill on the outskirts of Hammelburg, the small town (population 12,000) where Fell was born and raised, he built a log house that was powered entirely by renewables. The solar thermal panels on its southern-facing facade provided hot water, the vegetable-oil fired boiler in its basement, heating, and the 1.5 kilowatt photovoltaic system he installed in 1991 on its grass-covered roof, electricity.

In other aspects of his life, too, Fell found ways to demonstrate that it was possible to live sans fossil fuels. The flimsy-looking motorized tricycle he rode on the forty-mile round trip to and from the high school in nearby Schweinfurt where he taught was propelled by a combination of recharge-able batteries and pedal power. His other, more conventional vehicle, a VW sedan, ran on vegetable oil. In 1993, when Fell proposed drafting a local law based on the Aachen model to promote the take-up of solar energy, it would be the same thing. "Everyone told me that it was impossible to make such a law," he recalled happily. "So I created it, and it was possible."

Hammelburg nestles in a shallow valley in Franconia, a region of north-west Bavaria. A fire in the eighteenth century destroyed much of the old town, including most of its surrounding wall, but Hammelburg remains a picturesque melange of cobbled streets and twisted lanes lined with red-roofed two-story houses. It is the oldest wine-making municipality in Franconia. This the only part of Germany that bottles its wines in the stubby, round-shouldered *Bocksbeutel* rather than the familiar long, thin-necked bottle. (The shape derives, I was informed by the master of the town's wine cellar, from that of a goat's scrotum.) According to wine writer Hugh Johnson, Franconia is also the only place where better wine is made from the humble silvaner grape than the noble riesling. Dry rather than sweet, silvaner makes "one of the best German wines to drink with food."

Most people think of Bavaria as the land of lederhosen, oompah bands, and beer bacchanal, a strongly Catholic state that is traditional, conservative, and old-fashioned. And to be sure, the ruling right-wing Christian Social Union Party has been continuously in power in Bavaria longer than anywhere this side of Cuba. But although Bavarians still like to drink beer, the stereotype no longer holds true. Germany's largest state is home to global companies like BMW and Siemens. Its capital, Munich, encourages the kind of entrepreneurial culture in which high-tech start-ups can flourish.

Bavaria also boasts some of Germany's most progressive laws governing direct democracy. They make for spirited participation by citizens in local and municipal affairs. The Greens fare particularly well in southern German university towns and localities. Nuclear is the source of 70 percent of the state's electricity. Bavaria operates five of Germany's nineteen nuclear plants, more than any other state. Located in the south, Bavaria also happens to be the sunniest part of the country. As a result, anti-nuclear organizations and pro-solar groups are most numerous in the state. Between their memberships, there is much overlap.

Hans-Josef Fell was born on January 7, 1952. He is thus like Hermann Scheer, a sixty-eighter, a member of the generation that came of age during the radical student movement of late sixties. After graduating from the University of Würzburg, Fell taught physics and sports. He honed his competitive edge through volleyball, first as a player and subsequently as a coach. Environmentalism was part of his pedigree. "My father was a nature lover, he taught me how to think about environmental issues," the soft-spoken Fell told me. "I too am a nature lover; I want to protect nature, that's my intrinsic motivation."

His father was the mayor of Hammelburg and a member of the conservative Christian Democrats. On a hill outside Hammelburg is a big army base, which also happens to be the largest employer in the area. At one anti-war rally in the town, his father took up his place at the head of the soldiers, while Fell marched along with the demonstrators. The peace movement and the fight against what Fell called "right-wing radicalism" would remain key planks in his political platform. On the wall of his office in Berlin he keeps a picture of Mahatma Gandhi. The inscription reads: "There is no way to peace; peace is the way."

I meet Hans-Josef Fell twice: once at his office in 2007, then again the following year in Hammelburg. There, I take a terrifying uphill ride in his electric "Twike" with his charming and vivacious wife, Annemarie, to visit their grass-roofed, renewable-powered log cabin home. Balding, bearded, and bright-eyed, Fell is a cheerful man with an open, forthright manner. As befits a former volleyball player, he is tall and powerfully built. Fell exudes the kind of confidence that comes from twenty years of classroom teaching followed by ten years of successful politicking in the Bundestag.

The key characteristic with Fell, as with Hermann Scheer, is that he thinks for himself. Trained as a scientist, he can look at the facts of an issue and make up his own mind about what is right and what is not. Fell's radical thinking on energy dated back to the results of a 1980 federal government commission of inquiry. This listed renewables-only as one of five possible future energy scenarios. Admittedly it was the least likely scenario, and the government had chosen to go with the most likely, i.e., nuclear, but in Fell's mind the situation was quite clear. "I didn't like nuclear," he told me. "And, because I studied physics I can make my own decisions, I'm not dependent on what others tell me." Especially not on what Germany's big utilities were telling people during the eighties, that solar would never amount to anything more than a marginal technology.

Fell loves to tell the story of a project he used to assign his students. As preparation they measured how much solar energy a given area receives. "Then I gave the pupils the task: Go home and calculate how much energy the whole world receives, and compare it with the amount of energy the world consumes. And they would come back with the answer: Mr. Fell, the supply of solar is fifteen thousand times more than the total energy demand!" Then he would beam at them and say, "You see, the managers of the big utilities cannot perform such an easy experiment—they only produce disinformation!"

Politics was in Fell's blood. His father was mayor of Hammelburg for eighteen years, "so I learned in my youth how to make politics." Fell began his political career as an activist at the grassroots level. He worked with some of Bavaria's hundred-odd pro-solar NGOs, giving speeches at conferences and organizing local trade fairs. In 1990, following in his father's footsteps, he too was elected a town councillor. As such Fell automatically became a board member of the local utility. The smaller the utility, the easier it is to influence. In his case, Stadtwerke Hammelburg proved more amenable to persuasion than STAWAG. In late 1993 Fell drafted a law based on the rate-based, full-cost compensation Aachen model. The incentive rate would be 1.89DM per kilowatt-hour, the payback period twenty years, and the individual system limit 3 kilowatts. While Aachen was delayed by its legal battles, Hammelburg became the first place in Germany to implement such a measure.

With the law on the books, it was important to demonstrate its effectiveness in promoting the uptake of solar. Getting panels onto roofs was also crucial in winning acceptance for the Aachen model from the state regulatory commission, which had yet to rule on the matter. Fell decided to jump the gun, to present the authorities with a fait accompli. He launched a campaign to raise money to pay for PV systems, collecting two hundred thousand deutschmarks (about $120,000) from donors all over the country. Wolf von Fabeck was one of the first of seventy-odd individuals who contributed to the cause. Using this money, in 1994 Fell founded a company, Hammelburg Solar Power, with himself as CEO. Its mission was to install photovoltaic systems on the roofs of local residents, most of whom were also shareholders in the company. Whether the regulatory commission would allow utilities to pay premium rates to private generators was by no means a foregone conclusion. The pro-nuclear state government in Munich was vehemently opposed to renewables. "I had to tell everybody that we could not guarantee that they would get their money back," Fell said.

Hammelburg's law capped installations at just 15 kilowatts, a piddling quantity, enough to power only ten or so homes. But at least it was a start. "Today we laugh about this amount," Fell said, "but it was very important because we showed that 15 kilowatts could create a market." It represented more than 1 watt of solar power per person. Extrapolated nationwide, that would be 80 megawatts, a not insignificant amount. The learning curve was steep. Everything had to be done from scratch. "In order to install the photovoltaic systems, we had to create all the paperwork, make contracts with the utility and with insurance companies."

Meanwhile in Freising (population 42,000), near Munich, Fell's good friend Ernst Schrimpf, a professor of biology and coordinator of state-wide solar initiatives, was implementing a similar version of the Aachen model. The main difference was that the cap there was 100 kilowatts. As we have seen, North Rhine-Westphalia would be the first state to allow rate-based cost compensation. Once one state had green-lighted the model, it was relatively easy for Fell and his colleagues to put pressure on the Bavarian government by saying, Look at NRW—they're allowing it. In 1996 with big cities like Munich (population 1.3 million) and Nuremberg (population 500,000) also moving to adopt the Aachen model, Bavaria's regulators capitulated to the inevitable.

In 1998 Fell was elected to the Bundestag as a member of the first-ever Red-Green (Social Democrat-Green Party) coalition. It took over government from the Christian Democrats who had ruled Germany for the previous sixteen years. In Berlin, as we shall see in the next chapter, Fell would translate his invaluable expertise at the local level into the drafting of the federal government's epochal 2000 Renewable Energies Law. This would be the measure by which solar would at last be transformed from niche player to mainstream market. Once again, Fell would show nay-sayers that the impossible was possible.

17

AN ECOLOGICAL MASTERPIECE

*T*he "Red-Green" (Social Democrat-Green Party) coalition government took office in September 1998, committed to phasing out nuclear power and promoting renewable energy. Despite the adoption of solar rate-based cost-covering incentives in towns and cities across Germany during the previous four years, PV still needed all the help it could get. Other countries were forging ahead. Japan's New Sunshine Project was in full swing. The previous year, the Clinton administration had announced an ambitious-sounding Million Solar Roofs Program (see chapter 20).

One of the newly elected coalition's first actions was to sign the Kyoto Protocol. The world's sixth-largest producer of greenhouse gases, Germany pledged to reduce its emissions 21 percent from 1990 levels by 2008. One strategy for achieving this goal was to increase the portion of energy sourced from clean technologies. All parties realized that StrEG, the original feed-in tariff law, would have to be revised in order to stimulate the spread of solar, as the 1990 act had done so successfully for wind. Crafting the revisions would take time. Meanwhile, prompt action was required. "The photovoltaic sector was not doing too well," Hans-Josef Fell recalled. "We wanted to make a strong gesture in favor of renewable energies."

Hermann Scheer had first proposed his 100,000 Roofs Program back in 1993. Now he pushed for it to be dusted off and pressed into service as a stopgap measure. Its predecessor, the 1,000 Roofs Program, had been based on state and government subsidies. The successor would by contrast be financed by ten-year, interest-free loans from a government-owned development bank, Kreditanstalt für Wiederaufbau. This would be less of a direct burden on the taxpayer. It was, as Fell pointed out, "just a first step." For the sake of speed, the 100,000 Roofs Program did not take the form of a law, but of a decree enacted by a reluctant Ministry of Economics & Technology. The five-year program was scheduled to begin on January 1, 1999, with a budget of 1.1DM billion (about $660 million). In fact, the program was launched more than three months late as the ministry dragged its feet. Though considerable, the financial help the program provided was insufficient to engender a flourishing market. The target was to install 350 megawatts of PV by 2003. By the end of the program's first year, however, only 3,500 loans had been granted, representing a mere 9 megawatts in total.

In October 1999 moves to revise the feed-in law began in earnest. The Greens, led by Fell, their newly elected spokesman for research, wanted to transcribe local rate-based incentive laws into federal legislation. Scheer led the pro-renewables faction within the Social Democrats. For Fell, with his relatively small party of 47 deputies, persuading a majority was relatively straightforward. For Scheer, with his much bigger party, the task was far harder. Especially since the SDP had had a long and close relationship with the coal mining industry. But coal was palpably in decline. Scheer had to convince a majority of the SDP's 298 deputies to look to the future, not the past. Winning the backing of the unions was the turning point, he thought, in changing the attitudes of his fellow deputies in the Bundestag.

The Greens organized a coalition of supporters. In addition to various environmental groups, solar energy industry associations, and solar cell manufacturers, this included representatives from IG Metall, the influential metalworkers' trade union, and the Equipment and Machinery Producers, an industry association. The latter counted among its members some three thousand firms with a combined total of around a million employees. Its support demonstrated the increasingly broad-based legitimacy that renewables were beginning to enjoy in Germany.

For its part, the pro-renewables faction of the Social Democrats was also motivated by a strong interest in industrial policy. "They feared that the 1998 liberalization of the [electricity] market would lead to a long-term decline in the energy sector and in the associated capital goods industry which has always been a point of strength of German industry," wrote energy policy specialists Staffan Jacobsson and Volkmar Lauber. "Strong renewables legislation, these deputies argued, would put German industrial structure and employment on a more sustainable base both environmentally and economically." In other words, as well as being good for the climate, switching to renewables would also create new jobs.

Revising the feed-in law was nominally the responsibility of the federal economics minister, Werner Müller. A former electric power industry boss without party affiliation, he had been brought in by Social Democrat chancellor Gerhard Schroeder to beef up the government's pro-business credentials. The pro-solar parliamentarians believed they had reached an understanding with Müller regarding the nature of the revisions. Then the minister started making some worrying public statements. One concerned a renewable energy levy that the utilities would be able to impose voluntarily. Another, the idea of using a quota system to boost the amount of renewables in the energy mix. Soon, it was clear that Müller had no intention of respecting the understanding.

Faced with the minister's recalcitrance, the parliamentary groups led by Scheer and Fell seized the initiative. They decided that they would submit their own legislation. Fell, with his hands-on experience and knowledge of the issues, would himself draft what came to be known as the *Erneuerbare Energien Gesetz* (Renewable Energies Law), or EEG. What he wrote was a short document, only about four pages long. Its explicit aims were "to increase the share of renewable energies in the total energy supply to at least 12.5 percent by 2010 and to at least 20 percent by 2020, and to further develop technologies for the generation of electricity from renewable energies, thus contributing to the reduction in costs."

Unlike its predecessor, the Electricity Feed-In Law of 1990, which had been technology neutral, the EEG differentiated between generation technologies. It specified tariffs for each type of system including solar, calculated according to the formula that Harry Lehmann had worked out, based on the system cost plus a reasonable profit. The new law offered three justifications for requiring utilities to pay premium rates to generators of renewable energies. First, it referred to the "polluter pays" principle with regard to external costs. Second, it asserted that "conventional energy sources [coal and nuclear] still benefit from substantial government subsidies which keep their prices artificially low." Third, the law noted the need to break the vicious circle of high unit costs and low production volumes typical of technologies for the generation of electricity from renewable sources.

An innovation derived from the Aachen model and incorporated in the new law was that investors would be guaranteed the premium rates for twenty years. Incentives for wind were also increased to give investors greater security. But the increase in incentives for solar was far more dramatic. The new rate would be 0.99DM (51 eurocents, around 49 cents) per kilowatt-hour for roof-mounted PV systems of up to 5 kilowatts, capped at a cumulative total of 350 megawatts. This was more than three times the cost of conventionally generated electricity, then about 15 eurocents (around 14.5 cents) per kilowatt-hour. Another innovation that came from Aachen was what is known in the jargon as a "degression," literally, a stepwise reduction in the rate of compensation. In other words, to encourage buyers of PV systems not to postpone their purchases, and manufacturers to cut their production costs, the rate paid to new investors would decline by, initially, 5 percent per annum.

In the German political system it is highly unusual for legislation to come from the backbenches rather than the ministries of the governing party. In the case of renewables, however, expertise resided not with the bureaucrats, but with parliamentarians like Fell. The leadership of the SDP remained inwardly focused on decision making within parliament. Unbeknownst to them, Hermann Scheer had been surreptitiously gathering a large number of Social Democrat deputies behind him. I asked him how he had managed to pull off this remarkable political feat. He told me that he had built up support for his cause in his own inimitable way.

"Each parliamentarian wants to have in his constituency speeches by guests, where many people come," Sheer explained. "Normally, when they invite someone to speak on this or that, only twenty or thirty or forty people come, even if the speaker is a minister." Aware of his reputation as a charismatic speaker, Scheer's colleagues were keen to have him come to their constituencies and give a speech on renewable energy. The hosts could then bask in the reflected glory. Scheer would agree, on one condition: "You must support what I say," he told them, "because after I speak, your constituents will ask, What is your position—do you support what he says or not?" Thus Scheer won backing from many of the more open-minded among his colleagues.

Within their respective parties Scheer and Fell hammered out the details of the EEG. In order to chivvy the other members along, it was necessary from time to time to resort to what Fell described delicately as "tactics." "Sometimes I went to my group and told them, The Social Democrats agreed to this an hour ago, are you as Greens going to let them get ahead of you on this new policy? But in fact that was not correct, they had not agreed. Hermann Scheer would likewise go to his group and tell them, The Greens have already voted in favor of this—are you going to let them take the lead?" The two men developed a good working relationship. Scheer regarded Fell as his protégé. But he deferred to the younger man's hands-on experience with the legal and technological aspects of solar. For his part Fell respected Scheer's deep understanding, accumulated over two decades, of how to shepherd legislation through parliament: "Without him we could not have created this law," Fell said.

Scheer's tireless speech giving had won him many favors. On their own, however, these would not have sufficed to win a majority for the Renewable Energies Law. Out of party loyalty, deputies almost always toe the government line. The economics minister continued his unrelenting opposition to the bill. Then, fortuitously, something unexpected happened. Discussion and drafting of the EEG took place during October and November of 1999. The vote on the bill in the Bundestag was slated for February 2000. On December 9, just as the draft bill had been tabled, the news broke that Scheer had won the Right Livelihood Award, or the Alternative Nobel Prize as it is sometimes known. Like the Nobels this is a Swedish award, presented annually to honor those "working on practical and exemplary solutions to the most urgent challenges facing the world today." It covers such fields as environmental protection, human rights, sustainable development, health, education, and peace.

"The party was very proud of that," Scheer said. "The representatives of big energy went to [Chancellor Gerhardt] Schroeder and told him, You should block this bill! And he said to them, No, I will do nothing, I will let it pass. Because if I try to block it, I will have a conflict with Hermann Scheer, and at this moment I will lose this conflict!" Scheer recalled, laughing loudly. Winning the prize, he believed, represented a psychological breakthrough. It united his party behind him at exactly the right moment. "Half a year before it would have been much more difficult [to get the bill passed], or one year later too."

On February 25, 2000, in the magnificently refurbished debating chamber of the austere, battle-scarred Reichstag building, to which the Bundestag had returned the previous summer after an absence of sixty-six years, the EEG was voted into law. The tally ran 328 in favor, 217 against, with 5 abstentions. It was, according to one observer, an "ecological masterpiece . . . the coalition's single greatest accomplishment."

Combined with soft loans from the 100,000 Roofs Program, the revised feed-in law meant that solar became, for the first time, an attractive invest-ment. Especially for a new category of eligible system owners, commercial enterprises. Whereas an individual household might install up to 5 kilowatts, these new operators would put in much bigger systems, of 100 kilowatts and more. Prior to the passage of the EEG, the program had managed to under-write just 12 megawatts' worth of PV. Within a month of it becoming law, applications for a further 35 megawatts had been received. The year 2001 would be the most successful ever for installations of grid-connected solar.

By April 2002 some 34,000 permits had been issued under the program, and around 135 megawatts of PV installed. As applications for permits poured in, the administrators' phones rang off the hook. Assistants in Fell's office noted happily that on some days, up to 1,200 requests for permits had been recorded. In fact, things were moving too fast and the bureaucrats were soon overwhelmed. A three-month moratorium was imposed. Time and again the program had to be halted, and conditions for obtaining loans tight-ened. "I had such letters every month," Fell recalled, laughing. "Herr Fell, please help us, the money did not come! It was terrible for us, too."

In the parliamentary elections of 2002 the Red-Green coalition was reelected. Support for the Greens rose, while that for the Social Democrats declined. "[T]hus the Greens could claim a stronger position in government, and effectively secured the transfer of competency for renewable energy from the economics ministry (held by a Social Democrat), to the environ-ment ministry (held by a Green)." In 2002 the cap on the total amount to be installed under the 100,000 Roofs Program was tripled, from 350 megawatts to 1,000 megawatts. In the first six months of 2003 alone, permits for 150 megawatts were allocated. Then, having reached the target of 350 megawatts, the soft loans program ran out. The program had caused systems to be installed on more than 60,000 roofs, at a cost of 1.7 billion euros ($1.45 billion). During the time it had been in effect—1999–2003—the cost of solar systems dropped by around 24 percent.

In the second half of 2003 investment in PV systems plummeted. The reason was that, on their own, the premium rates stipulated by the EEG were not sufficient to persuade would-be owners to invest. To sustain the momentum a revision to the law was urgently required. Once again Hermann Scheer proposed upping the ante, in the shape of a new and more ambitious target: a million roofs by 2010. This time, however, the Greens objected. Fell was against further subsidy programs because he wanted to wean solar off dependence on taxpayer funds. Eliminating subsidies also meant the bureaucrats would no longer be able to drag their feet. However, he and Scheer did agree on another crucial objective: the removal of the cap on the amount of PV that could be installed.

In the struggle to revise the law, Wolfgang Clement replaced Müller as "super-minister" of economics and labor. Clement had supported the introduction of premium feed-in tariffs for solar. He was also closely aligned with coal interests in his home state of North Rhine-Westphalia. By 2003 the minister had come to question the underlying principle of feed-in tariffs. He argued that the premium rates paid were excessive: the coal industry had to be protected. Clement was joined by critics from the opposition conservative parties, notably Angela Merkel, the future chancellor. She attacked the revised EEG, arguing that its subsidies would represent a burden on the budget. This was patently false. Since feed-in tariffs are paid directly by electricity consumers, they do not figure in the budget.

Fierce opposition to feed-in tariffs also came from the energy sector, most notably the four big investor-owned utilities that dominate the German electricity market. They were still reeling from the decision, taken by the Red-Green coalition in 2002, to phase out nuclear power, all nineteen of the country's reactors to be shut down by 2030. Now, with the proposed revisions to the EEG, the utilities faced the very real prospect of losing over time at least some of "their" market. The main strategy of the power companies, dating back to 1996, had been to attack feed-in tariffs in the courts of the European Commission on the grounds that they violated rules against member states providing aid to specific industries. In 2001 their case was finally thrown out.

At the same time support for feed-in tariffs was growing. In 2003 the ranks of the pro-renewables consortium were swelled by the German Confederation of Small- and Medium-Sized Enterprises, representing two-thirds of the country's employers, and the service workers union, representing 2.2 million employees. Nor was support restricted to the left of the political spectrum. Farmers had begun to recognize that solar represented an attractive new business opportunity for them. They could make money by selling electricity from PV systems installed on their barns and in their fields. The farmers' lobby put pressure on the Christian Democrats to support the revision.

By 2003 it was becoming clear that wind and solar were creating thousands of new jobs, many of them in the manufacturing sector. Opinion polls showed that close to 80 percent of Germans supported the government's policy of promoting renewables. It was also understood that feed-in tariffs were not imposing a huge burden on electricity ratepayers. In fact, at around a tenth of a cent per kilowatt-hour, the increase was too small for most consumers to notice. The annual cost for an average family was just eight euros (around $7) a year. This was, according to the somewhat unfortunate analogy used by environment minister Jürgen Trittin, "the equivalent of two packs of cigarettes (for all those who still smoke)."

The negotiations over the revisions to the EEG coincided with a clash between economic and environment ministers over reductions to Germany's carbon emissions. Trittin wanted to implement major cuts; Clement, backed by industry, argued in favor of less drastic measures. In the resultant compromise the Greens, to their surprise, got more or less exactly the solar feed-in tariffs they had requested, while the SDP won a watered-down emissions trading system. In 2004 the revised EEG passed, becoming law on August 1 that year. Tariffs paid for solar electricity were increased to the point where they were commercially attractive without additional subsidies from government. Tariffs were also differentiated according to system size. The rate for residential rooftop PV was raised to 57 eurocents (68 cents) per kilowatt-hour.

Germany's introduction of the enhanced new feed-in tariffs represents the pivotal moment in solar energy's long struggle for widespread adoption. The engine was already revving at full throttle. Now the chocks were yanked from in front of the wheels. With rates increased and caps removed, sales of PV systems in Germany took off. Six years later, despite turbulence from the global financial crisis, they were still rising.

In 2004 German installations of solar systems shot up, quadrupling to 600 megawatts from 150 megawatts the previous year, outstripping in one bound the cumulative total of all previous installations (405 megawatts). Germany's nascent solar industry was soon employing 25,000 people, racking up between them sales of 3 billion euros. Four years later PV-related employment had spurted to over 40,000, outstripping the coal industry's 35,000. Many of the new jobs were in the formerly communist east, where unemployment was double that in the rest of the country. Sales of PV had reached 7 billion euros (around $9.9 billion). System costs had tumbled by around a third. The solar sector mustered more than fifty leading players, including seventeen listed companies. Analysts joked that the TecDax, the stock index that tracks the performance of the thirty largest German companies in the technology sector, should be renamed the SunDax, because of the predominance of solar firms. The sector had created hundreds of start-ups and mobilized an astonishing amount of private capital. Joseph Schumpeter, who coined the word *Unternehmergeist* ("entrepreneur-spirit") while teaching economics in Bonn, would doubtless have been delighted.

By 2008 solar systems were clamped to almost half a million German roofs. That year the country purchased around one and a half giga (billion) watts of PV, representing 50 percent of all solar installations in the world. In 2009, a recession year for almost all other industries, Germany installed over 2.3 gigawatts. A study showed that one in five German homeowners was considering buying a PV system.

Why has the German feed-in tariff model been so successful? In large part, because it is simple to understand. Especially compared to the confusing melange of bureaucratic mechanisms employed elsewhere. With feed-in tariffs the utilities are responsible for paying for the provision of clean energy. The role of government is merely to ensure that producers of clean energy are appropriately compensated. The paperwork is minimal. Not a penny of taxpayer money is involved. Feed-in tariffs are based on a pragmatic view of the world. They appeal not just to people who want to save the planet, but also to people who want to make money. "These are people who take a pencil, make the calculation, and say, What's in it for me?" Eicke Weber said. "Are there more people who have a green heart or who want to make money?" Weber returned to Germany in 2006 to become director of the Fraunhofer Institute of Solar Energy after twenty-three years teaching materials science at the University of California at Berkeley. "Anyone who can calculate finds out that feed-in tariffs are an incredible thing, because you can finance your investment with a loan from the bank, and you can pay back the loan from the income from your PV system," Weber told me. "Then, in a few years, when you have paid your loan back, you start getting a check each month."

Feed-in tariffs were superior precisely because of this profit-making ability. With net metering, by contrast, it was only possible to bring your electricity bill to zero. Weber recalled a Swiss colleague of his at Berkeley who, having zeroed his bill, went out looking for new appliances to buy in order to absorb the excess electricity he was generating rather than feed it into the grid for no return. "The net metering concept, which basically implies that you cannot make money, is completely stupid." Feed-in tariffs were also, Weber believed, superior to the quota systems—also known as renewable portfolio standards—adopted in the US and the UK. These prescribe that a given percentage of power should be derived from a particular renewable technology. Such prescriptive systems are perforce weighed down with rules and regulations. By contrast the German system uses the market mechanism to determine which technology is most efficient at producing clean energy within the established framework.

Germany has been accused by die-hard free marketers of picking winners. They ask why solar should be so highly subsidized, criticizing the government for its "Stalinist" approach. But in fact, feed-in tariffs are not limited to solar. They also apply to wind, biomass, and other forms of renewable energy. And the amounts paid to generators of solar energy are peanuts by comparison with the subsidies that have been lavished, over many decades, on the coal and nuclear industries.

Dozens of other countries and states around the world have paid Germany the ultimate compliment of introducing their own feed-in tariffs. They accounted for over 90 percent of solar installations made in 2008. However, in designing their systems, regulators often omit or modify some aspect of the mechanism, with unfortunate consequences. For example, sunny Spain instituted a feed-in tariff regime that locked in high rates with no built-in differentiation between small rooftop systems and huge ground-based ones, or capacity for stepwise reductions. Lacking the ability to react, the system boomed spectacularly in 2008, then promptly went bust. From 2.7 gigawatts, representing almost half of the world market, Spain capped its market at 500 megawatts (an amount still larger than all newly installed PV in the US that year). Also in 2008 South Korea emerged as the fourth-largest PV market in the world, rivaling Japan as Asia's largest, thanks to its system of feed-in tariffs. But the South Korean government tried to cover the cost of the tariffs out of its own budget rather than from electricity rates, a variation that quickly proved unsustainable.

Meanwhile, back in Germany, though feed-in tariffs now enjoy support from all the major political parties, that does not stop an annual battle taking place in the Bundestag over the amount of the degression, the annual reduction in rates. Free marketeers argue that the prices paid by ratepayers are too high, and that manufacturers do not have enough incentive to reduce their prices. Also, that high prices benefit Chinese equipment suppliers at the expense of domestic ones. Supporters counter that, in order to attract investment and build up a robust and sustainable industry, high profit margins are required, at least in the early years. *Sie müssen speck ansetzen*, the Germans say, "You must let them get fat."

The degression was originally set at 5 percent per annum for rooftop systems. In 2008 the Christian Democrats, now governing in coalition with the SDP following the election of 2005, proposed increasing the drop to 30 percent that year. They argued that the law had been too successful for its own good. The amount of solar energy was now increasing much faster than anticipated. Though electricity consumers were currently paying a surcharge of just 1.01 euros ($1.69) a month, by 2015 they could be paying as much as eight euros ($12.32). That would work out to an estimated 120 billion euros ($184 billion) in total. Public support for solar would be lost.

Experts like Eicke Weber countered that the estimate was too high because it did not take into account the rising price of conventional electricity. The actual amount would not be more that 60 billion euros, he claimed. That was just a third of what the state was paying to subsidize its aging coal industry. "If we're willing to burden the population with 180 billion euros of support for a dying industry, why do we worry about taking one-third of this to make Germany the leader in photovoltaic technology?" he demanded. After a contentious political debate, the Bundestag voted to cut solar price by between 8 and 10 percent in each of the following three years. The industry's verdict: tough, but fair. Solar subsidies would continue to be reduced. But just as regulations create markets, so markets create constituencies. It seemed likely that Germany's increasingly powerful clean-tech lobby would prevent the new government from doing anything too drastic. State support for the solar industry would continue for the foreseeable future. (For the debate over reductions to the feed-in tariff following the 2009 German federal election, see epilogue.)

On April 5, 2006, a party was held in Munich to mark a momentous milestone, with "live music, good food, and"—appropriately, in the city that hosted the Oktoberfest—"enough beer." For the first time anywhere, the ratio of electricity fed into the public grid from photovoltaic sources in Bavaria had topped 1 percent. That was just the average: on sunny afternoons, the state was getting as much as 10 percent of its power from PV. By 2010 the figure was more like 25 percent.

By the time I visited Bavaria two years later, the average had exceeded

2 percent and was still rising. Elsewhere in Germany you see the odd roof-mounting PV here and there. In Bavaria, you cannot miss them—there are panel-covered roofs everywhere. The state musters around half of all Germany's PV installations, more than the US and Japan put together, more than anywhere in the world. In part this is simply because there is more sun in the south than in other parts of Germany. But it is also because Bavaria has more places to put PV. Large spaces capable of supporting 20- to 50-kilowatt arrays are available in the form of the roofs of farmers' barns. The locals call them "energy farmers." Exposed to competition from low-cost exporters like Brazil, they have learned that selling power is more profitable than raising pigs and more predictable, too. Energy farmers find it relatively easy to get cheap loans that enable them to install field-size systems.

I visit Adelsdorf, a village (population 7,000) in northern Bavaria, whose best-known feature is a castle that dates back to the twelfth century. Adelsdorf is the headquarters of Sunset Energietechnik, one of the country's oldest installers (founded 1979) of solar systems. On what was, incongruously, a misty, drizzly November morning, I am picked up at my hotel by the company's founder and CEO, Olaf Fleck. In his (Bavarian-made) BMW he drives me around Adelsdorf, pointing out PV installations, of which there were many, not all of them made by his company. At every turn solar roofs emerge from the mist, big systems and small ones, crystalline and thin-film, on houses, schools, factories, municipal offices, and the town hall.

As you would expect, individual purchasers of solar are mostly middle-class citizens. Some work in the nearby city of Erlangen, where Siemens runs several large plants. Fleck tells me that, these days, the average size of a PV installation on an individual house is upward of 5 kilowatts. Naturally, almost all systems are installed on south-facing roofs, to catch the maximum sunlight. But he also shows me several examples of systems clamped on north-facing roofs, by owners who are so happy with the returns from their existing south-facing systems that they figure, What the heck—why not go ahead and install some more?

Perhaps the most impressive installation is on a small factory, whose roof is completely covered with dark-colored thin-film PV panels. Had it not been pointed out to me, I would never have guessed that it was a solar roof. This is how PV will be in the future, I think: systems will be so integrated with buildings that they will become invisible. Adelsdorf is an eye-opening demonstration of the extent to which solar has been accepted in German society. But even here, just one in four or five houses is equipped with PV. "We are only at the beginning," Fleck admits.

From Adelsdorf, I move on to Hammelburg, where I am met by Hans-Josef Fell. He drives me to nearby Arnstein to see what was, until recently, one of the largest PV plants in the world. Completed in 2005, it is a breathtaking sight. On what was a farm where once sheep grazed are now several fields full of photovoltaic, receding in rows almost as far as the eye can see. The plant consists of 1,500 individual units. Each unit is a giant composite array consisting of twelve large panels that present fifty square meters of PV to the heavens. The arrays are mounted on concrete plinths equipped with gears that enable them automatically to tilt and rotate in order to track the sun. They produce a combined total of 12 megawatts. Though enormous, this installation has since been eclipsed by even larger solar plants.

"In matters where persistency and unremitting industry are called for, the Germans are to be praised for being able to advance things further than other peoples." So wrote the philosopher Immanuel Kant in 1783. Through their persistency and unremitting industry, Wolf von Fabeck, Hermann Scheer, Hans-Josef Fell, and hundreds, perhaps thousands, of their colleagues and fellow activists have been able to advance the solar cause further—far further—than anyone else. Before we leave Germany, however, we must meet one more important set of actors in the drama: German solar entrepreneurs.

18

MAKING OF SOLAR VALLEY

*A*nton Milner's head was reeling. The time was November 1999, the place, Berlin. For two hours three crazy Germans had been bombarding the Englishman, a former McKinsey consultant, with information about this company they wanted to start. It had something to do with photovoltaic, a word that Milner was not even sure he knew how to spell. Finally, he could not take in any more. "Look," Milner pleaded, "can you just summarize what this is about, and what you want from me?"

"OK," Reiner Lemoine, the leader of the crazies explained, "it's like this. Point one, renewable technologies will come, they have to come for the sake of the planet, and there will be a market for them, as you can see from the subsidy programs that the new coalition government is laying out. Point two, there will be changes to the technologies which will make them a little bit better than they are today. Point three, we have a business plan, we've identified a site, and if you believe what the plan says, we might actually be able to make some money on this. But you have to understand that nobody has ever made any money from solar, throughout the history of the industry they've only had losses. So you can write business plans and you can get the right numbers, but nobody's ever managed to actually accomplish them. Point four, we need financing, but we are two scientists and one engineer— we can't communicate with the banks and, point five, even if we could, they wouldn't believe us." Then Lemoine paused, lit a cigarette, and looked Milner in the eye. "Oh and by the way, there's a point six," he said with a mischievous grin. "We can't afford to pay you anything and we don't have any money."

It didn't sound very promising. Nonetheless, Milner was sufficiently intrigued to take their business plan home with him. He read it overnight. The plan had obviously been written by technologists, it was not something that you could raise any financing on, but its potential screamed out at him.

Next day Milner went back for a second meeting. "Look, Reiner," he told Lemoine, "the very last thing you need is a consultant who comes in and tells you what to do. What you need is someone who comes in and helps you build a company. I'd like to do it, so if you fancy it, let's do it." To cement the partnership, they shook hands. On Monday, November 22, 1999, the four partners set up a company called Q-Cells. It had paid-in capital of just 60,000 euros (about $59,000) and no employees. Within ten years, Q-Cells would be the largest manufacturer of photovoltaic, not just in Germany, but in the whole world.

It was an unlikely coupling, the soft-spoken Englishman and the outspoken German, the philosophical, egalitarian engineer and the "horrible, nasty McKinsey capitalist," as Milner jokingly described himself. Despite their political and ideological differences, the two men liked each other from the outset. Born in Sheffield in 1961, Milner was the younger of the two. He was himself a quarter German, his paternal grandfather having been born in Usedom, a village in the east of the country. Milner's father had gone to live in Germany as a teenager, but had been evacuated to England in 1939 and thrown into an internment camp. He had subsequently cut off all his ties to the country. Milner graduated in chemical engineering, then joined Royal Dutch Shell, where he was involved in oil trading and risk management. After a spell working in Hamburg, he went back to London to do an MBA, then in 1993 joined McKinsey, the world's largest consultancy. He moved to Berlin because his partner was German. In early 1999 Milner left McKinsey, intending to take a year off. By the time a friend called to introduce Lemoine and his partners, Milner was awfully bored. He desperately needed a new challenge.

"Anton's almost unique skill is his can-do sort of attitude," said David Hogg, CEO of CSG Solar, a sometime Q-Cells affiliate. "He's a combination of a clever strategist with a huge amount of drive. And he's able to bring people along with him, that's what's driven the growth of that company."

Reiner Lemoine was born in Berlin in 1949, during the blockade, when the Russians tried to wrest control of the city from the Allies. In 1968, the year that university campuses erupted in protest, he was an idealistic youngster of nineteen. In later years, Lemoine would look back fondly on his time manning the barricades as a student radical. "He was very socialist-communist oriented," Milner said. "Not especially from political motivation, but from philosophy." In addition to political activity, the late sixties and early seventies was a time of intense all-night philosophical discussions among German students. Much of the discussion focused on nuclear war. "For us in Germany sitting at the [Berlin] Wall, we were probably at the place where the Third World War would happen," Harry Lehmann said. "This led people to ask each other, What are the consequences of our activities as a physicist or an engineer? This in turn led to the fact that the people studying these subjects were affected by ethical or environmental ideas, which they took over and used later to help them find their future and their career."

"We made a lot of politics in university," agreed Günther Cramer, CEO of inverter maker SMA Solar Technology. "Renewable energy gave us as electronic engineers the chance to do what we loved to do, to make new technology. With this new technology, we could change the politics and make a different world. And that was fascinating for us." There was also perhaps another, particularly German, motivation for this altruistic attitude. Namely, guilt for the horrors perpetrated by their Nazi parents during World War II imbued idealistic young Germans with a strong desire to do good for humanity.

At the Technical University of Berlin, Lemoine studied aerospace engineering. He was not a compliant student. He found rote learning tiresome and swotting for exams repugnant. His fellow students called him "Professor Drop-out: "Drop-out" because with his long hair and wire-rimmed glasses he looked like John Lennon, "Professor" because he made bold speeches that outraged his teachers. But Lemoine had no respect for the academics, with their antiquated gowns and outdated ideas. "Why do drudge work?" he demanded. "We should think about the future."

It was a time when young people were investigating all kinds of alternatives. From early on Lemoine was fascinated with renewable energy. His commitment had a strong social and political component. Renewables would enable energy autonomy, thus wean the country off dependence on imports. By lessening the likelihood of conflict over scarce resources, as well as protecting the environment, renewables would also serve the cause of peace. While the struggle against nuclear power raged, Lemoine turned to technology. It was not enough to be anti-nuclear, he felt, you also had to try to make a better world. Upon graduating in 1978, together with some like-minded fellow students, he founded a company. Actually, Wuseltronik (roughly "Bustle-tronic") was not so much a company, more a socialist engineering collective. "You couldn't call them entrepreneurs," commented Olaf

Fleck. "They were more like artists in their technology. They made things, but not very professionally."

Wuseltronik was based in Berlin's bohemian Kreuzberg district. Enclosed on three sides by the Berlin Wall, this low-rent area was a magnet for students, artists, and immigrants. Kreuzberg became famous for alternative lifestyles and radical politics. A rundown, slightly wild place, it was home to squatters and leather-clad punk rockers, who would frequently clash with the police. Lemoine and his partners built measuring devices for use in wind and solar applications. They gave their machines whimsical names, like Wumm ("Whoompf"). On principle, they refused to accept contract work from military institutions like the army. During breaks, they played table tennis and beat on drums. Lemoine himself was fond of the strategic board game Go. Sitting cross-legged he would play it for hours at a time, oblivious to the outside world. All workers in the enterprise participated in decision making; all were paid the same wage. Not that they made much money. That was fine with Lemoine, who despised profit making. His motto was always: "Commercialism be damned. Let's do what's right."

"What Reiner and others did in the early days, they worked for long years, for nearly nothing," Cramer said. "There is a word for this in German, we call it 'self-exploitation.'" For many years Lemoine's radiologist wife kept the family afloat.

In 1996 Lemoine cofounded Solon, a company whose mission was the manufacture of PV modules. (I was slow to realize that the name has nothing to do with lawmakers. It is of course a compound of two words, "Sol" and "on," perhaps cocking a snook at the giant German electricity utility E.on.) In 1998 Solon became Germany's first solar company to go public. Some flavor of the firm's in-your-face attitude may be derived from its corporate slogan: "Don't leave the planet to the stupid." As a module assembler, Solon was dependent on outside suppliers for its solar cells. But the quality of the cells that the company could procure was annoyingly variable. Originally, Solon proposed to build its own small plant to produce cells in Berlin. But the company could not raise the capital, and the project had to be abandoned. Lemoine and his colleagues, Paul Grunow and Holger Feist, decided to leave and set up their own specialist cell-maker.

The trio drew up a blueprint for a firm whose rationale would be the production of reliable high-performance solar cells for independent module manufacturers like Solon. They named the company Q-Cells, the "Q" standing for "quality." Now what they needed was someone who understood how to start up and run a company. Enter Anton Milner. "With their Green perspective, they were thinking, We've got to get this to market, get the cost and prices down as much as possible, but we don't really want to earn money," Milner told me. "And, as a technology freak, Reiner wanted to create a new technology." This was very different to Milner's "We've-got-to-build-a-business-and-you've-got-to-do-certain-things-to-do-this-successfully" pragmatic capitalist approach. "As you can imagine, it took us quite a long time to align our values in what we do and how we think about the world."

The company's mission, the objective that united all four founders, was to accelerate the development of photovoltaic technologies as quickly as possible. That way solar would become a major source of energy, replacing less advantageous sources. To reach that goal they would use commercial and technical mechanisms. "What we did not want to do—we were all of the same opinion—was build just another company, with a hierarchy, telling people what to do, how to do it, all that sort of stuff," Milner said. "We wanted to build a very different type of organization, one where people could really develop their potential, grow into this thing, be themselves, express themselves."

Lemoine and Milner were not the first to have such thoughts. The formation of Q-Cells together with a slew of other German solar start-ups brought to mind another, similar outburst of entrepreneurial activity that had occurred several decades earlier, in what had formerly been apricot orchards on a certain peninsula south of San Francisco. As with photovoltaic, the focus of that activity had also been the fabrication from silicon of semiconductor devices. No surprise then that the site where Q-Cells chose to locate its factory should in time come to be known as Solar Valley.

Solar Valley was not actually a valley, it was in fact flat as a pancake. "Solar Plain" would have been a more accurate description. Nor was there a world-class university like Stanford in the vicinity. It was a greenfield site, near a hamlet called Thalheim, just outside the small town of Bitterfeld, about a hundred miles southwest of Berlin, in what had formerly been East Germany. During the communist era Bitterfeld—literally "bitter field"—was a grim, sullen place, notorious as the most polluted city in Europe. Many chemical companies, notably the photographic film maker Agfa, had located in the town to gain access to its large deposits of brown coal. This was dug up from an open-cut mine, a huge hole in the landscape that ultimately would stretch four miles across.

Brown coal is the filthiest of the fossil fuels. Tons of fly ash from the smokestacks of a dozen brown-coal-fired power plants used to fall on Bitterfeld every day, coating its buildings with soot. Fumes pouring from obsolete chemical dye and pesticide works meant that a foul smell hung perpetually in the air. So poor was the environmental quality, a study found that life expectancy for the wretched inhabitants of Bitterfeld was years less than elsewhere. Following reunification in 1989, the obsolete power plants and factories were shut down, leaving behind a devastated industrial wasteland. Unemployment soared above 20 percent. Young people left in droves.

When Bitterfeld mayor Manfred Kressin heard that Reiner Lemoine and his partners were looking for somewhere to locate their new factory, he received them with open arms, rolling out the red carpet, bombarding them with reasons why they should choose his town for their base. Foremost among them was a lack of the red tape that had bedeviled them elsewhere. "Had we tried to build this factory in Berlin or Brandenburg, we would probably still be in the approval phase," Lemoine commented later. Whereas in Thalheim, the mayor told them, "You can start immediately." Decisions would be made quickly, Kressin assured them. If need be, he himself would hop on his bicycle and round up the local council. They would have an answer within a week.

Wide-open greenfield sites were available where they could build. There was no shortage of labor, plenty of well-qualified locals available who had formerly worked in the photographic film industry. The locals were used to shift-work and more flexible than their unionized counterparts in the west. Another incentive was generous investment grants from the state government of Saxony-Anhalt for companies that promised to employ local people. "The state said, Look, we need to build infrastructure and provide jobs: if you come here, we will give you a loan you can pay back later so that you've actually got the capital to start," Milner said. " I think we would have spent a long time raising that kind of money without a track record."

Seed financing for the start-up came from an angel investor, Immo Ströher. A scion of the family that owned Wella, a leading German maker of hair-care products, Ströher had caught the solar bug. He had been an early-stage investor in Solon, becoming that company's largest shareholder with a stake of 20 percent. It was therefore natural that Ströher should be the first person that Lemoine and Milner approached. "I said to him in my best consultant-like manner, I've got all these business plans I can give you, I can come over and talk you through them," Milner recalled. "Immo looked at Reiner, and he said, Reiner, I don't have time to read business plans: Should I do this, is it a good one? And Reiner said, Yes, you should do it. Then Immo said, OK, I'll do it. That was our first investment round: it was all very much people- and trust-based."

There was no slam-dunk security in making such an investment: a leap of faith was required. To be sure, the government's 100,000 Roofs Program had been formulated, but it was still under discussion by the politicians in Bonn. "You could see the demand, but there was as yet no real market," said Milner. "It was a very, very high-risk enterprise." Financing closed in December 2000. For the next seven months everyone worked flat-out. The following July, the fledgling firm—which by now had nineteen employees—produced its first cells. "It was the quickest greenfield build of a cell factory—particularly through four beginners who had never done this before—that the industry has ever seen," Milner boasted. Speed would continue to be the company's trademark.

In 2002 Q-Cells produced 9.3 megawatts' worth of solar cells, not enough to power a decent-sized village. That year Milner and his partners went looking for more funding to finance their second-phase expansion. But the dot-com bubble had just burst and Germany's conservative banks weren't lending, certainly not to dodgy start-ups in an industry that financiers didn't understand. Once again they were fortunate in finding a sympathetic investor. Good Energies is a Basel-based private firm that specializes in funding renewable energy start-ups. It was founded in 2001 by Marcel Brenninkmeijer, a member of one of the richest families in Europe.

In 1861 two brothers—Clemens and August Brenninkmeijer—started selling clothes to cash-strapped Dutch farmers. This grew into the C&A chain of downmarket retail clothing stores, a familiar brand on many European high streets. After two decades working in the family business, Marcel decided that he was more interested in environmental issues. In 1989, after attempting to green C&A, he saw an opportunity to foster more sensible environmental practices through investment. "Good Energies has been a really terrific investor," Milner said. "Marcel did what the textbooks say that venture capitalists should do," agreed David Hogg of CSG Solar, another start-up funded by Good Energies. "That is, forget about trying to pick the technology winner, choose the people you think can drive the business and back them."

In 2003, having built a new factory twice the size of its first one, Q-Cells tripled its production. The following year output swelled eightfold, then doubled again in 2005. That October Q-Cells went public, its initial share offering oversubscribed forty times. By 2006 Q-Cells was the fastest-growing company in Germany. In the five-year period from 2002 through 2007 Q-Cells' shipments grew at a compound annual rate of 112 percent. In 2007 sales revenue broke through the billion-dollar mark for the first time. The company was running seven production lines, including one in Malaysia, and employing around 2,500 people.

In December 2006 Reiner Lemoine died of a brain tumor. Tragically, he did not live to see the firm he cofounded overtake Sharp of Japan the following year to become the world's number one maker of solar cells. Shortly before his death, Lemoine set up a foundation to support young researchers in the field of renewable energies. In August 2008 Q-Cells named its new fifty-million-euro R&D center in his honor. "He was a wonderful guy," said Milner, "and a big inspiration."

A small sign announces "Solar Valley" on a turn-off from the north-south autobahn that bypasses Bitterfeld on its way from Dessau, home of the Bauhaus, to Leipzig, where Johann Sebastian Bach lived and worked. On the bleak November day I visit the valley, the weather is miserable, a light drizzle turning into snow. On the way I pass through several small villages where drab brick houses line the tidy streets. Braving the cold, old ladies are vigorously scrubbing their front windows. If there are any solar panels on the red-tiled roofs, I do not see them. The countryside is littered with abandoned factories and derelict power stations, their blackened concrete cooling towers blotting the landscape. I encounter, coming the other way, a solitary Trabant, the iconic compact car whose poor performance and ugly appearance made it a symbol of life under communism. Its appearance is proof positive that I am in the former East Germany.

Arriving at Thalheim is like entering another world. The scene that confronts me is one of ceaseless activity. Looking down *Sonnenallee* ("Sun Alley"), Solar Valley's main street, I can see no fewer than five giant new factories under construction. Beside them white silos containing industrial gases rise tall. Cranes dominate the skyline; heavy trucks hauling sand rumble past. Off to one side is the original Q-Cells factory, a squat six-story cube. Its distinctive southern face is covered with blue polycrystalline solar cells on which is superimposed a huge silver Q. Dwarfing the cube is the facility next door. With its tall tapered smokestack, it looks like a relic from the bad old days, but is in fact, Milner assures me, a relatively recent float-glass plant. Only one of the new factories belongs to Q-Cells itself, he says, the rest are joint ventures and subsidiaries.

The core strategy of Q-Cells was originally to concentrate on one link in the value chain. The company would make only solar cells, leaving assembly of the cells into panels to customers like Solon. This was a departure from the traditional solar industry model of vertical integration, established back in the midseventies by the likes of ARCO Solar, which combines wafer manufacture, cell production, and module assembly under one roof.

Other German start-ups like Q-Cells' exact contemporary, Bonn-based SolarWorld, took a more traditional route. Founded in 1998 by "five crazy guys who people thought were on drugs," SolarWorld began as a distributer of solar modules. Fearful of becoming dependent on its suppliers, the company gradually worked its way up the value chain, producing everything from ingots of silicon, wafers—Q-Cells was one of their first clients—all the way to cells and finished modules. In February 2006 SolarWorld purchased ARCO Solar's venerable plant in Camarillo, California. Two years later it established a second, much larger facility in Hillsboro, Oregon. Today, SolarWorld is the largest manufacturer of PV in the US.

By summer 2007 Germany was the undisputed world leader in photovoltaics. The industry had grown there faster than anyone could have imagined, outperforming even the wildest predictions of the optimists. The country's $5 billion solar industry mustered more than fifty companies all the way up the value chain from silicon and wafers, to cells and thin films, from module assemblers like Solon to fully integrated manufacturers like Solar-World. Germany was home to the world's largest concentration of solar manufacturing plants, three-quarters of them located in the former communist east.

In addition to the cell and module business, there was also another, lesser-known but equally extraordinary, success story in the shape of inverter maker SMA Solar Technology, a company based in the central German university town of Kassel, home of the Brothers Grimm. (The initials stand for *System Mess und Anlagen*, "system measurement and power plant" technologies.) Inverters are electronic devices that take the direct current output by solar panels and turn it into grid-friendly alternating current. To use an analogy from personal computers, if solar cells are memory chips, inverters are microprocessors. They contain smarts that will become increasingly important as controllers in the coming era of hybrid renewable-conventional grids. In the early days, however, inverters were PV's Achilles' heel. They were forever breaking down or burning out and having to be replaced, much to the annoyance of the customer and the despair of the installer who had to drive out with the replacement. Then in 1995 SMA launched the Sunny Boy.

Not only was this product more efficient at converting DC to AC, the Sunny Boy was also a simpler design, meaning that it was easier to install. In Long Beach, California, installer Pat Redgate recalled, "Our market was about ready to die because the inverters were so poor. Then everybody discovered the Sunny Boy and started using it, and as a result the business just kept growing because slowly people—buyers and sellers—became more confident."

Today, the firm dominates the market—worth $2.8 billion in 2009—with a share of more than 40 percent. Though inverters are high tech, it was not so much the technical sophistication of the Sunny Boy that enabled SMA to win the lion's share. Rather, as CEO Günther Cramer explained, service was the secret of their success. "If our customers had a problem they could just call us, and within twenty-four hours we would send them a new inverter." Cramer and his two partners were originally academics from the University of Kessel. "We never expected anything like this," he tells me bemusedly when I meet the soft-spoken chief executive in the company's beautiful new ultra-modern boardroom.

I was curious to know whether SMA had emulated any corporate models. It turned out that they had made everything up as they went along. "Our only idea was to do things like we were used to working at the university, where we had no clear rules, and we could create everything on our own. But we didn't speak about it, we just did it. After a while, when we became fifty, a hundred people, we had to say, How can we communicate, what are the rules for this? Then we started to discuss what we were doing, to find out what it is, and what is the difference to the normal way? So now we do know very well how it works, but in the very beginning it was something like naturally grown."

Astonishingly the company, which was founded in 1981, has never needed to depend on external sources of funding. "We always financed by ourselves from working capital," Cramer said proudly. Initially, this was out of necessity. When they started it was hard persuading financiers to lend them even paltry sums. One bank went so far as to insist that the three partners take out life insurance policies as security. Building good relationships with customers was one part of SMA's corporate culture. The other, internal part was to involve employees in the information and decision-making processes. "So they can participate in creating this company, so they are really involved, not only with the brain and the heart, but also with the wallet. They participate in the profits of the company, this can be a lot of money, and that's the reason they are highly identified with the company, so we don't have any turnover, even with the engineers."

The company has virtually taken over the suburb of Kassel where its headquarters and manufacturing facilities are located. I count at least twenty-one buildings, plus a huge new factory under construction. A big sign outside the front gate proudly informs passersby that the new plant will be carbon neutral: this is a company that practices what it preaches. Today, in developing its products, SMA emphasizes ease of installation. Previously, it took a qualified specialist to install PV systems. Now, with plug-and-play inverters, almost anyone can do the job. The result is a big reduction in the cost of ownership.

I ask Cramer, who is also president of the German Solar Industry Association, about the future. He tells me that a recent get-together of solar CEOs at the European Photovoltaic Industry Association in Brussels had established a new target. "We want to reach 12 percent of European electrical energy by 2020. The old target was 3 percent, but that's nothing—all the effort and all the money that has been invested in PV doesn't make sense if we only reach 3 percent by 2020." Attaining 12 percent would mean a growth rate of between 30 and 40 percent. "But that is not so much, we did it in all of the last ten years, in fact more than that."

I ask Anton Milner the same question. "The roadmap's quite clear," he responds. "I always tell my guys that everything we're doing now is just warming up. One of two things will happen in this industry: either it explodes in scale, or it dies. The reason is quite simple: if we get our costs down quickly enough, you have a crossover point where you're suddenly competitive against grid electricity pricing. Even with all of the tremendous growth that this industry's had, we're still only producing about the equivalent of one or two power stations a year. Suddenly, you go from a gigawatt market to a terawatt market. (One terawatt equals one trillion, a thousand billion [giga] watts.) Then you're talking about growth the like of which we've never seen. That's the prize, because at that stage you can really start to make a huge market for photovoltaic products as a major energy supply tool. So I think we're talking about extremely rapid growth over the next few years—we're going to become a multi-multi-multi-billion-dollar industry."

When would the crossover point come, the holy grail of "grid parity" be reached, when the cost of electricity from the roof drops below the cost of electricity from the socket? "By 2015," Milner replies. These were brave words when they were spoken, in early 2008, before the onset of the global financial crisis. The downturn hit the PV industry hard. Unsold inventory swelled, the price of solar panels plummeted up to 40 percent below 2008 levels. In 2009 Q-Cells recorded its first-ever loss, a stunning $1.9 billion, forcing the company to slash five hundred jobs, nearly a fifth of its workforce, close down its older lines in Germany, and move production to its lower-cost Malaysian plants.

The unique business model of focusing exclusively on cell manufacture had led to Q-Cells being hit harder than its vertically integrated rivals. Now it was time to "de-risk" the model and move to a more conventional strategy that encompassed module production and system integration. Taking responsibility for the financial markets' loss of confidence in the company, in March 2010 Milner resigned as CEO.

Overall, the slump in the first half of 2009 was a shocking time for the solar industry, which for almost a decade had been accustomed to continuous double-digit growth. For observers of the semiconductor industry, however, the scenario seemed quite familiar. Booms are always followed by busts. Some companies would no doubt go under, others would be taken over. The survivors would emerge from the ordeal stronger and ready to resume their rapid growth. Meanwhile, though bad for producers, plunging prices were good news for potential customers of PV systems. They represented a giant step on the long march toward grid parity.

Other parts of the world were keen to emulate the German experience with solar. But as we shall see in part 4, California, by far the most active player as well as the largest market for solar in the US, would have problems in trying to replicate Germany's spectacular growth.

Bill Yerkes outside Camarillo factory, California, November 2009. (*Bob Johnstone*)

David Freeman at Los Angeles Department of Water & Power, November 2009. *(Bob Johnstone)*

Markus Real at his chalet, Schwyz, Switzerland, November 2007. *(Bob Johnstone)*

Don Osborn *(right)* and Dave Collier *(left)* under the 540-kilowatt solar system at the Sacramento State Exposition Fairground parking lot, November 2008. *(Bob Johnstone)*

Wolf von Fabeck at the offices of Solarenergie-Förderverein, Aachen, November 2008. *(Bob Johnstone)*

Hermann Scheer, Berlin, November 2007. *(Bob Johnstone)*

Hans-Josef Fell, Berlin, November 2007. *(Bob Johnstone)*

Hans-Josef Fell at a solar farm near Hammelburg, Bavaria, November 2008. *(Bob Johnstone)*

David Hochschild *(left)* and Adam Browning *(right)* of Vote Solar, Anaheim, October 2009. *(Bob Johnstone)*

(Left to right) Hermann Scheer, Arnold Schwarzenegger, and Bernadette Del Chiaro, Sacramento, 2005. *(Courtesy of Environment California)*

Randy Schultz of Southern California Edison on warehouse roof with First Solar panels, Chino, October 2009. *(Bob Johnstone)*

Richard Swanson of SunPower, San Jose, November 2008. *(Bob Johnstone)*

Martin Green at ARC Photovoltaics Centre of Excellence, University of New South Wales, 2007. *(Courtesy of Martin Green)*

Part 4:
The Second Coming

PUSHING PV IN CALIFORNIA, BOTTOM UP AND
TOP DOWN

19

CATCHING RAYS IN FOG TOWN

*E*ven if Adam Browning and David Hochschild had tried, they could hardly have picked a more auspicious time to launch their San Francisco–based, pro-solar initiative than January 2001. That month, California lost control of its energy destiny. The partial deregulation of the state's electricity market was forcing its giant investor-owned utilities to pay five times more for wholesale power than they were allowed to charge their retail customers. As a result of having to borrow to buy electricity, which they then had to sell at a loss, PG&E and Southern California Edison were rapidly running out of cash. Deregulation was proving, Governor Gray Davis lamented, a "colossal and dangerous failure."

On January 17, to prevent the grid crashing and the lights going out, rolling blackouts were imposed on the state for the first time since World War II. The blackouts affected several hundred thousand consumers in Northern and Central California. Next day Davis declared a state of emergency, authorizing the use of taxpayer funds to buy power on behalf of the beleaguered utilities. The following week, to minimize the number of television sets switched on simultaneously, football fans were encouraged to congregate in groups to watch Super Bowl XXXV.

Things kept getting worse. In mid-March blackouts were extended to Southern California, affecting over 1.5 million consumers there. On April 6 Davis proposed record rate hikes, averaging 37 percent. Next day PG&E filed for bankruptcy. The situation, the governor confessed on *Frontline*, was "a hell of a mess." State historian Kevin Starr called it "one of the most dramatic crises ever to afflict California." To avoid grid collapse during the summer months, when people would be turning on their air conditioners en mass, Davis imposed a drastic energy conservation program. To lead the program he named "that wily old fellow" David Freeman as his chief energy advisor.

Still the blackouts continued. The incoming Bush administration made it clear that its sympathies lay with energy traders like Houston-based Enron and not with California-based rate-payers. Vice President Dick Cheney went out of his way to mock energy conservation as a way of dealing with electricity shortages. Not until the following year would the extent to which the out-of-state shysters had gamed—that is to say, manipulated—the system to their advantage emerge. California had suffered, Freeman told a Senate subcommittee, "the largest transfer of money out of consumers' pockets in utility history."

Energy prices did not stabilize until September 2001. A few weeks later, on November 7, San Franciscans voted on Proposition B, a ballot measure authorizing $100 million worth of revenue bonds to pay for the installation of 10 megawatts' worth of solar panels on city-owned buildings. They approved it by a whopping 73 percent. Proposition B was popular because it promised to deliver the city's long-suffering residents at least some degree of independence from the vicissitudes of the energy market. The proposition represented, according to its proponents, the largest purchase of solar energy in the US, allowing a single city to install as many solar panels as the entire nation did each year.

Proposition B was the creation of the Vote Solar Initiative, a grassroots advocacy group founded by two old college friends, Adam Browning and David Hochschild. After graduating from Swarthmore in the early nineties, both had headed for Africa. Browning, who had been a champion swimmer, worked as a Peace Corps volunteer in Guinea-Bissau, a small west African nation. Hochschild participated in Nelson Mandela's youth program in a black township in South Africa's Eastern Cape province. There he saw, and was intrigued by, solar panels used to power lights to illuminate post-office mail boxes, an off-grid application.

On returning to his native San Francisco, Hochschild talked his parents into putting solar panels on the roof of their house. "They basically said, If you can figure out how to do it, and find someone to do the installation, we'll pay for it." It was not the first rooftop solar system in San Francisco, but it was one of the first. Hochschild subsequently installed solar panels on his own house. Now he was more than just a believer in solar, he was also a consumer. Personal experience with PV would lend him credence in his attempts to persuade others of solar's merits: "Hey, this technology really works, it's reliable and simple, and it'll bring your electricity bill down."

After South Africa, Hochschild did a master's degree in public policy at Harvard's Kennedy School of Government. In summer 1998 he worked as a White House intern. It was a frustrating experience: everyone was distracted by the Monica Lewinsky scandal. Hochschild felt detached from the impact of what he was doing. Instead of working for the government in Washington as he had planned, he ended up heading home to San Francisco. There he got a job as an aide to Mayor Willie Brown. In his two years at City Hall, Hochschild learned a great deal about how to get things done.

Having recently installed solar panels on his own house, it was natural that Hochschild should think of putting PV on the roof of City Hall as well. "I just didn't understand," he told me, "if we have a way to make energy that's clean and quiet, really durable, can go on a rooftop, and doesn't require drilling or invading other countries to get, why are we not putting this up everywhere? As I started to look into it, it became clear to me what the main obstacles to solar were—policy barriers and the higher cost."

The key to overcoming the higher cost was financing. Hochschild knew how cities raise money: they issue revenue bonds. Bonds gave them access to long-term, low-cost financing. It should be possible, he thought, to issue bonds that would enable solar panels to be put on city-owned buildings. The clean power the panels generated would replace conventionally sourced electricity that the city would otherwise have to buy. Thus the PV would pay for itself, with no need to increase taxes.

Hochschild told his buddy Adam Browning, who was immediately taken with the idea. Since his return from Africa in 1995, Browning had been running the Toxics Release Inventory Program for the Environmental Protection Agency's San Francisco office. "I had a lot to do with enforcement, going out and inspecting facilities, finding people who were doing illegal things, gathering evidence, then suing the hell out of them," he told me. "It gave me a good introduction to how you control things 'at the end of the pipe.' Solar just seemed like this way of avoiding the pipe altogether. It was a completely new way of doing things that instead of trying to control the problem circumvented it."

The pair teamed up, meeting once a week for coffee and chat with a group of like-minded others, such as Randy Hayes, who back in 1985 had founded the Rainforest Action Network. They were an effective combo: Hochschild, a born enthusiast, had an appealing attitude; Browning had a mellifluous voice that made him an attractive advocate. Together, they cooked up the idea for what became Proposition B. To promote it, they founded the Vote Solar Initiative as a nonprofit organization. To get a proposition on the November ballot they needed a city legislator to sponsor it. Since his appointment in 1998 as a supervisor—as San Francisco calls its councillors—Mark Leno had made a name for himself supporting equal access and gay rights issues. Leno enthusiastically embraced the pro-PV proposition, telling local reporters that he wanted foggy San Francisco to overtake Sacramento as the country's number one municipal producer of solar power.

An excited Hochschild quit his job, Browning scaled back his hours, and the pair threw themselves into campaigning. They set up shop in a little office on the top floor of a nondescript building in the city's boho Mission District. "We raised about $125,000 and recruited about two hundred volunteers," Browning recalled, "people who really wanted to be part of something larger than themselves." They won endorsements from green groups and celebrity environmentalists like Bonnie Raitt and Robert Redford. They plastered copies of a brightly colored poster all over the city that depicted a big yellow sun whose rays filled the sky above the Golden Gate Bridge and the fog bank beyond it. "At the time, so much of environmentalism was about saying no to this, no to that, stop this, stop that," Browning said. "We ran a very different campaign based on yes on B! Clean air, clean energy! It was all about the positive alternatives."

Meanwhile, the state's energy crisis dragged on. "There were more blackouts after we had begun the campaign and introduced the legislation for the solar bond," Hochschild said. "Businesses were getting shut down. There was a feeling of, OK, this is not working, we have to make a big change. Suddenly, solar was getting a lot of attention. In the face of all that malfeasance, you had Gray Davis essentially demanding that the feds step in. But the feds, directed by George Bush, were refusing. There was a sense of powerlessness. Solar represented a more democratic path to electricity generation—you no longer had to wait for politicians and utilities to make the right choices, you could just do it yourself. So we were already on our way, it was like a snowball going downhill." It was all great fun, so fulfilling, such exhilaration. "In the final days of the campaign we had all these flyers, with armies of volunteers handing them out throughout the city at all the transit stations and subway stops," Browning recalled. Campaigning appealed to his combative instincts: "To do this kind of work, identifying goals and hacking through until you win, you have to enjoy the scrap a bit, be a scrapper."

Following the passage of Proposition B, the first municipal edifice to get a solar rooftop was the Moscone Convention Center in downtown San Francisco, the venue that each year hosts Macworld. In early 2003 a five-thousand-panel, 675-kilowatt system was installed on its roof at a cost of $7.4 million. The project was on track to pay for itself within eight years.

Vote Solar's campaign caught the attention of national media. "We got all these calls from other cities around the country saying, Hey, we want to do this too, how should we go about going solar on our own, can you help?" Browning said. Clearly, the pair were onto something. For the next four years they focused on building a grassroots advocacy movement and extending it beyond San Francisco. Eventually membership of Vote Solar would reach around 60,000. Like others before them, the activists recognized that if they could create a large enough market for solar, then costs would come down until it would be competitive with conventional forms of energy. "Our whole mantra was that with scale the PV industry would grow to the point where it would no longer need subsidies, it would be at grid parity," Browning said. "At that point, our job would be over."

They set themselves a goal: ten cities in two years, replicating the model that they'd established in San Francisco. Their focus would be municipalities, large energy users that could borrow money cheaply through bonds. It would, they thought, be easy. It was not. "We quickly learned that nothing is cookie-cutter, every place has its own particular dynamics," Browning admitted. "We realized that it's not how much energy you generate, it's how much that electricity is worth. In the US the variability of the cost of energy is incredible. Up in the Pacific Northwest, where hydropower is plentiful, they pay only a few cents per kilowatt-hour. It made sense to concentrate on Californian cities, where we pay up to ten times as much." As their first target, they chose San Diego.

During California's energy crisis, San Diego had been the canary in the coal mine. San Diego Gas & Electric was the first utility in the state to sell off its generators. Within a year, as wholesale prices soared, consumers had seen their electricity bills double, then triple. San Diego is known as a navy town. Historically its political culture has leaned to the right. In 2002 the mayor, Dick Murphy, was a Republican. Though fiscally conservative, Murphy was a moderate. He also had a daughter who was an environmentalist. She was forever pushing him to embrace green issues. Besides, solar, as Browning liked to remind people, was a nonpartisan issue. Determined to insulate San Diego from the whims of the power market, Murphy made achieving energy independence a priority for the city. He invited the Vote Solar folks down to make a presentation about San Francisco's solar bond. For added muscle, they brought with them Ed Smeloff, formerly of SMUD, now in charge of power policy at the San Francisco Public Utilities Commission. The mayor seemed to like what he heard, especially the bit about going solar not costing the city any money. The question was, how to turn approbation into action?

Having the mayor on board was a good start. Now what they needed was what Browning called "an internal champion, someone who takes an issue on, makes it their own, and bulldogs it through to completion." Donna Frye is not your typical city council member. She was in fact an outsider, a surf-shop owner who got into politics campaigning against poor water quality on popular San Diego city beaches. When some of her customers picked up viral infections from sewage-tainted seawater, she founded an organization, Surfers Tired Of Pollution (STOP). Her campaign manager and senior policy advisor, Nicole Capretz, had worked for the Environmental Health Coalition, founded in 1980, one of the country's oldest green groups. Frye and Capretz got together with Vote Solar's third person, former Greenpeace activist J. P. Ross, to cook up something concrete to put on the table.

"We said, What's a target that's high, but not totally pie-in-the-sky?" Capretz told me. "We came up with a goal of 50 megawatts within ten years. That became our rallying point." For the time it was an aggressive figure. Sacramento had deployed around 8 megawatts of PV, San Francisco had pledged to do 10. But 50 megawatts exceeded the cap on the percentage of peak demand that could be net metered, as officials at San Diego Gas & Electricity were quick to point out. However, once the goal had been officially endorsed by Mayor Murphy at a press conference held in September 2003—"this city needs to enthusiastically embrace the pursuit of sustainable, reliable energy," he told reporters—the utility quickly folded.

It helped that, in addition to the environmentalists, Mike Turk, a Republican and a prominent local developer who had been building energy-efficient homes since the midseventies, also rallied round the solar flag. The city's plan mandated that all new or refurbished public buildings should henceforth derive at least 10 percent of their energy from renewable sources. But if San Diego was to meet this ambitious goal, the private sector would also have to be persuaded to join in. Since in the developer world time is money, Turk suggested that the city's permitting process should be expedited for homebuilders intending to take advantage of the program. That would create an incentive for the builders to go solar.

"Mike was at the city's development services department all the time," Capretz explained. "He'd been very frustrated by the lengthy process needed to put up solar panels on his projects. He wanted to figure out a way to streamline that process, to make it easier and more appealing for builders." Great idea in theory; in practice, however, the bureaucrats who ran the city's development services department refused to buy in. "It's been a carrot, but it hasn't drawn enough of the private sector into the program," Capretz said. "Now we're exploring mandates for the private sector, too."

By 2009 San Diego was around halfway to meeting its 50-megawatt goal. The city had installed large solar arrays on the roofs of its police head-quarters, water treatment works, trash truck maintenance center, state fair-grounds, and environmental services building. As a result, it had cut hundreds of thousands of dollars from its annual energy bill. San Diego was also leading the nation in solar rooftop systems, with 2,262, up from essentially none in 2001, compared to 1,350 in San Francisco. But for the majority of its power San Diego continued to depend on imports. To bring in electricity from elsewhere SDG&E proposed to build a new, large-scale transmission line—dubbed Sunrise Powerlink—at a cost of $7 billion. Dick Murphy was no longer mayor, but the city was considering updating its solar goal to make it more aggressive. As Donna Frye wrote in a 2007 op-ed, "[R]educing reliance on imports and increasing reliance on local distributed generation is the smarter choice for San Diego's energy future." The scrap continues.

Vote Solar's campaign to increase the deployment of PV in California was predicated on two little-noticed temporary measures. Both had been hurriedly enacted by Governor Davis in early 2001 during the depths of the energy crisis to encourage the introduction of new, privately owned distributed-generation systems. The idea was that these would help reduce load on the grid, especially at peak hours.

One measure was the Self-Generation Incentive Program, introduced March 27, 2001. This was an initiative that would pay up to half of the cost of solar and other renewable systems large enough to provide on-site generation to supply all or part of an organization's energy needs. The program dished out rebates of $125 million a year, the money coming from charges paid by the utilities' commercial customers. Most recipients were small businesses, but the scheme also funded the large-scale local government installations in San Francisco, including the rooftop solar system on the Moscone Center. Though successful, the program was set to expire at the end of 2003. This would have stopped Vote Solar's campaign in its tracks. Meanwhile, pro-solar politician Mark Leno had been elected to the state assembly. At the urging of Browning and Hochschild, one of the first pieces of legislation Leno got passed was a four-year reauthorization of the scheme. As we shall see in the next chapter, the lowly Self-Generation Incentive Program would become the template for the vastly larger California Solar Initiative.

The other, complementary measure was a temporary increase in the system cap for net metering, from 10 kilowatts to 1 megawatt, introduced on April 19, 2001. Net metering, readers will recall, allows system owners to gain credit for the surplus energy they generate. The increase meant that as well as households, farmers, small businesses, and local governments could also save on their electricity bills. The new measure had an immediate impact. The number of large solar installations soared. By mid-2002 some 2,200 customers in California were taking advantage of net metering, up from around 500 two years previously. Another 700 applications were pending.

As we saw in chapter 10, in 1995 California's investor-owned utilities had allowed the state legislature to pass the nation's first net metering act. At the time they were distracted, having bigger fish to fry, namely, deregulation. By now, however, the power companies were well aware of the threat to their revenue that net metering represented. They geared up to fight it tooth and nail. The expanded net metering provision was due to sunset in August 2002, threatening PV's newfound momentum.

"We realized that if net metering went away," Browning said, "so would all our city projects." Now it was all hands on deck to persuade legislators in Sacramento to reauthorize the bill. Vote Solar took out a full-page ad in the West Coast edition of the *New York Times*. "Don't Let the Utilities Turn the Lights Out on Solar Energy in California" was the headline. The activists also managed to place op-ed pieces in three major California daily papers. They sent out an e-mail signed by faithful supporters Bonnie Raitt and

Robert Redford (plus Graham Nash and David Crosby) urging voters to lobby their representatives. About 2,500 people responded by sending protests to the California Public Utilities Commission.

The big utilities fought hard to stop the reauthorization. Southern California Edison's chief lobbyist, Catherine Hackney, argued that the current net metering provisions were "indefensible" because they supported a "fundamental inequity" between solar producers and other consumers. The few PV system owners were being subsidized by the many regular, nonsolar customers. In fact, the utilities' own figures showed that the financial burden on rate-payers amounted to less than a penny a month. Eventually, realizing that they could not stop the reauthorization bill, the power companies sought instead to hijack it. Hackney got an amendment inserted that reduced net metering credits for large solar producers by up to 50 percent, thus doubling the payback period. The amendment also required customers to install an additional meter at their own expense, unnecessarily complicating the administrative process. It undermined the main virtue of net metering, its simplicity. The revised bill, which Governor Davis signed into law, begged the question put by Kari Smith of Berkeley-based PowerLight, a system integrator that specialized in large-scale projects like the Moscone Center: "Does California want to support solar or not?"

The lengths to which big utilities were prepared to go to discourage the spread of net metered solar is well illustrated by the case of Fred Adelman. Admittedly, Adelman was not your average solar system owner. A retired entrepreneur who had made millions selling his company to Nokia, Adelman and his wife, Gabrielle, were environmentalists who happened to like flying their private planes and helicopter. To assuage their guilt about the damage their hobby was doing to the environment, they had bought several electric cars. Charging the batteries of these vehicles sent their electricity bills soaring.

Hearing that the 10-kilowatt system cap on net metering was going to be lifted and that attractive new incentives would be available, Adelman decided that the time had come to go solar. In the grounds of their house in the hills just south of Santa Cruz, he built what at 30 kilowatts was said to be the largest residential solar system in California. It cost $360,000, minus a $135,000 rebate under the Self-Generation Incentive Program.

On April 20, 2001, the day after the new net metering law went into effect, Adelman submitted his application for interconnection to PG&E. Several weeks later, he received an e-mail from the utility. It explained that before his system could connect to the grid, an engineering impact study would have to be performed—at the customer's expense, of course. A month passed with no further information forthcoming. Eventually, losing patience, Adelman called PG&E. He was told that the utility proposed to charge him $605,000 for the upgrades to the company's local distribution network that interconnection of his system would necessitate.

It was, Adelman felt, a bizarre situation. "Pacific Gas & Electric was bankrupt from having to pay outrageous wholesale prices for peak-time power," he wrote on his website. "Here I was, willing to give them peak-time power in exchange for kilowatts that they would deliver to me at 2 AM to charge my cars. They were acting not just against the best interests of the public, but seemingly against their own best interest!"

Adelman went public with the story in the local media. The utility quickly caved in and approved his system. Next day, however, they changed their mind, instructing Adelman to disconnect his system from the grid, forcibly cutting him off when he refused. It took almost a year of costly legal dickering to reach a settlement. Eventually, the Public Utilities Ccommission reduced the amount Adelman would have to pay for upgrades from $605,000 to just $11,000.

California's energy crisis would give solar one final, totally unexpected boost. In the recall election of late 2003, the state's voters turfed Gray Davis out of office, in large part for what they perceived as his costly mismanagement of the crisis. In his stead, they chose a most unlikely successor as governor. How Arnold Schwarzenegger became America's most enthusiastic political champion of PV since Jimmy Carter, and the ambitious pro-solar program he initiated in the Golden State, is the subject of the next chapter.

20

SOLAR'S NOT FOR SISSIES

*O*n September 21, 2003, a few weeks after announcing his candidacy for the governorship in California's recall election to Jay Leno on *The Tonight Show*, Arnold Schwarzenegger posted a visionary environmental action plan on his website. Its goals included reducing the state's energy consumption by 20 percent within two years. This would be achieved, among other measures, by equipping 50 percent of new homes in the state with solar panels.

After Schwarzenegger was elected governor in mid-November, the challenge for planners at California's Environmental Protection Agency was to concoct an initiative that would live up to his campaign promises, something that was really *Arnold-sized*. In response, they proposed a scheme that called for solar systems to be deployed on the roofs of a million homes throughout the state. A million sounded like a nice, big number.

The planners eventually got the chance to pitch their proposal to the great man in person at his Sacramento office. Their presentation went spectacularly well. The slides flowed nicely. The governor would ask a good question then, *tap!*—up would come the answer on the very next slide. When the presentation ended, Schwarzenegger seemed impressed. Turning to the gaggle of veteran aides—chief of staff, policy advisors, energy experts—who ran his legislative agenda, he asked them, "What do you guys think?"

The functionaries took turns tearing the proposal to pieces. Impossible, they said, too expensive, not going to work. Like a poker player sizing up his opponents, the governor listened impassively. Suddenly, he picked up the proposal then slammed it back down on his desk. "Dammit," he snapped, "this is a great proposal!" Schwarzenegger proceeded to shoot down their criticisms one by one. He pointed out that there were always going to be naysayers who said things were not possible, or too expensive, or whatever. "I want to see this go forward," he instructed. Then, adding that he was late for his flight to Santa Monica, he stood up, pivoted on his lizard-skin cowboy boots, and walked out.

It felt, according to one participant, like a scene from a B-grade movie, where the knight in shining armor—the last action hero, you might say—comes riding in to save the day. Except that, in this movie, the star was very much A-list.

"There are a lot of things I disagree with the governor on," said Adam Browning of Vote Solar. "But I will say that solar in this state would never have happened in the way it did without his true leadership. He set higher goals than anyone thought politically possible, and stood by them."

Knowing that he was a cigar-smoking, Hummer-driving Republican, many Californians were bemused to discover that Arnold Schwarzenegger was also an environmentalist. They would not have been had they heard their new governor talk, as he often did, about the nasty shock he got on arriving in Los Angeles in 1968, at the age of twenty-one. As the young bodybuilder pumped iron on Muscle Beach, the smog would sting his eyes and burn his lungs. Schwarzenegger had expected a pristine seafront. What he found was polluted shorelines and garbage-littered sidewalks. Growing up in postwar Austria where electricity was rationed had made Schwarzenegger a self-confessed "conservation fanatic," unable to walk out of a room without turning off the lights. His 1986 marriage to Maria Shriver brought him into contact with her cousin, Robert Kennedy Jr., a leading environmental lawyer with an impressive record of suing corporate polluters. Over a family dinner a couple of days after Schwarzenegger declared his candidacy, Kennedy advised him that, since most Californians cared about the environment, he had better work up a robust environmental plan. To fill him in on the details, Kennedy suggested the candidate should call his friend Terry Tamminen, with whom he had worked for many years in the Water-keeper Alliance, an anti-pollution advocacy group.

Terry Tamminen was the possessor of what one observer described as "an almost comically varied résumé." Born in Wisconsin to a Finnish-American family, he had spent his high school years in Australia, after which he had lived and worked in Europe and Africa. Tamminen had been employed in a bewildering variety of capacities, including mariculture in the Gulf States, livestock disease control in Minnesota, recreational services, even real estate management. He spoke German, Dutch, and Spanish, was a licensed ship's captain, an avid airplane and helicopter pilot, and the author of a best-selling series of guidebooks on pools and hot spas. When the call from the candidate came, he was running a foundation called Environment Now.

Schwarzenegger invited Tamminen to drop by his office in Santa Monica for lunch. Over the next few days they met several times over pizza, gradually formulating what would became the candidate's environmental action plan, an eight-page document listing all the things he pledged to do if he became governor. "It was the only detailed platform he actually issued during the recall campaign," Tamminen told me. When Schwarzenegger was elected Tamminen was one of his first appointees, initially as secretary of the California EPA then, in December 2004, as chief policy advisor.

"Terry is just an extraordinary individual," said Drew Bohan, one of Tamminen's colleagues from those heady days. "He captured the governor's imagination and he had his ear."

However, credit for the idea of putting solar on the roof of every new home in California belongs not to Terry Tamminen, but to Bernadette Del Chiaro of Environment California, a public advocacy group. Northern California born and Berkeley educated, Del Chiaro was an activist who had cut her teeth fighting a proposed radioactive nuclear waste dump out in the California desert in the midnineties. From there she had moved on to a campaign to clean up Connecticut's "filthy five" coal power plants.

Returning to Sacramento in 2002 to direct Environment California's clean energy program, Del Chiaro realized that solar was being overlooked. None of the other environmental organizations were taking PV seriously. "We just saw it as a huge missed opportunity," she told me. "Solar roofs were one of the most tangible ways that Californians could do something about our energy and environmental problems." Vote Solar's success in San Francisco had shown that voters embraced PV. It was time to step things up to another level. "The solar initiative in San Francisco was great," Del Chiaro said, "but even that was pretty small in terms of watts installed, compared to what California should be doing. We decided to launch a bigger campaign." In the early 2000s the housing market was booming. "We basically said, Look, the vision is that solar needs to be as mainstream as double-paned windows, it just has to be part of the basic fabric of our built environment." Pushing for solar on new homes, Del Chiaro thought, was "the smartest, most cost-effective way to go."

Senate Bill 289, introduced to the California legislature under Environment California's sponsorship in May 2003, required that half of all new homes constructed in the state after 2006 be built with solar panels on their roofs as a standard feature. In addition to making the state more self-reliant, the measure would also allow individuals to be more energy independent. "Given the blackouts we have had," the bill's author, Senator Kevin Murray, said, "I think a lot of people would like to know that they can operate for at least some portion of the day, in case the whole electricity grid goes down." Unfortunately, any reminder of the energy crisis was exactly what Governor Gray Davis was eager to avoid. The solar homes bill was quietly killed at the committee stage. Though unsuccessful, the bill was noticed by a lot of people, notably Terry Tamminen. It would serve as the basis for Arnold Schwarzenegger's pro-solar platform.

As a candidate, Schwarzenegger had been big on solar. During his first months in office, however, the governor failed to follow through on his commitment. Lacking his endorsement, a revised version of the bill looked likely to meet the same fate as its predecessor. It was necessary, Del Chiaro decided, somehow to capture the governor's attention. In summer 2004 Environment California bought a bunch of life-size cardboard cut-outs of Schwarzenegger in his signature role as the Terminator. Volunteers took the cut-outs onto the streets of Californian cities, snapping pictures of ordinary people standing next to "the Solarnator" with messages of support for the solar homes bill. Del Chiaro mailed hundreds of these snaps to the governor's office.

Environment California also tapped the Internet for donations. In just two days, from over four hundred donors, they received $25,000. That was enough to pay for a half-page ad that ran in the August 2, 2004, edition of the *Los Angeles Times*. It featured a cartoon depicting the Solarnator as superhero. "Governor, keep your promise to build solar homes," urged the text. It was in direct response to this mounting political pressure that Schwarzenegger called the meeting at which the EPA planners pitched their Million Solar Homes proposal. There would be no doubting his administration's commitment to solar from then on.

When Tamminen and his assistants at the EPA released a draft of their Million Solar Homes initiative in early August 2004, they were unaware of the existence of a similarly named Department of Energy program, the Million Solar Roofs initiative. This had been launched by President Bill Clinton in June 1997, in a speech on climate change at the United Nations. Prior to that, despite having Al Gore as his vice president, Clinton had shown no interest in renewable energy. "We could not even get it on his radar screen," the still-frustrated solar architect Don Aitken recalled. "He wouldn't touch it, it was terribly disappointing."

President Clinton's new initiative proved little more than smoke and mirrors. Its budget was minuscule, especially when compared to the huge sums routinely doled out to subsidize conventional energy programs. Over ten years, the federal government invested just $16 million in the Million Solar Roofs initiative. What little money there was went mostly in the form of grants of up to $50,000 to some ninety cities, which in return pledged to install a certain amount of solar by 2010. "The idea was to get these local governments to sign on as Million Solar Roofs communities," Heather Mulligan, the program's coordinator, told me. "We would try to get them the technical assistance they needed from [national laboratories]. . . . To be honest, it was a lot of community organizing, going to a city and bringing together different groups who could play a role in promoting solar there."

Within the Department of Energy itself, nobody took the initiative seriously. "We were relegated to a corner of a room," Mulligan recalled. "Everybody kind of laughed and said, Oh, there's those solar nutcases again." Solar was still stigmatized, especially by Republicans, as being for weird-beards only. When Clinton proposed his initiative, it drew a contemptuous retort from Trent Lott, then the Senate majority leader. "I'm sure you petroleum folks understand that solar power will solve all our problems," Lott sneered in a speech to the Independent Petroleum Association of America. "How much money have we blown on that? This is the hippies' program from the seventies and they're still pushing this stuff."

When George W. Bush took office in 2001, things got even worse. It was like the Carter-Reagan transition all over again, except on a much smaller scale. "The Bush people wanted *nothing* to do with a Clinton administration grassroots hippie program," Mulligan said. "From the outset, they had the program in their crosshairs, it was going to go one way or another." By 2003, when the California solar homes program was first mooted, the US PV industry was in bad shape. That year shipments decreased 14 percent, to just over 100 megawatts. Leading companies like BP Solar were cutting back production, while AstroPower, the nation's second-largest domestic producer of solar cells, went bankrupt.

In getting the Million Solar Homes initiative off the ground, its proponents had first to overcome the prevailing perception, especially among state Republicans and business types, that PV was only for granola-eating, Birkenstock-wearing, long-haired freaks. Real men ran giant power plants that spun big turbines fueled by coal or natural gas. Solar, by contrast, was goofy stuff, a fringe thing that hippies put on their cabins out in the back-woods. Or, it was green bling for environmentally conscious elites in Holly-wood. Attempting to address this image problem, Vote Solar came up with the slogan "Hip, Not Hippie." But when solar won Schwarzenegger's endorsement it was, ipso facto, no longer for sissies.

To bolster their case, the EPA planners assembled some compelling fig-ures. Solar was going to be good for the state, it would generate new busi-nesses, and that meant jobs. It helped that well-known Japanese PV makers like Sharp were coming on strong. The success of the large-scale Japanese and German government subsidy programs was becoming hard to ignore, especially in a state that had once led the world in solar. As a native speaker of German, Schwarzenegger was well aware of Germany's prowess in PV. He and Tamminen had talked about the fact that cloudy Germany had taken the lead in solar deployments. "Arnold is a competitive bodybuilder," Tam-minen said. "If you can do ten pounds, he can do a hundred. It was an embar-rassment to him that a state with so much sun was lagging behind." If Ger-many could do a hundred thousand roofs, then surely California could do a million.

The turbocharger for solar in Germany was, as we have seen, the feed-in tariff. But in California in late 2003, the efficacy of feed-in tariffs was not as obvious as it would become after the revision of the Renewable Energies Law the following year. "I don't think in the beginning we ever seriously considered a feed-in tariff," Tamminen told me. "It would have been much more difficult to legislate, much more complicated politically to maneuver." Americans were not accustomed to taking their lead from Germany: Japan was still the policy exemplar. California's planners decided to emulate the familiar, tried-and-tested route the Japanese had taken. Upfront rebates were seen as the best way to encourage people to buy solar systems. Happily, a rebate scheme—the modest Self-Generation Incentives Program that the Davis administration had introduced as a temporary measure—was already in place. All they had to do was super-size it.

In its initial form the draft bill proposed that, within ten years, half of all new homes built in the state should run at least in part on solar energy. To this end $100 million a year would be allocated in the form of rebates to builders of new homes and to homeowners for retrofits on existing houses. The money would come out of a new ratepayer surcharge of about 25 to 30 cents a month per household. The cap on net metering provisions would be lifted, initially to 2 percent and subsequently to 5 percent of peak energy demand. Homeowners would be able to sell their excess energy to the utilities. They would save, so the planners claimed, more money in reduced electricity charges than the monthly payments on their solar systems. Large developers would be obliged to equip a percentage of new houses in a subdivision with solar. It would start at 5 percent the first year, increasing annually until the target of 50 percent of the 150,000-odd new homes built in the state annually was reached in 2020. But, deeply resenting the prospect of being told what to do, the California Building Industry Association baulked. The builders used their considerable clout with Republican lawmakers to get the bill killed at the committee stage.

Six months later, in February 2005, an undeterred governor resurrected the legislation. "In California, where we are famous for the sun," Schwarzenegger insisted, "we are going to put the positive benefits of the sun to good use." Senate Bill 1 was a bipartisan effort coauthored by state senators Kevin Murray, an African American Democrat from Los Angeles who two years previously had authored the original solar homes bill, and John Campbell, a conservative Republican from Orange County. The bill had been chosen as the top priority for the senate that year. Throwing extra weight behind SB 1 were various organizations including Shell Solar, the American Lung Association, a group of nearly three hundred churches, and a cohort of celebrity greens led by Robert Redford. "We've put together the broadest coalition of any bill that I can think of in history," Murray boasted.

The revised legislation did away with the mandatory requirements, focusing instead on ensuring long-term market stability for solar. It would guarantee rebates of around $3 billion to help fund 3 gigawatts of solar power over ten years. Three gigawatts was equivalent to 1 million homes equipped with 3 kilowatts each. However, much of this PV would be installed on commercial and municipal buildings in the form of large systems of up to 1 megawatt. The bill, its backers asserted, would make California the world's number one PV market. "This is the best piece of solar legislation ever introduced in the US," enthused David Hochschild of Vote Solar.

In March 2005 Environment California brought Hermann Scheer, the father of Germany's 100,000 Solar Roofs Program, to Sacramento to confer his blessing on the bill. He met with the governor and his chief advisor, Terry Tamminen. The fact that all three spoke German helped them hit it off, despite the fact that Scheer and Schwarzenegger came from opposite ends of the political spectrum. SB 1 was, Scheer announced, a signal to other countries and states that a global "Olympic competition of renewable energy" had commenced. It was also a signal to German PV firms that the Californian market was open for business.

In April Hochschild persuaded the mayors of the six largest cities in the state—Los Angeles, Oakland, Sacramento, San Diego, San Jose, and San Francisco—to sign an open letter addressed to the state legislature. It endorsed the bill as an idea whose time had come. For its part Environment California went into overdrive, hiring hundreds of organizers to hit the streets of every city in the state, getting citizens to sign petitions, make phone calls, donate money, and attend district meetings with local legislators. Thanks largely to the group's concerted efforts, the bill attracted more public support than gay marriage, legalization of marijuana, driver's licenses for immigrants, or any other hot-button issue. One contemporary poll found that more than three-quarters of Californians favored increasing the state's investment in solar energy.

In June 2005 SB 1 sailed through the state senate by a majority of thirty to five. Then, in the larger (eighty-member) assembly, the bill ran into trouble. This time, it was largely Democrats who were to blame. The powerful International Brotherhood of Electrical Workers decided that they would support the bill only if all the work that resulted from its passage was limited to its highly qualified (and highly paid) members. This was foolish: solar installers were mostly specialists who knew their business much better than IBEW members. Besides, there was more to the job than just its electrical component: you also had to know roofing, how to flash and waterproof an installation. The real damage had been done behind the scenes, by the state's giant investor-owned utilities. They feared the bill because of the potential threat it posed to their monopoly. Their lobbyists picked off the Republicans and business-friendly Democrats. That left the bill's passage dependent on left-leaning Democrat senators sensitive to the influence of organized labor.

Ratepayer advocates like the public utility commission–funded group The Utility Reform Network (TURN) fretted about the $3 billion price tag.

They vigorously opposed *any* increase in electricity rates, no matter how worthy the goal. And they objected to the idea that residential customers should subsidize commercial installations.

The impasse also had a personal dimension. Tensions between the Democrats and Schwarzenegger were running high. "He has gone out of his way to antagonize us," complained Mark Leno, who had previously led the way in sponsoring solar legislation. One of the first openly gay members to be elected to the state assembly, Leno was particularly upset that the governor had refused to apologize for calling lawmakers "girlie men" the previous year. Leno and his colleagues were in no mood to hand Schwarzenegger a big victory. When the legislature ended its 2005 session on September 9, the bill died a second death.

Having come this far, having, as he put it, "pushed and pushed and pushed," Schwarzenegger was not about to give up. In early fall 2005, when it had become clear that the legislature was not going to pass his plan, the governor summoned Michael Peevey, the president of California's Public Utilities Commission, to meet him in Sacramento. "Can you help do this at the PUC?" he asked Peevey. The request did not come as a surprise. The first time Peevey met Schwarzenegger, the governor had expressed a strong interest in solar, telling him about how Jim Cameron, director of the *Terminator* movies, had installed PV all over his very large homestead. Though a Democrat, appointed under the Davis administration, Peevey was won over by Schwarzenegger. "He's upbeat and optimistic and charming," the commissioner gushed to a reporter. "He makes you feel good when you meet him."

After what Terry Tamminen called "a meeting of minds," Peevey drafted a plan that was closely modeled on the Million Solar Homes bill. Because it would be ratepayer based—that is, funded by a small increase in electricity rates—the CPUC had the state constitution–sanctioned authority to implement the scheme by administrative action rather than legislation. Still, pushing through the CPUC's plan hinged on rallying a broad base. That fall environmental and public policy advocacy groups sent out an alert to their memberships, urging them to contact the commission to express their support. The most instrumental organization was Moveon.org, whose e-mail list in California was around 500,000, larger than any other group's by an order of magnitude. Their office happened to be right next door to Vote Solar. Fifty thousand e-mails flooded into the CPUC, many of them from Moveon members, shutting down its server for a day. Never in its hundred-year history, Peevey confessed, had the commission received so much interest from the public.

On January 12, 2006, the CPUC announced that it had approved by a three-to-one majority a subsidy plan, to be known as the California Solar Initiative. The commission would direct the state's investor-owned utilities to spend an estimated $3.2 billion over eleven years to encourage business and residential customers to install 3 gigawatts of solar electricity. This would be the equivalent of six medium-sized coal-fired power plants or forty gas-fired "peaker" plants. The program would be paid for using existing funds already earmarked for solar, plus a small surcharge—around a dollar a month—to rate payers. The largest solar program in the US would become, the commissioners hoped, a model for other states. It was, thought Del Chiaro, who was present when Peevey made the announcement, "a real watershed moment in the story of solar power in California."

Presented with a fait accompli, the state legislature fell meekly into line, passing SB 1 to include municipal utilities in the solar initiative as well as the CPUC-regulated IOUs (investor-owned utilities). The bill also increased the cap on net metering from 0.5 to 2.5 percent, enough for roughly half a million new solar system owners to join the program. In addition, it required that all new homes built after 2011 should come with solar power as a standard feature. On August 21, a jubilant Schwarzenegger signed the bill into law. "This is really terrific," the governor enthused at the signing ceremony. "This is what we wanted from the start, a system where families across our state can take advantage of clean solar power, and so that every rooftop can become a clean solar plant."

Was it a coincidence that just two weeks after the CPUC announced the California Solar Initiative, President George W. Bush in his January 2006 State of the Union address proposed an Advanced Energy Initiative, a key component of which would be a Solar America Initiative? The following month Bush became the first president since Jimmy Carter to visit the National Renewable Energy Laboratory in Colorado. Unfortunately, as a result of the most recent round of cuts to NREL's budget, thirty-two employees at the lab had just been laid off. The day before Bush arrived, an embarrassed Department of Energy hurriedly transferred sufficient funds to enable the lab to restore the axed jobs.

Bush visiting NREL was a bit like Nixon going to China, thought veteran solar cell researcher Larry Kazmerski. Had the president at long last grasped the need for renewables? "The vision for solar," Bush said during a panel discussion at the lab, "is one day each home becomes a little power unit unto itself, that photovoltaic processes will enable you to become a little power generator, and that if you generate more power than you use, you can feed it back into the grid." The president was using weasel words. By "one day," he meant some indeterminate point in the future, certainly not on his watch.

That October at the National Renewable Energy Conference in St. Louis, Bush told the audience he believed that "with the proper amount of research, whether it be public or private, we will have solar roofs that will enable the American family to be able to generate their own electricity." Here, Bush was rehashing the erroneous but widely held view that breakthroughs were still needed before PV could be deployed commercially. Solar technology remained a tentative possibility for tomorrow, not a well-established fact for today. However qualified, Bush's endorsement did indicate that solar was at last being taken seriously by decision makers at the highest level.

The new Solar America Initiative was the old DoE Million Solar Roofs program on steroids. Its stated aim was to generate from 5 to 10 gigawatts of solar power by 2015, enough (according to the department's estimate) to power between 1 and 2 million homes. To fund this ambitious-sounding program, in his fiscal 2007 budget the president requested just $148 million. It was a piddling amount compared with the vast sums Bush had signed into law the previous year in his energy bill, which dished out $6 billion in subsidies to oil and gas, $9 billion to coal, and $12 billion to nuclear.

The most important contribution the federal government made to the spread of solar was a 30 percent investment tax credit. This provision was introduced in 2005 as part of the Bush administration's Energy Policy Act. It was due to expire in 2008, but the incoming Speaker of the House of Representatives, Nancy Pelosi, made extending the credit her number one priority. A religious person, Pelosi believed that achieving energy independence was a moral issue: the future of the planet was at stake. She muscled the bill through, against the opposition not just of Republicans but also Democrats from coal states.

The extension of the tax credits was, thought David Hochschild, "the biggest solar policy victory in the US" because it gave the solar industry certainty that there would be consistent support for its products, at least until 2016, by which time it was assumed that solar energy would be one of the least expensive sources of electricity for consumers. The tax credit could be combined with state and city rebates, notably the upfront incentive from the California Solar Initiative. Cumbersomely entitled the Expected Performance-Based Buydown, this incentive was based on an estimate of a solar system's future output. Like feed-in tariffs, the payment would decline in steps as capacity targets were met. The incentive began at $2.50 a watt, meaning that the purchaser of a typical domestic 3-kilowatt solar system would get a rebate of $7,500, or somewhere between a quarter and a third of the total cost. With net metering the system could pay for itself in as little as a decade.

It seemed that the CSI would get off to a flying start. After all, in 2006, even without rebates, more than 7,000 Californian householders had applied to install solar systems, up from 4,000 in each of the two previous years. Unfortunately, however, the new legislation contained a "time-of-use" provision that forced applicants for the rebates to sign up for a costly pricing plan. Under this scheme, electricity delivered by the utilities during peak periods would be the most expensive. Potential customers whose roofs could not accommodate a solar system large enough to make them self-sufficient would have ended up paying *more* for their electricity, not less. In effect, they would have been punished for going solar. As red-faced state officials scrambled to fix the glitch in the legislation, applications plummeted.

To get an idea of what was happening in the field, it was instructive to talk to an installer of PV systems. Pat Redgate of Long Beach–based Ameco had been in the solar business for more than thirty years. By September 2007, his ten-employee firm was close to being smothered by the red tape the rebate program generated. For Redgate, the CSI was "a big pain." The root of the problem as he saw it was that, although the CPUC had created the program, the commission had left it to the utilities to run as they saw fit. Formerly it had been possible to get rebates paid within a couple of months. Now getting the money took seven or eight months. The paperwork had ballooned from a simple, one-page application to a forty-nine-page document. "I don't know if there are that many pages involved when you join the military," Redgate said sarcastically. At his company, more man-hours went into processing paperwork and coordinating with various officials for site inspections than into the actual installation of the system.

On top of the byzantine CSI bureaucracy, there was also the net metering agreement to contend with. This had bloated out to the point where customers were required to indemnify the power company. Customers had to prove they had insurance, Redgate explained, "so that the utility could go after their insurance company if their wicked, evil solar system is going to, y'know, kill any babies and stuff." Redgate was unsure whether such provisions were a conspiracy perpetrated by the utilities to make life difficult for firms such as his. "But I can tell you that they're not in it to make our life any easier," he said. He was growing tired of the seemingly never-ending battle against the power companies. "What it amounts to is purely an economic thing. [Southern California] Edison, PG&E, and all these other utilities, they're for solar, they want solar, but they want to own and run the solar systems themselves. They don't want people like me helping customers get off the grid."

But if the California Solar Initiative were to achieve its goals, then companies like Ameco would have to succeed. "I'm a small business, but I'm the paradigm of the American economy," Redgate said. "You want to see a new industry take off, a lot of companies like me growing and competing and trying to make things cheaper and better and more economic, instead of big companies with all their built-in profit margins because they're part of the utility structure. I'm all for solar, and if the only way we can have solar is for the utilities to do it then, OK, fine—just don't kill me, don't put me out of business, just make a law that says I can't do it anymore. We're making a living, we're doing OK," he concluded ruefully, "it's just a bit like *Alice in Wonderland*."

In 2008 California added 156 megawatts of new solar capacity, almost double the 2007 increase of 81 megawatts. The state now boasted more than 50,000 solar installations, representing over two-thirds the US national total. Their combined output was more than 500 megawatts, about the same as one coal-fired power plant, with the California Solar Initiative responsible for around half of the total. The customer base for solar was expanding beyond tree-huggers and do-it-yourself tinkerers, to hands-off types whose concerns were at least partially economic rather than environmental.

"California is making major moves," a determined-sounding Schwarzenegger said in late 2008 at the launch of a new solar manufacturing facility in Bakersfield. "We know that Germany is number one in solar power, but let me tell you something: We're going to catch up with them very quickly the way we are going, and I'm very happy about that. . . . My vision is that, when I fly up and down the state of California, I see . . . every available space blanketed with solar, if it's parking lots, if it's on top of buildings, on top of prisons, universities, government buildings, hospitals—solar, solar, solar, that is my goal."

In truth, solar still had a very long way to go in order to reach that goal. Even with all the new capacity, PV was still a pipsqueak among energy sources. It represented around a quarter of 1 percent of California's total energy capacity, little more than a rounding error. While California was adding 150 megawatts a year, Germany, with a population just twice as large but only half as much sun, was putting on 1.5 *gigawatts* a year, ten times as much. (The ratio remains constant. In July 2010 the CPUC reported that, since January, the CSI program had received a record of nearly 300 megawatts of new applications. Meanwhile, analysts were predicting that Germany would install 6.5 gigawatts of solar in 2010.)

By 2008 it was already clear that the California Solar Initiative on its own was not going to be sufficient to bring about the necessary paradigm shift in energy generation. Especially since it was based on a finite amount of money, which would run out sooner than expected, leaving the solar tax credit as the key driver. Other efforts were also required. In California as in Germany, innovation, as we shall see in the next chapter, was bubbling up from the local level in two very different cities.

21

A TALE OF TWO CITIES

*I*t was doubtless predictable that Berkeley, a city that championed environmental consciousness, where municipal trucks run on recycled vegetable oil harvested from local restaurants, should come up with an innovative model for making rooftop solar affordable to its residents. It was, however, utterly unexpected that Palm Desert, a city that epitomized environmental excess, where the palm trees that line its main streets are wrapped with decorative lights that shine all night, should be first to implement such a model and blaze the trail for other Californian cities to follow.

Like Arnold Schwarzenegger, Berkeley's mayor Tom Bates traced his environmental awareness to a frugal upbringing. The son of working-class parents, Bates came to UC Berkeley on a football scholarship. During the sixties he drove a Volkswagen van decorated with a green stars and stripes sticker. In twenty years as a state assembly member he had commuted to and from Sacramento in a carpool. Bates had even signed his name on the many bills he had authored using a green felt-tip pen, until he discovered that over time green ink fades. At home, in preference to using an energy-guzzling dryer, he and his wife hung their washing on a clothesline.

In 2006 Bates installed solar panels on the roof of his brown-shingled Craftsman house. That November, when the city council asked local residents to vote in a referendum on Measure G, a Kyoto Protocol–inspired plan to cut Berkeley's greenhouse gas emissions 80 percent by 2050, their overwhelming response—four out of five—was yes. Bates himself was a true believer in the pressing need to tackle climate change. He told people that it was the responsibility of elected officials at every level of government to lead. His chief of staff was Cisco DeVries, an intense-looking thirty-something policy wonk, whom the mayor hailed as one of the best aides he had had in thirty years of public service. Bates tasked DeVries with translating Berkeley's greenhouse gas plan into action at the community level.

The younger man decided it behooved him to follow his boss's lead and put some PV on the roof of his house in Oakland. "I got two or three bids, looked at what it would take, and had the same sort of realization that everybody else had. Which was, that's a really big check you have to write." DeVries had a home equity line of credit, which meant he could have afforded to buy a system. But, daunted by the scale of the commitment and, having realized based on a spreadsheet analysis of his last three years of PG&E bills that there was no way the investment was going to work (especially not within the time he expected to live in his house), he didn't. Instead, DeVries began contemplating ways of circumventing the biggest obstacle to owning solar, the high upfront cost.

At the time DeVries happened to be peripherally involved with a neighborhood in Berkeley that was "undergrounding" its power lines. Burying cables not only improves the esthetics of streets, it also reduces risk from earthquake and fire. To fund the construction work—digging trenches, laying cables, backfilling, and so on—the neighborhood had formed what is known as an "underground utility district." "Underground utility" is one variety of what are more broadly referred to as "land-secured financing districts." For more than a century expanding municipalities have used this mechanism to pay for public infrastructure like sewer systems. The bonds are paid off, over many years, via an extra line item on the annual property tax bills of all residents who benefit.

DeVries did not have a single Eureka! moment. It just gradually dawned on him that adapting this financing mechanism to underwrite the installation of solar systems on the roofs of homes might be possible. San Francisco, as we have seen, had used municipal bonds to put panels on the roofs of city buildings. Utilities like SMUD had used "on-bill financing" to repay loans for purchasing appliances like energy-efficient refrigerators. But no one had previously thought of using property tax bills to finance rooftop PV. It was a genuine innovation. "There are certain things about the property tax bill that are frankly brilliant for this," DeVries told me. "It's extremely secure: if ever there's a foreclosure, the property taxes get paid back first, ahead of any mortgage. It's a bill that everybody pays, because there are huge consequences if you don't. The delinquency rate in Berkeley of people who fail to pay their property taxes on time is less than 2 percent."

DeVries wrote a memo outlining his idea and sent it to Bates, attaching the question, Do you think we could do this? "To the mayor's immense credit, instead of saying no, which is what most busy people say when confronted with things that seem crazy, he replied maybe." To help DeVries flesh out his concept, Bates assigned him a lawyer and a financial advisor. The program the team developed—immodestly entitled Berkeley FIRST (Financing Initiative for Renewable and Solar Technology)—would enable property owners to borrow money from the city to install solar systems. The amount—typically between twenty and thirty thousand dollars—would then be repaid, over twenty years, through a special tax on the property. In addition to solving the problem of high upfront cost, the program also dealt with another major disincentive to potential solar buyers. Homeowners often hesitate to purchase PV because they worry that they will not remain in the house long enough to have the investment pay for itself. Under the Berkeley model, since the solar system stays with the property, so does the tax obligation. If the home were sold, any remaining taxes would simply be paid by the new owners.

In November 2007 the city council unanimously approved the concept. National media picked up the story. Later that month when Bates and DcVries flew up to Seattle for a mayors conference, they were bombarded with questions. "It was like the most interesting thing that people had heard of, because it was novel," Bates said. On their return nearly two dozen cities called indicating that they wanted to follow Berkeley's lead. "I've never been part of something like this where the power of an idea has grabbed so many people so quickly," a dazed DeVries told the *New York Times*. Most cities, however, were cautious. They would wait to see how successful Berkeley was. "Everybody loved this idea, but nobody knew how to do it," DeVries said. "They wanted some way that it could be done in their community, without any muss or fuss."

DeVries subsequently left local government. He went on to found a start-up called Renewable Funding. Among other things, the firm would line up financing for the Berkeley program. The program's pilot phase was limited to $1.5 million, enough to pay for forty installations. The initial slots were snapped up in just nine minutes. Such rapid acceptance was not surprising in a city that already boasted California's highest per-capita deployment of PV, with solar systems deployed on almost four hundred roofs. On a typical $23,000 system, homeowners would in theory repay roughly $180 a month. But the amount would be less when the CSI rebate and federal tax credit were factored in.

Across the bay, in March 2009 San Francisco announced a plan to disburse between $20 million and $30 million in a similar property tax–based program. "We're going to take Berkeley's idea to a whole 'nother level," Mayor Gavin Newsom boasted. San Francisco hired Renewable Funding to get the program going. San Diego was likewise pushing a privately financed tax-based program, also aided and abetted by DeVries's firm. City officials anticipated that when the program launched in fall 2009, it would attract five hundred participants. Los Angeles County followed suit, announcing an $11 million investment in federal grant funds. By the end of 2010, 23 counties and 184 cities in the state were scheduled to have initiatives in place. Around the country seventeen states passed legislation to enable what had come to be known as PACE—for Property-Assessed Clean Energy—programs. (As of late 2010, however, momentum had been slowed by opposition from Fannie Mae and Freddie Mac. The government-chartered mortgage giants were concerned about losing out to property tax assessments if homeowners default or go into foreclosure.)

The Berkeley program's second phase was also due to begin in late 2009. The city's ultimate goal was to sign up a quarter of property owners to its solar financing plan, reducing Berkeley's greenhouse gas emissions by two thousand tons a year. But with a population of 105,000 there was, as DeVries acknowledged, "still a long way to go." In October 2009 Vice President Joe Biden announced that the Obama administration would adopt the Berkeley financing model on a national level as a central feature of its Recovery Through Retrofit program. A delighted DeVries was invited to the White House to meet Biden. "This is a remarkable vindication of what Berkeley did," DeVries said. "For an idea that started in Berkeley, it's proven to be very nonideological."

Berkeley is a charter city, meaning that it was free to opt out of California's state government code and go its own way. There are, however, limits to what the state is willing to accept. Too radical a departure from accepted practice tends to spook the bond market, which is highly risk averse, especially when the economy as a whole is perilously close to meltdown. Berkeley required that private capital should provide all the financing for its solar program. As a consequence, the city was not able to approve its first solar loans until February 2009.

Palm Desert is also a charter city. Rather than go straight to the bond market, however, Palm Desert opted to change state law first, to explicitly allow land-secured financing. The bill the city drafted sailed through the state legislature, winning a two-thirds majority in both houses, the governor signing it into law in July 2008. Within a month, Palm Desert was issuing its first solar loans.

Other than both being charter cities, Berkeley and Palm Desert had little else in common. The first, located in the north of the state, is known as an intellectual powerhouse and bastion of progressivism. The latter, located 125 miles east of Los Angeles in the south of the state, is a wealthy conservative resort community. Palm Desert is populated mostly by aging baby boomers and, during the winter months, by "snowbirds" who fly down from colder climates. The city nestles near Joshua Tree National Park in the picturesque Coachella Valley, which contains nine other municipalities including Palm Springs, Rancho Mirage, and Indian Wells. The city's population is 50,000; the valley's, close to half a million. Tourism is the primary industry; Palm Desert alone boasts twenty-eight golf courses and more than a dozen hotels. Residents are encouraged to trundle around in golf carts via special lanes set aside on city streets.

Berkeley got into PV out of environmental consciousness. The motivation for Palm Desert was by contrast largely economic. The arid city is one of the hottest in the US. In summer temperatures can soar above 120 degrees F (49 degrees C) and stay there for days at a time. In such scorching heat fights sometimes erupt over who gets the shady spots in parking lots. During blackouts, people die. Utility rates are among the highest in the nation. Running air conditioners day and night means that many residents must pay more than a thousand dollars a month for their electricity.

Jim Ferguson, the prime mover on Palm Desert's council and a two-term mayor, was politically the antithesis of Tom Bates. Ferguson had been a Republican his whole life, beginning his career as press secretary to the conservative libertarian senator Barry Goldwater, later becoming a political appointee during the second Reagan administration. But Ferguson and Bates did share some values. He may not have been a tree-hugging tofu eater, but Ferguson was, he confessed, "an unwitting environmentalist." He had grown up in Arizona, a state blessed with many breathtaking vistas, the Grand Canyon and Monument Valley to name just two.

Ferguson had been embarrassed by President Bush's refusal to sign the Kyoto Protocol. When in 2005 Palm Desert was asked to endorse a conference of mayors resolution calling for local governments to take action to reduce their greenhouse gas emissions, the city voted it down. Ferguson drafted an alternative that sounded less like Al Gore: "If there are ways to keep from throwing crap in the air, then we support them." It passed unanimously. "I'm a Republican," he told me, "but I live in a beautiful valley, with beautiful mountains, and I like seeing those mountains. I like clean air, and I'm all for doing things that will keep us from becoming the next Los Angeles."

In August 2001, in the depths of California's energy crisis, Palm Desert received an executive order from Governor Gray Davis instructing the city to cut its electricity consumption by 20 percent for the following month. Ferguson doubted that this would be possible. However, by virtue of implementing small behavioral changes—such as shifting work schedules so that city employees could come in and leave earlier, and relaxing dress codes to allow the wearing of shorts—within two weeks Palm Desert had shaved the requisite 20 percent off its municipal load.

"It became normal for people to turn off their computer monitors if they were leaving the room, whereas before they could have cared less," Ferguson said. He liked to think that during this period, "people changed the way that they thought about electricity: it was no longer a right that they could simply flip on with a switch, it was a commodity for which they were spending money, and they had to make a decision about whether they wanted to spend that money or not." Then the governor's executive order expired, the crisis went away, and the sense of urgency was lost. By that winter most people had reverted to their wasteful ways. But rates kept going up, and that bothered Ferguson. He reckoned he was shelling out around $1,500 a month for his electricity, more than any other household expense except his mortgage. To make matters worse, despite having to pay through the teeth for his energy, he didn't even get to choose his provider: Southern California Edison had a monopoly on the supply. Ferguson saw himself as an unwilling partner in a shotgun marriage arranged by the California Public Utilities Commission.

In 2005 Ferguson was invited to a retreat in Sweden, a country known for having some of Europe's most energy efficient dwellings. The agenda included a sightseeing trip to the Estonian capital, Tallinn. The cabins on the overnight ferry back to Stockholm were tiny. As soon as the ship left port, Ferguson headed for the cocktail bar. There he found CPUC commissioner Mike Peevey with senior executives from Southern California Edison.

The commissioner had long been a proponent of energy efficiency. His mantra was that saving energy was always going to be cheaper than building new power plants or importing electricity from out of state. Palm Desert had demonstrated that it could save 20 percent on its energy bills. Knowing this, Ferguson boldly proposed that if the CPUC would provide funding, his city would undertake to reduce its energy consumption by 30 percent. The commissioner was skeptical. "You'll never make it," Ferguson recalled Peevey scoffing, "you guys wrap your trees in lights." It was true, desert cities were indeed notorious electricity hogs. But as Ferguson was quick to riposte, Peevey's hometown of La Cañada, an upmarket hillside community on LA's northern rim, was not exactly known for being poorly lit at night, either. Touché.

The cut and thrust continued through the night as they bobbed across the Baltic. From this exotic encounter emerged a draft resolution that became known, on account of its point of origin, as the Estonia Protocol. The CPUC agreed to give Palm Desert $7 million a year for five years to implement a demonstration program. In return, the city would develop a roadmap that other cities could follow. If profligate Palm Desert could reduce its energy by 30 percent, or even come close, then so could most other Californian communities.

When Ferguson got home, he appointed a city official named Pat Conlon to implement the energy management program. Conlon was not impressed. "It was a crazy goal," he told me. "If I had been on that boat I would have told them to put down their drinks and start talking real." To get to 10, 15 percent was easy-peasy, you could do that sitting on your hands, in city buildings and at home, too. But 30 percent? That was going to involve more than behavioral changes, that meant installing new equipment, plus a financing program to lend people the money to buy that equipment.

Initially, equipment meant more efficient appliances, in particular modern replacements for energy guzzlers like air conditioners and pool pumps. But trading in appliances would only take the city so far. The program began in January 2007. By the end of the first year, savings had reached just 12 percent. Pretty soon, it became clear that the only way Palm Desert was going to get to 30 percent was for homeowners, especially those living in older houses, to start producing their own energy. "We needed to have solar," Ferguson said. To him, the notion of self-generation also had its own intrinsic appeal. It was an opportunity to look at energy, "not as something that Big Brother or the government is going to give you, but as something that you can generate in your own backyard." The bill Palm Desert sponsored allowed cities to loan residents the money to equip their homes with solar systems (and other energy-saving devices). Technically designated AB 811, in Palm Desert it was dubbed the Energy Independence Act. Officially, that meant independence from foreign oil; unofficially, (partial) independence from the utility. Ferguson was delighted: Southern California Edison's monopoly on energy generation had finally been broken. Invited to invest in the city's solar loan program, Edison declined.

Despite its well-deserved reputation as a citadel of the looney left, Berkeley had been extremely fiscally conservative in implementing its program. Only private financing could be used, with no obligation to the city's general fund. "The rules I was given were clear—no subsidy, no liability, no risk," DeVries told me. "If those rules weren't met: no program." Berkeley has one of the highest bond ratings in the state. Though suffering through budget cuts, it maintained millions of dollars in a reserve account for rainy days. Try as he might to get the city to let him use those funds for the program, DeVries had zero success. Hence the delay in implementation.

Though politically conservative, Palm Desert took a much more liberal approach toward implementing its program. The city was willing to put its own dollars at risk, tapping its general fund to issue the initial low-interest loans. In the first phase of the pilot program, launched in August 2008, $2.5 million was snapped up in eighteen days. A further tranche of $5 million from the fund was fully subscribed within four weeks. By March 2009 nearly a hundred Palm Desert households had been approved for PV panels. Installations were going, as one local solar entrepreneur infelicitously put it, "through the roof." On July 6, 2009, officials opened the doors at City Hall to find a line of fifty residents and developers waiting outside, eagerly clutching applications. The $1.25 million the city had reserved for new solar loans went within twenty minutes. The queue of residents waiting for solar was long. "Based on the demand we see we have the potential here in Palm Desert to use $15 million to $20 million for energy loans," Conlon said. Unleashing such large amounts of money would, the city hoped, trigger a gold rush of solar companies setting up shop to take advantage of this new marketplace. "The Coachella Valley is poised to become a new solar Silicon Valley," Ferguson boasted.

Other places were eager to follow Palm Desert's example. "We've sent out information to over sixty-five cities, fifty-two of them in California," Conlon told me. "There's no doubt that this loan program could easily be replicated in other cities." Fifteen states have since passed versions of AB

811. Within California, the concept has been adopted by over twenty counties and close to a hundred cities. "Palm Desert really has done a great job," Cisco DeVries commented admiringly.

The crucial question for cash-strapped cities struggling in the aftermath of the financial crisis was: Where would the money come from? The Berkeley model stipulated that only private capital should be used. Declining rebates from the California Solar Initiative would only go so far. "The CSI is helpful, no doubt about it," Conlon said, "but the incentives are dropping too fast and the application process is an absolute nightmare. I mean, I'm government, I know how to fill out forms, how to follow instructions, that's what I do. I filled out the application for our 80-kilowatt system here at City Hall. I could not believe what a pain that was, it was absolutely horrible, incredibly inefficient—it took us ten months to get our CSI rebate."

Ferguson was even more scathing in his criticism of the rebate program. "A lot of people are gung-ho about the CSI," Ferguson said, "but candidly it's been a colossal failure, the incentives it has created have just not been great enough—Arnold Schwarzenegger's Million Solar Roofs have not materialized."

Palm Desert was not done yet making pro-solar laws. The California Solar Initiative would only pay for a certain number of kilowatts, based on the predicted annual usage for a given site. That meant homeowners had no incentive to install a system with the capacity to produce more than they needed to cover their personal load. "If I generate excess electricity, then all my energy credits get flushed down the toilet at the end of the year," Ferguson complained. "Or, stated another way, Edison gets those credits for free. Our thought was, people should get paid for the excess capacity that they're currently putting into the grid for free." What Palm Desert's PV producers needed, in short, was a feed-in tariff. A tariff that required the utility to pay top-tier prices for solar electricity would complement the loan program nicely. "If you look at it in a business sense," Ferguson explained, "a feed-in tariff really reduces your capital cost return from about fourteen or fifteen years down to six or seven years. That makes it much more financeable for your average homeowner." Ultimately, he saw a system of feed-in tariffs supplanting the CSI, which would only cover about 13 percent of the cost of installing a solar system.

In October 2008 Ferguson journeyed to Spain to witness that country's feed-in tariff in action. He chose Spain rather than Germany, where solar feed-in tariffs originated, because, "Spain was kind of tomorrow's Germany . . . it just seemed like a logical place to go. Our council has always been of the view that the only way to see something is to go kick the tires, meet with the local people, with the federal government people, ask them what the benefits were; if they had to do it over again, what they would do different." That year, thanks to lucrative tariffs the Spanish PV market exploded, adding more than 2.5 gigawatts, making it the world's largest. It was an astonishing and, as it turned out, unsustainable boom. At the time of his visit, however, Ferguson was dazzled by what he called "an unbridled free-market response to a subsidized program."

In March 2009, at Ferguson's urging, assemblyman Brian Nestande (R-Palm Desert) introduced AB 432 into the state legislature. The bill's modest proposal was to set up a pilot solar feed-in tariff program in Palm Desert that would oblige Southern California Edison to pay a premium rate for excess generation. The tariff would be set by the CPUC. A rate of between 26 and 32 cents per kilowatt-hour would be about right, Ferguson thought. The following month, however, before its first hearing, Nestande pulled the bill. The proposal had run into opposition from, as he put it, "some industry folks." Meaning, in particular, Southern California Edison. AB 432 was unnecessary, according to SCE director of public affairs Catherine Hackney: California already had a feed-in tariff. Which was true, although misleading, because the feed-in tariff introduced by the CPUC the previous year was designed for large industrial installations rather than small residential ones, and the rate it paid was as low as eight cents per kilowatt-hour. Besides, Edison argued, why should it be the only utility obliged to accept contracts from Palm Desert, and why should Palm Desert be the only city authorized to issue such contracts?

Furious at the utility's backstabbing, Ferguson resolved that Palm Desert would reintroduce its bill in 2010. Similar proposals from other Californian municipalities, notably Santa Monica, were also in the pipeline. But as we shall see in the next chapter, another US city had already managed to implement a solar feed-in tariff, a German-style one at that. Gainesville was located not in California, but on the opposite side of the country, in Florida, the self-styled Sunshine State. Another, crucial difference was that, unlike Palm Desert, Gainesville was served not by an investor-owned utility, but a municipal one.

22

FIT USA?

Gainesville is a small city in northern Florida that prides itself on the prowess of its top-ranking college football team, the Fighting Gators. As befits the home of the University of Florida, the nation's third-largest institute of higher learning, Gainesville also prides itself on its smarts. "We've learned to spray all the utility poles up and down University Avenue with [cooking oil] after a national championship," Ed Regan joked, "because that keeps the kids from climbing up them and getting killed."

Regan works for Gainesville Regional Utilities, which in addition to supplying its 90,000-plus customers with electricity, natural gas, water, waste water, and telecommunications, also provides the city with almost 40 percent of its revenue. An environmental engineer, Regan was hired in 1979 in the wake of the oil embargoes to set up energy conservation programs. In 1989 he became GRU's manager of strategic planning. The planners' main job was to determine the cheapest fuel from which to generate electricity. Mostly, this meant coal: "We burn to earn" was GRU's unofficial motto. In 2003, faced with a growing population and the need to replace obsolete generators, the utility came up with the obvious least-cost solution. It proposed to build another coal-fired power plant. Meantime, however, the community had begun to fret about climate change. Municipal utilities are governed by elected officials: they tend to reflect community interests. GRU was tasked to respond to the Kyoto Protocol by reducing its carbon emissions 7 percent from their 1990 level. The utility's primary mechanisms to achieve this goal were urging people to use less energy and adopt renewables.

Like its peers around the US, GRU offered cash incentives to encourage residents to buy solar systems. To enable them to reduce their electricity bills, it had introduced net metering. There were problems with both of these programs. The utility disbursed rebates—$300,000 of ratepayer dollars each year—to allow homeowners to purchase equipment for their roofs. But when

Regan drove around the city, he would often see solar systems that had been installed under trees, where shade prevented them from performing well. "My community invested in that," he would think gloomily. (Heaven forbid that you should suggest chopping down trees in leafy Gainesville, a two-time winner of Tree City USA awards.) To make sure that GRU was getting a bang for its bucks, utility employees were obliged to become "solar police," a role with which they were not comfortable. Ideally, solar systems should have been installed on the roofs of large buildings like supermarkets or over open spaces like parking lots. With net metering that would not happen. Under a scheme where the best you can do is zero out your electricity bill at the end of the month, investors had no incentive to put in large systems whose output far exceeded their on-site needs. Especially since most businesses in Gainesville operated out of leased premises.

The public was invited to participate in the debate over future energy supply and climate change. At one such meeting, when renewable energy policy was under discussion, a man from the audience stood up to have his say. Regan recognized him. Harald "Harry" Kegelmann had migrated to the Sunshine State because he found the weather in his native Bavaria so miserable. After teaching computer science at Florida State, Kegelmann had set up one of Gainesville's first Internet service providers. On vacations back home, he had witnessed the PV revolution taking place with his own eyes. A single German soccer stadium was mounted with more solar panels than the whole of Florida. "You should do what they do in Germany," Kegelmann told the board, adding, "in Germany, they pay 60 cents a kilowatt-hour for solar electricity." Regan was taken aback. "*How in the world* do they justify doing that?" he wondered. The going rate in Gainesville was less than 14 cents. "Is this guy pulling our leg or what?" Shortly afterward, in June 2008, Regan had a chance to go to Germany, along with thirty other decision makers from utilities around the US, to find out for himself. It was, in his words, "a mind-boggling experience."

Over five days the Americans toured German municipal and investor-owned utilities, solar manufacturers, and research institutes. They saw large-scale roof- and ground-mounted PV installations. The trip culminated in a visit to Europe's largest solar trade show, where attendees numbered in the tens of thousands. For the visitors, it was a sobering experience to encounter the full-on spirit of optimism about solar that prevailed in Germany; also, the thousands of jobs that solar had created there. Especially since some of those jobs were at transplanted US manufacturers like First Solar. The basic solar technologies had almost all been invented in America, but now here they were, flourishing on foreign soil, thanks to a more attractive incentive model.

Regan was struck by how his German counterparts seemed not to be thinking along traditional, least-cost utility lines. Their concern was rather with energy independence and economic development. Germany relied on imports for three-quarters of its energy. Making the most of renewables was thus strategically vital for the Germans. Could not the same be said of the US? In many states, including Florida, decisions on energy investment were driven by renewable energy credits. A form of derivative, these are certificates issued by governments to generators, who then sell the credits to buyers, normally utilities looking to fulfill their government-mandated quota. The trading system allows utilities to buy credits from sellers who offer the lowest prices, which favors operators of large-scale facilities over rooftop generators. Everywhere he went in Germany, Regan kept asking the same question: Who gets the credits for the generation of renewable energy? His hosts invariably looked puzzled. That is not how we do it here, they told him politely. We have this other system, called "feed-in tariffs."

Regan returned to Gainesville preaching the gospel of German-style incentives. His evangelism soon converted the city commissioners. They immediately grasped the economic implications of this new energy policy. Especially appealing was the fact that feed-in tariffs would involve the whole community. Local residents would make investments with loans from local banks in systems installed by local firms. In October 2008 GRU held a public workshop to explain the incentive mechanism; on March 1, 2009, Gainesville became the first US utility to introduce a German-style feed-in tariff.

As a planner Regan marveled at how easy it had been to design the tariff. He was used to grappling with complicated forecasts based on predictions of fluctuations in the price of fossil fuels. Here, instead of having to decide what method of generation would be cheapest in the future, the costs were known. It boiled down to a simple question: What was the appropriate rate of return? The commission picked 5 percent, after taxes. GRU would pay 32 cents a kilowatt-hour for solar electricity, the amount fixed for twenty years. After two years, the rate would decline by 5 percent. The program was capped at 4 megawatts per annum, the utility agreeing to dish out around $1.3 million. This it was estimated would add about 70 cents to the average homeowners' monthly electricity bill. It was, as Regan pointed out, the equivalent of a can of soft drink, considerably less than the adjustments to rates necessitated by variations in the cost of fossil fuel.

The utility anticipated there would be demand for 2 megawatts the first year and 4 the second. Two megawatts would in itself have doubled the amount of PV then deployed in all of Florida, a pathetic total for the so-called Sunshine State. In fact, GRU was overwhelmed, as applications for *40* megawatts poured in. The program operated on a first-come, first-served basis. The initial allocation was snapped up within days. The utility set up a queue for future allocations for would-be system owners. GRU soon received enough applications to meet its program caps through 2016. Already, two large solar farms were under construction and a 2-megawatt system was planned for the roof of Gainesville's largest shopping center by the end of 2010.

Citing the speed with which the queue had filled up, critics charged Gainesville with having set an overgenerous rate. But while signing up was easy, obtaining financing and permits for projects proved harder. "Most of the delays have been for financing," Regan told me, "which would suggest that the money isn't that lucrative." He estimated that 4 megawatts represented over $30 million worth of investment. "Some of it is going into equipment that's going to be brought in from elsewhere," Regan said, "but a big part of the system cost is roofers, electricians, solar contractors, engineers, and designers." A vibrant, competitive market for solar energy was springing up in the community, creating jobs and investment opportunities.

The idea that anybody could now become an energy producer drove traditionalists at GRU nuts. But at least utility employees no longer had to play solar police: if systems did not generate electricity, that was their owners' lookout. In addition to reduced administrative costs, a bonus for the utility was the valuable renewable energy credits that, as part of the feed-in tariff deal, they got to keep.

Barry Cinnamon was another American who, bemused by Germany's success with PV, had flown there to find out for himself what was going on. Cinnamon's interest in solar dated back to 1977, when as a nineteen-year-old he had been inspired by Jimmy Carter's speech about winning independence from imported energy being "the moral equivalent of war." A student of mechanical engineering, Cinnamon worked on passive solar houses at MIT. He wrote his thesis on the design of a solar thermal panel. After graduating, Cinnamon worked for small solar firms in the Boston area until the Reagan administration axed Carter's tax credits, sending the industry into hibernation. Cinnamon went to business school, then moved to the Bay Area. There, he spent the next fifteen years toiling, as he put it, "in the wilderness of the software industry."

In early 2001, as we have seen, California's energy crisis sent electricity prices soaring. Sick of running his Internet company, Cinnamon decided it was a good time to stick some solar panels on his roof. He called local installers for quotes, asking them what the optimal system size was and what the payback period would be. Unsatisfied with their replies, Cinnamon designed a website that collated information on solar systems. Much to his surprise, he started getting e-mails from people who had visited the site and, assuming that he was an installer, wanted to buy a system from him. The demand was obviously there, but before he could exploit it, Cinnamon had to learn the basics. He took a course, then installed his first solar system, on his own house. In May 2001 the state government increased solar incentives by 50 percent. Overnight the economic rationale for solar had shifted. The payback period dropped from fifteen or twenty years to around ten. Cinnamon founded a firm, Akeena, in Los Gatos, a city in the foothills of the Santa Cruz mountains. He joked that his was a Silicon Valley start-up that didn't start *in* a garage, it started *on* a garage, because that's where he had put his panels.

In 2001 the solar industry was still dominated by mom-and-pop shops, mostly old-timers who had been in business since the seventies. As a tie-dyed T-shirt–wearing Grateful Dead fan, Cinnamon fitted in well with this laid-back community. But unlike the hippies, he had an engineering degree from MIT, an MBA from Wharton, and an entrepreneur's drive. Cinnamon did some marketing, hired some salespeople to track down potential leads, some installers to put in the panels, some administrative people to keep up with the paperwork, then did some more marketing. "I kept turning that crank and it just grew," he told me. Pretty soon, Akeena was one of California's leading solar installers.

But while Akeena's sales were climbing steadily, those of its German counterparts were soaring toward the stratosphere. This was odd, because the Germans were installing the same equipment, in some cases actually paying more for their panels than Cinnamon was for his. Somehow, their prices were significantly cheaper. What was their secret? Puzzled, in 2006 he set off to discover the answer. A lucrative feed-in tariff was the most obvious explanation for the difference. But Cinnamon also identified two other, more subtle reasons for the German success. One was that, while in California his customers had to scratch around to find financing for their solar systems, in Germany PV purchases were underwritten by automatic low-interest loans from KfW, a government-owned development bank. (Kreditanstalt für Wiederaufbau—literally, Credit Institute for Reconstruction—was formed after World War II as part of the Marshall Plan.)

The other reason was the paperwork or rather, in the German case, the lack of it. In California the bureaucracy had been bad enough in 2001, when Cinnamon installed his first system on his garage. Back then it amounted to some 50 pages. With the advent of the California Solar Initiative rebates, it had since bloated out to a mind-blowing *148 pages* for a standard residential system. As a result, while equipment and installation costs had almost halved since 2001, administrative costs had tripled.

Would-be solar system owners were obliged to clear three hurdles. Two were with their utility, an agreement allowing owners to connect to its grid, and an application for net metering. The third was with the California Energy Commission, to receive the CSI rebate. To qualify, in addition to simple facts like number of panels and output, owners also had to provide a shading analysis of their roofs, an energy audit (for conservation purposes), and a calculation of the estimated production of their systems (rebates were based on annual on-site usage). In some places like, at one point, San Francisco, it could take up to six months to get a permit for a solar installation approved. In San Diego a permit cost more than five hundred dollars.

The installer would typically fill in forms on behalf of the customer, print them out, then mail a hard copy to the utility and the commission. There, astonishingly in twenty-first-century California, the information would be laboriously reentered by hand. Installers objected that this process could be much more efficiently accomplished via a website. The stuffy official rejoinder was that a "wet-ink" signature was needed before authorization could be granted. Akeena employed an entire department of clerical staff dedicated to dealing with the mountains of paperwork that accumulated.

Cinnamon's salespeople got more training in filling out forms than they did in selling solar systems. After an installation was completed, at least four inspections were required: by the utility, the building department, the fire department, and the rebate administrators. Each necessitated the presence of a representative from the installer. On top of which, despite the fact that solar panels were almost bulletproof—they routinely survived hailstorms that wrecked roof tiles—system owners also had to comply with burdensome requirements from unsympathetic insurers.

"Ludicrous" was the word that Cinnamon used to describe this duplicative, byzantine, *Alice-in-Wonderland* process. In Germany, he kept prompting his hosts to show him their documentation package. The result was a failure to communicate that had nothing to do with lack of language. "After three or four days I finally figured out that the reason I always got a blank stare when I asked about the paperwork was . . . there wasn't any!" Well, almost none: merely a simple three-page form consisting of an application, a permit, and an invoice. In Germany the boot was on the other foot: instead of politely asking for permission, German homeowners simply informed their utility that they would be connecting a PV system. The utility was obliged, by law, to accept the connection. Streamlined administration procedures were thus the main reason why installing PV in Germany was so much cheaper. In the US, as one bemused German observed, it was easier to buy a gun than to install a solar system on your roof.

Cinnamon embarked on a crusade whose aim was not to reduce paperwork but, as near as possible, to eliminate it entirely. Disposing of bureaucracy would be the quickest path to grid parity. After all, he argued, you didn't need to ask anyone's permission to plug in a hairdryer that drew 1.6 kilowatts; why should you need a permit to connect a 3-kilowatt solar system? Soon big-box stores would be stocking plug-and-play kits like the Andalay system his company had developed. How would authorities deal with customers who bought some solar panels, took them home, and hooked them up? Arrest them as criminals? Americans, Cinnamon said pointedly, did not take kindly to prohibitions.

When Germany introduced its epochal Renewable Energies Law in 2000, few in the US even knew what the phrase "feed-in tariff" meant. To suggest that such an obscure stratagem was worth considering as a serious policy alternative was to invite ridicule. However, following the passage of the law and the subsequent, spectacular growth of solar in Germany, it was no longer possible to pooh-pooh what was obviously a highly effective mechanism. In early 2005, after his meeting with Governor Schwarzenegger, Hermann Scheer had declared California open for business. Thus encouraged, German companies began to take an active interest in the US as a potential market. The German-American Chamber of Commerce led the way, organizing a series of workshops around the country to explain the benefits of feed-in tariffs to leery locals.

While it was difficult to deny the success of the German approach, in California there was an understandable reluctance to consider adopting feed-in tariffs, at least in the short term. Policy makers and PV manufacturers had after all just spent three years fighting tooth and nail to force the California Solar Initiative through the legislature. Pro-solar advocates like David Hochschild felt that it would be a mistake, not to mention politically difficult, to revamp the program just as it was getting started. Admittedly, the patchwork quilt of federal tax credits, state rebates, local incentives, and net metering that had been stitched together was bewilderingly complex compared with what everyone agreed was the elegant simplicity of the German model. But Americans, it was argued, were familiar with tax credits and rebates, whereas feed-in tariffs were an exotic idea, one that was moreover Not Invented Here. Besides, wasn't Germany a notorious nanny state? Dirigisme—economic control and planning by the state—was alien to the individualistic, market-led American way. US policy makers were not accustomed to taking their lead from Germany. Just because a policy worked in one country, that did not mean it would work in another. The United States was unique, different from the rest of the world. It was not Germany.

The criticism most often voiced by Americans about feed-in tariffs was that they were exorbitantly expensive. The German government was handing out "bags of money" to owners of solar systems. "People think that they want a feed-in tariff, but what they really want is those bags of money," argued Vote Solar's Adam Browning. Though electricity was certainly more expensive in Germany, the amount feed-in tariffs added to its price was remarkably small. The commonly used analogy for the burden imposed each month on the average German electricity ratepayer was a loaf of bread.

Not surprisingly, the main opposition to feed-in tariffs in the US came from utilities like PG&E concerned about the threat posed to their monopoly. Buying electricity at retail rates from customers would not be cost-effective, the utilities insisted, nor fair to their other, nonsolar customers. As time went by, however, it became increasingly clear that California was falling behind in its efforts to meet its self-imposed renewable energy target. In December 2007 the California Energy Commission, the state's primary energy policy and planning agency, recommended in a report that, in order to reach Governor Schwarzenegger's ambitious goal of deriving 33 percent of energy from renewables by 2020, "California must move to a new system, such as the expanded use of feed-in tariffs." Terry Tamminen, the governor's chief policy advisor, added his endorsement. Feed-in tariffs, he said, "turn homes, farms, and businesses into entrepreneurs who will accelerate our path to clean energy."

Two months later, in February 2008, responding to pressure from Sacramento, the California Public Utilities Commission approved a feed-in tariff. But it was very limited in scope, applying mostly to renewable energy systems of up to 1.5 megawatts, at public water and waste water facilities that had excess generation. The rates it offered were very low, being calculated based not, as in Germany, on the cost of generation but on the wholesale price the utility would otherwise have to pay for the power. This was not the kind of incentive that was likely to produce a dramatic increase in participation. Indeed, in the scheme's first six months, there were no takers.

In the nation's capital, charmed by what he described as the "solid, take-it-to-the bank security of feed-in tariffs," Jay Inslee, a Democrat from Washington State, announced his intention to introduce legislation that would create European-style solar tariffs at the federal level. Feed-in tariffs had spurred the development of the solar industry in Germany, an industry that the US had once dominated but in which it was now only a minor player. "It's time for us to get back in the game," the congressman told reporters. In just a few months, the term "feed-in tariff" had gone from being an arcane piece of jargon known only to energy policy wonks to a buzzword that almost everyone seemed to be using (if not necessarily understanding). Europe was showing the US the way. Of the twenty-seven member countries making up the European Union, twenty had already implemented feed-in tariffs. Between them these countries accounted for 80 percent of all PV installed in 2008, versus just 6 percent by the US. In addition to Germany,

France and Italy had also adopted versions of the mechanism. But the country most often cited as a glowing exemplar of the astonishing growth that tariffs could engender was Spain. In fact, however, Spain would prove to be a cautionary tale of how *not* to implement feed-in tariffs.

Spain had been eager to emulate Germany's stunning success. In 2007 Madrid enacted a similar set of incentives. In their haste, however, the bureaucrats overlooked the important fact that Spain gets twice as much sun as Germany. The result was a feed-in tariff that offered investors an outrageously generous rate of return. Gifted an opportunity to mint money, speculators flooded into the market. Unlike in Germany, where farmers and individual homeowners made up the bulk of solar purchases, in Spain it was mostly companies building large, ground-based solar farms, many of them poorly designed. Spanish planners drafted complicated administrative procedures, excluding residential investors. With ordinary citizens effectively prevented from participating in what one critic described as "a drunken orgy of speculation," feed-in tariffs became unpopular with the people. The Spanish government's aim had been to achieve 400 megawatts of solar capacity within three years. The target was reached within three months. Planners were taken by surprise as the market exploded to 2.7 gigawatts, representing almost half of all solar installations made in the world in 2008.

Canny German planners had instituted a flexible market threshold. This could be varied to keep growth within a well-defined corridor: failure to reach the threshold meant that incentives needed to be increased; too rapid growth meant that they needed to be reduced. The Spanish by contrast had omitted to build in a control mechanism to throttle back on tariffs when targets were exceeded. Worse, the peculiarities of the Spanish electricity market meant that the additional costs could not be passed onto ratepayers, leaving the government to foot a bill that was estimated at more than $26 billion. With the market out of control, in September 2008 the Spanish government slammed on the brakes. It slashed feed-in tariffs and imposed an annual cap of 500 megawatts. The local PV industry, which had invested aggressively in ramping up production to meet the surging demand, collapsed.

Spain's disastrous solar bubble provided ammunition for critics in the

US who saw the incentives as an example of how government policies intended to promote renewable energy could lead to unsustainable outcomes. "Spain's Solar Market Crash Offers a Cautionary Tale about Feed-In Tariffs," ran the headline of one *New York Times* article. Nonetheless, evidence that feed-in tariffs were a more effective support mechanism than renewable energy credits continued to mount. During 2009 the National Renewable Energy Laboratory in Colorado issued a series of technical reports on the topic. "We deal with data and the evidence is very clear," NREL analyst Toby Couture told a reporter, "Feed-in tariffs have consistently proven to be cheaper for consumers. That's the bottom line."

In late 2008 the European Photovoltaic Industry Association announced an aggressive goal: to supply up to 12 percent of European electricity demand with PV by 2020. This would represent a massive 400 gigawatts of installed capacity. The association calculated what would be the cost of the feed-in tariff required to support the market development to achieve this goal. The calculation assumed an internal rate of return to investors of 6 to 8 percent and a continuing decline in the price of systems. "The cost will not represent more than 2.2 percent on the total electricity bill during the full feed-in tariff period," said Adel El Gammal, the association's secretary general. "It's peanuts," he added.

Meanwhile, back in California, support for feed-in tariffs continued to mount. In July 2009, following Gainesville's lead, the Sacramento Municipal Utility District announced that it would introduce a feed-in tariff for homeowners. The program, which was capped at 100 megawatts, would pay up to 30 cents a kilowatt-hour for electricity, but only during peak periods. The utility had fielded lots of inquires from customers interested in building systems that produced more energy than they needed, and thus did not qualify for net metering. "If [the feed-in tariff program] prevents us from having to build more remotely sited generation plants or transmission resources, then we think that's a good thing," said John Bertolino, a superintendent at SMUD. "State-wide you're going to see California move very rapidly toward feed-in tariffs," Bertolino added. When its program took effect in January 2010, in the first week the utility received applications that exceeded the 100 megawatt cap.

In October 2009 Arnold Schwarzenegger signed two important pro-solar bills. One obliged utilities to pay owners of small systems for the excess electricity they generated under net metering. At the same time, the governor secured a commitment from PG&E to increase the amount of net metered electricity allowed in its territory from 2.5 percent to 3.5 percent of peak demand. In a press release he committed himself to introducing legislation that would "permanently eliminate all caps on net metering in California so there are no arbitrary limits on the amount of solar we can install, then number of jobs we can create, and the amount of energy we can save." (In March 2010 the state legislature raised California's cap on net metering to 5 percent.)

The other bill raised the limit on the size of systems that qualified for California's existing feed-in tariff, from 1.5 megawatts to 3 megawatts. It also increased the rate to reflect the value of the electricity generated. In an accompanying letter to the state senate, Schwarzenegger emphasized that it was necessary "to use all of the tools available" under existing programs to reach the goal of 33 percent renewables by 2020. Noting that the California Public Utilities Commission was currently exploring an expanded feed-in tariff that would cover small- to medium-size systems, he encouraged the CPUC to continue its work "so that we can take advantage of the new renewable energy capacity that a robust [feed-in tariff] program can provide."

Observers like John Geesman, a former California Energy Commissioner, agreed. "It's not clear to me that the scalability objectives of government policy will be accomplished without some form of feed-in tariff," Geesman said. By the end of 2008, two years into its solar initiative rebate program, California had managed to install just over 50,000 solar systems, far short of its million roofs goal, with a total capacity of around half a gigawatt. By mid-2010 almost three-quarters of a gigawatt of solar was in operation or in the pipeline.

In July 2010 UC Berkeley's Renewable and Appropriate Energy Laboratory published a study examining the economic benefits of a comprehensive feed-in tariff. The analysis showed that enacting a robust feed-in tariff in California to achieve the state's renewable portfolio standard goal would create over the next decade 280,000 new jobs and stimulate up to $50 billion in new private investment.

Commenting on the halting progress toward a state-wide feed-in tariff, Barry Cinnamon liked to quote Winston Churchill: "Americans can always be expected to do the right thing—after first exploring all the other alternatives."

Boris Klebensberger is president of SolarWorld USA, the largest manufacturer of PV in the US. Like many Europeans, Klebensberger believed that the US was a sleeping giant. Sooner or later, America would wake up and become the world's largest market for PV.

Accordingly, Bonn-based SolarWorld had located more than half of its gigawatt-plus production capacity in the US, in two separate facilities. One was the Camarillo factory north of Los Angeles that Bill Yerkes had originally built for ARCO Solar almost twenty years previously, the other a much larger plant up in Hillsboro, Oregon. In November 2009 I visited the Camarillo factory with Yerkes. Together, we watched solar panels being assembled—one every minute—by a troupe of exquisitely choreographed robots. Afterward I was surprised to learn that all these panels were destined, not for the local market, but for export, mostly to Germany.

There was, Klebensberger explained, a major difference between the US and German PV markets. "The US is mainly driven by large installations, by utilities and big corporations. Almost all of them have to be financed, but when financing becomes scarce like it is now, it's not that easy. There have been a lot of announcements, but on the execution side it's not flying as quickly as everybody had anticipated. In Germany, by contrast, the reason why the German market is doing so well and dominating in 2009 is that it is not based on big corporations, or on big financing, but on the belief of Germans in PV and on the genius of the feed-in tariff." Klebensberger was bemused by the fact that, in the land of the free, regulations on the production of energy were so onerous. "The feed-in tariff in Germany uses everything that you would normally find in a free market: competition, quality, self-determination," he told me. "That's not the case in most of the American schemes, because you're depending on the regulation, on someone else, to decide whether it's good or bad, and that's driving the bureaucracy."

While red tape hampered market growth, California's mishmash of solar incentives had nonetheless induced responses. Some, as in Germany, came from grassroots groups determined to have solar; others from a combination of what Jimmy Carter thirty years previously had called "private enterprise, individual initiative, and the inventive genius of the United States." As we shall see in the next chapter, in and around Silicon Valley, a rash of solar start-ups was popping up. Their focus was on the installation end of the value chain, which accounted for half the total cost. Run by determined young entrepreneurs, their goal was to take the hassle out of going solar for the homeowner and, in so doing, enable the long-awaited exponential growth to occur in the US.

23

EXIT HIPPIES,
ENTER STANFORD MBAs

*A*l Gore's climate change documentary, *An Inconvenient Truth*, which began its US screenings in May 2006, concludes with a call to action. Among those who responded most vigorously to the call were residents of Silicon Valley. About twenty members of a homeowners association in Portola Valley, a little town nestled in the foothills behind the Stanford campus, organized a trip to see the movie. Afterward the group assembled at the home of one of their members, Armand Neukermans, to discuss what action they could take. One member had an idea: Why not club together to buy solar? A bulk purchase would enable residents to reduce the daunting upfront cost. Having four years previously put a system on his roof, the Belgian-born Neukermans knew a bit about solar. It was natural he should volunteer to find out whether any of the local solar firms would be prepared to offer a group discount. There were about five or six installers in the area. Most were run by old-timers who had been in the business many years. Neukermans took his proposal to a solar exhibition in San Jose. "It turned out that none of them were interested, with one exception, SolarCity, a small start-up," he told me. "I talked to the young guy that runs it, and he immediately saw the potential. It got started from there."

The homeowners association organized an evening meeting for interested members. Much to everyone's surprise a hundred people showed up. Neukermans gave an evangelical talk about the benefits of solar power. He began by pointing out that if people wanted to reduce their carbon emissions, the first thing they should do was conserve energy, not install solar. However, he conceded, there was undoubtedly something magical about PV, exploiting energy pumped out by the sun, that attracted people: no need to give up the little electrical luxuries they were used to. If members could commit to buy 175 kilowatts' worth of solar systems, SolarCity would give them a discount of 30 percent. That would knock around $6,000 off the cost of an average-size system. After two further meetings held at local church halls and six weeks of neighborhood canvassing, Portola Valley reached its target. By December 2006, thirty-eight homes had signed up to the program. Eventually the number of solar installations would top seventy. Getting all those panels on their roofs was, Neukermans thought, a great source of community pride.

The collective power program had been a great success. But was the model replicable? Portola Valley was after all hardly your typical community. With a median annual income of over $150,000, it was the sixth-wealthiest town in California. Portola Valley was also an environmentally conscious place with strict regulations aimed at preserving the prevailing pastoral quiet. Its residents were, as Neukermans put it, "sort of unusual." He himself was a semi-retired physicist and high-tech entrepreneur who in 2001 had been voted Inventor of the Year by the Silicon Valley Patent Lawyers Association.

Located in the heart of Silicon Valley a few miles east of Portola Valley, Mountain View is a very different sort of community. Its median income is around half that of its affluent neighbor. Many Mountain View residents are Stanford graduates. It's the kind of place where a lot of smart, tech-savvy people start their careers. For more than two decades Mountain View had also been the home of Bruce Karney. A Stanford math grad who specialized in the hybrid discipline of knowledge management, Karney had taken early retirement in 2005 after a twenty-four-year career at Hewlett-Packard. He and his wife, Twana, a chemical engineer who worked for an oil company, were members of a movie club. In May 2006 they attended a sneak preview of *An Inconvenient Truth*. The documentary was a turning point in their lives. They had not previously been involved with the environmental movement. Karney emerged from the movie convinced that "there was the equivalent of a war on climate change going on." Though he had not served in the army, this was a war for which he wanted to volunteer. He and Twana would sign up, learn new skills, then sally forth to fight the good fight.

Their first step was to modify their own behavior, reducing the amount of greenhouse gases they were adding to the atmosphere. They bought a hybrid car, insulated their walls, strung out a clothesline in their backyard. The next step was to help others. Bruce joined a program whose purpose was to perform energy audits on people's homes. In January 2007 he visited Armand Neukermans. The Belgian happened to mention to Bruce the Portola Valley community program, which had closed the previous week, and how many people it had signed up. Group discount sounded like a brilliant idea to Karney. He immediately wanted to copy it. He called SolarCity: If he could put together enough homeowners and small businesses in Mountain View, would they give him the same deal as Portola Valley? The answer was yes.

Karney used e-mail lists to spread the good word. An article by the supportive editor of the local weekly led to coverage in the *San Jose Mercury News*. The first public meeting was held in the hall of the Mountain View Senior Center on February 10, 2007, a Saturday afternoon. Two hundred and fifty people showed up, well beyond anyone's expectations. A hundred more attended two follow-up meetings. To qualify for the discount, Mountain View needed to recruit 60 households. Within a month of the first meeting, they had received 25 pledges; by the end of April, the number had swelled to over 100. Eventually 119 homeowners would commit, making it the largest-ever group purchase of solar systems in the US. By the end of the year, little Mountain View (population 75,000) had become one of the leading adopters of residential solar in the state, trailing only San Diego and San Jose, both of which have populations of over a million.

Thus far Karney had worked pro bono, for nothing other than personal satisfaction. Now, recognizing his value as a solar champion, SolarCity hired him as marketing manager for community programs. Over the next eighteen months Karney ran twenty programs in cities and towns throughout California. Success depended on identifying an enthusiastic local volunteer to play the role he himself had played, a classic case of setting a thief to catch a thief. One of the most successful—and to Karney, most surprising—programs ran in Clovis, a suburb of Fresno, located in the San Joaquin Valley midway between San Francisco and Los Angeles. Clovis was a solidly Republican town, a far cry from the liberal likes of Mountain View. "People there have big houses on big lots, and the climate is very hot," Karney explained. "Solar just makes financial sense to them." Property-based financing programs like Palm Desert's were still at least a year away. "In Clovis, people *bought* the damn things, they wrote checks and took out loans."

Thanks to its community-based business model, SolarCity would complete during 2007 and 2008 more than eight hundred installations, 4.5 megawatts in aggregate. As a result the company would become the largest provider of residential solar power in the state; and, since California still accounts for around two-thirds of all US residential installations, the nation.

The seed for what would become SolarCity was sown in 2004 at Burning Man, the boho gathering that happens every summer in the Nevada desert, during a conversation between two precocious South-African-born cousins. One was thirty-three-year-old Elon Musk, an engineer-entrepreneur who had made serious money out of high-tech start-ups, notably PayPal, which he cofounded and sold to eBay in 2002 for $1.5 billion. In addition to exploring space Musk's passions included helping to bring about the transition to a sustainable energy economy. He was lead investor, chairman, and CEO of Tesla Motors, a Silicon Valley–based manufacturer of electric cars. Musk had also put a solar system on the roof of his Bel Air mansion. The experience had been unsatisfactory. Installation was still essentially a cottage industry. "It was all mom-and-pop contractors," he told a reporter, adding, "they basically suck." Finding an installer who could do a professional job had proved impossible. Musk had no confidence that the company that installed his system would still be around in a few years to perform maintenance and repair. The mom-and-pop approach struck him as expensive and inefficient. No effort had been made to hone installation practices, to eliminate excess parts and labor, or to achieve economies of scale by buying panels in bulk. Musk was thus especially receptive to a proposal put to him by his cousin, a smart, largely self-educated twenty-eight-year-old with piercing eyes named Lyndon Rive.

By the time he was a gangly teenager growing up in suburban Johannesburg, Rive was already manifesting the traits that mark the classic entrepreneur. At fourteen he discovered underwater hockey. Also known as octopush, this bizarre sport involves teams of six players equipped with masks, snorkels, fins, and a short hand-stick. The objective is to push a lead puck across the bottom of a swimming pool into goals located at either end. Since individual players cannot hold their breath forever, success depends on focus, strategy, and teamwork—just like a successful start-up.

At seventeen Rive founded his first company, to distribute a homeopathic remedy that soothed his grandmother's arthritis. The firm that made the product only sold locally. Rive approached them with an offer to distribute nationally, throughout South Africa. They laughed at him, telling the youngster to come back when he had grown up and got a degree. Rive persisted, persuading them to back his company. Running the business forced him to cut classes. The principal of his high school threatened to expel him. Rive responded by showing his financial statements. He argued that it made no sense to give up his company to finish school. Realizing that his pupil was making more money than he was, the principal agreed to let Rive take his final exams and graduate without attending classes.

At eighteen Rive came to the US as a member of the South African men's underwater hockey team, to participate in a tournament in San Jose. He fell in love with Silicon Valley. His older brother Peter, a computer science graduate, was already working there. With funding from their cousin Elon, the pair started their own remote monitoring software company, Ever-Dream. It was 1999, the peak of the dot-com bubble. But by 2006, after eight years of eighty-hour weeks, the thrill of doing enterprise software had gone. They quit the company, selling it to Dell. For their next venture, the brothers wanted something more fulfilling. A keen outdoor sportsman, Lyndon was also an enthusiastic environmentalist. It made sense to investigate opportunities in renewable energy, the new new thing for the valley's venture capitalists. That July the brothers started SolarCity with $10 million seed funding from Musk. For the company's headquarters, they chose an unassuming office park in Foster City, near San Francisco International Airport. Peter the technologist would handle operations, design, and engineering; Lyndon, the more outgoing of the pair, sales and marketing, and what he self-deprecatingly described as "all the fluff."

At Burning Man, aware that his cousin was seeking a new challenge, Musk had suggested solar as an area where Rive could really make a difference. It would mean thinking big—something that Musk liked to do—leveraging economies of scale, establishing a nationally recognized brand, transforming the installation of solar from a complex nightmare into something much simpler. The Rive brothers vetted the idea. Before taking the plunge, they quizzed existing solar companies. The quintessential description of such companies is "two hippies and a van" or, if successful, "four hippies and two vans." The brothers asked installers questions like where they saw their business headed and how they intended to reduce costs. The answers were unsatisfactory. Scaling up by dribs and drabs was not, Rive saw, going to solve the environmental challenge facing the nation. "If you want to make a true difference," he said, "you need to install in millions of homes." You had to build an infrastructure that could accommodate adoption on such a large scale, and do it right away, not in five or ten years. There was no time to lose.

The company started small, acquiring two existing installers. But the name Rive chose indicated the scope of his ambition: the goal was to equip not individual houses, but entire *cities* with solar. Where would such demand come from? The answer in the first instance was, much to Rive's surprise, from concerned communities like Portola Valley and Mountain View. Soon, for Bay Area residents, it was common to see SolarCity's snappy green vans speeding off to yet another installation. A brand had been born.

Community aggregation programs involving hundreds of homes were a great way to reach economies of scale. But organizing events like solar education seminars and teams of volunteers was a lot of work, and hiring halls typically cost hundreds of dollars. Initially, SolarCity's competitors did not know how to react, conceding the field to their upstart rival. As time went by, however, local installers regrouped, dropping their prices to match the discounts SolarCity was offering. They would hand out flyers as people walked into meetings, or stick them under attendees' windshield wipers in parking lots.

In April 2008 SolarCity launched a second major innovation: solar leases. It was a solution to what had always been the biggest barrier to adoption, solar's high upfront cost. Rive's key insight was that most people didn't care whether they owned the solar panels, what they really wanted was clean energy. Instead of asking customers to dig deep into their pockets to pay for a system, why not do what the auto industry did with cars and simply lease them the equipment? Under SolarCity's solar lease agreement customers would not have to put *any* money down. They would sign up for fifteen years, paying around $100 a month on a system for a typical three-bedroom Bay Area home. With net metering, their electricity would end up costing them 10 to 15 percent less. Integrating financing through leasing thus turned solar into a way of *saving* money. This was an important motivator, especially in times of financial crisis. In addition to installation, SolarCity also agreed to maintain the systems, guaranteeing the production of power. There was, Rive argued, no downside. His biggest challenge now was overcoming customer skepticism. The lease deal sounded too good to be true.

"The big idea of leasing was so much more powerful than community programs," Bruce Karney told me. "It was more scalable, you didn't have to bring people together in a church basement and show them PowerPoint slides for an hour to get them to schedule a site visit. You could do it all over the phone, people would sign up and say, Yes, I want to reduce my energy expenditures." Karney himself moved from the company's community programs group to its lease team. Solar leases were an instant success. Within three months of their introduction, SolarCity's sales leapt 300 percent. By late 2008 the company was installing 130 systems a month, three-quarters of them leased. During the fifteen-year lease period SolarCity would retain ownership of the systems. The company was thus able to pocket the lucrative federal commercial tax credits and depreciation benefits that were not available to homeowners. It would sell the credits to investment banks like Morgan Stanley, which had a seemingly insatiable "tax appetite."

The global financial crisis caused the supply of tax equity investors to dry up. SolarCity was forced to impose a temporary freeze on hiring. By late 2009, however, having secured new sources of financing, including Pacific Ventures, a subsidiary of PG&E, the company was again taking on new people. In mid-2010 SolarCity had 630 employees, up from just two in 2006. In addition to California, it had expanded its activities to eight other states and was planning to go national. The company claimed to have installed solar panels for over seven thousand customers. "My goal is to get to tens and then hundreds of thousands of homes," Rive said. "Once we've hit that point and start crossing over to millions of homes, we're really moving the need when it comes to environmental impact." The US solar market, he confidently predicted, was on track to become the world's number one.

Danny Kennedy had been an environmental activist since he was a schoolboy. Born in Los Angeles in 1971 to Australian parents, Kennedy grew up in Sydney. Though he cut his teeth on forestry issues, his passion was clean energy. In Houston in 1997 Kennedy had been arrested and spent three days in a Texan jail for hanging a huge banner reading "Houston we have a problem: No new oil exploration" off a crane opposite the World Petroleum Conference. In San Francisco in 2001 as a coordinator for Greenpeace, Kennedy had run a successful campaign to help Vote Solar get Proposition B passed. His work as an activist brought Kennedy into contact with a lot of solar industry people. He saw the way the solar industry was developing. "I was telling them, Guys, you're getting it all wrong!" In his view the industry was fixated on the upstream end of the value chain, on honing mass production. "Nobody was focused on the customer relationship, the brand, the sales, shipping, logistics, all of that stuff," Kennedy told me. But downstream was where in his view the real business opportunity lay.

Kennedy had a friend, Alec Guettel, whom he had met twenty years earlier when both were youth delegates attending an environmental conference. After a stint working for the EPA during the Clinton administration, Guettel had gone to Stanford, subsequently becoming an entrepreneur, with two start-ups to his credit. Guettel kept telling Kennedy that he would be better off running a business than campaigns. The friends brainstormed the kind of company they could start together. Another friend, Andrew Birch, a Scottish banker who had formerly been a business development manager for BP Solar, came up with the idea. A downstream solar business would leverage the power of the Internet. It would transform a complex sale into a simple transaction, the same trick that Dell had pulled off for personal computers.

On April 1, 2008, the trio launched Sungevity, based in Berkeley, with $2.7 million of seed money that came mostly from angel investors and friends like the actress Cate Blanchett. Its business plan was based on what Kennedy liked to call the company's "secret sauce." One key ingredient was Microsoft's Virtual Earth visualization program. This combined top-down imagery from satellites with arial photographs taken from planes equipped with special cameras which could shoot a building from four angles simultaneously. From these multiple views, it was possible to construct a three-dimensional visual model of a home that incorporated accurate dimensions of the roof area and its slope. GPS coordinates would give the solar exposure, a database of local electricity rates, and rebates. All a potential customer had to do was go to Sungevity's website and type in his address and monthly electricity usage. Within twenty-four hours the company would reply with an e-mail containing a detailed estimate, including the cost of several standard systems of various sizes, and a simulated picture of what the home would look like with solar panels on its roof. As with any other Internet purchase, the customer could pay online. Once an order was placed, a crew from Sungevity could be on site within days to perform the installation.

In its first week Sungevity processed three hundred quotes. Using the conventional method—sending out a salesperson equipped with a tape measure in a truck—drafting that many quotes would have taken a year. Though the company still had to dispatch a crew to perform actual installations, the huge savings in pre-sale labor costs allowed Sungevity to offer its systems 10 percent cheaper than rival installers. In its first six months with just three sales staff and two crews, Sungevity racked up over 120 sales. Each sale was exuberantly celebrated at its cramped second-story offices with whoops and a vigorous shaking of maracas.

Included with Sungevity's original quote came an estimate of how much money the system would save homeowners during its working life. Also, the prospective increase in the value of their home as a result of having installed that system. Return on investment was something that Kennedy felt the industry had failed to explain properly. Not only was solar "an incredible technology which takes free fuel from the sun and creates cold beer and hot showers," it was also "an economic asset which makes money for customers over twenty-five years." For a Californian homeowner in 2009, Kennedy reckoned, it was hard to imagine a better investment than solar.

For most people, however, going solar remained daunting. In 2006 Sylvia Ventura, a Parisienne advertising executive specializing in biotech, and her husband, Dan Barahona, an engineer and high-tech start-up strategist, decided to buy a solar system for their home in San Francisco's Panhandle district. They discovered firsthand how confusing, intimidating, and arduous the process could be. "We went through the pains of learning about the technology and the rebates and the net metering, and all that good stuff," Ventura told a reporter. Once their system was installed, friends wanting to follow suit asked their advice. Determined to help others traverse the information minefield, the couple set up a how-to website to share their hard-earned knowledge. In return for a referral fee for successful sales, they passed on leads from the site to Dave Llorens, the young man who had sold them their system. Llorens and Barahona thought along similar lines. The two soon became good friends.

After graduating from the University of Illinois with a degree in electrical engineering, the Chicago-born Llorens had headed west. Like many others, he had fallen in love with San Francisco. He got a job as a salesman with NextEnergy, a Bay Area installer. He would drive out to potential customers' homes, make a presentation in their living room, take measurements of their roof, then draw up a design and make a proposal. The whole process would take three or four hours. It was a frustrating experience. "For a lot of these people, solar was just such a ridiculous slam-dunk," Llorens told me. "It was like four times better than prepaying their mortgage. But I couldn't sell it to anyone, I would go to ten homes and maybe one would buy it." Part of the problem, he realized, was that people simply did not trust salesmen. The salesman's goal after all was to sell systems at the highest possible price. To prevent being duped, potential customers would get quotes from a couple of other installers. But buying a solar system was not like buying a kitchen cabinet. Solar was a complicated technology—what exactly *was* the difference between kilowatts and kilowatt-hours? Evaluating the various proposals was hard. Confused, many potential purchasers would simply give up.

With their website, Ventura and Barahona had recognized the need for impartial advice. Even more important was the issue of how to get the best price. You could band together and demand a discount, as the community programs run by the likes of SolarCity had demonstrated. But what was it a discount *from*? The solar industry had done its best to obfuscate pricing, so you could not be sure, especially not when buying from a single vendor. There was thus, Llorens concluded, a need for an organization that could not only aggregate potential customers, but that could also act as a neutral middleman between those customers and the solar installers. To create market competition, the organization would issue a request for proposals (RFP), then have its experts vet the companies that responded. That way, they could be sure that the vendors were kosher, capable not only of installing systems but also of servicing warranties down the track. After checking out the proposals and negotiating aggressively on price, the organization would recommend the best one overall. People would feel more comfortable about taking the plunge if they knew that hundreds of their neighbors were getting the exact same deal. Safety in numbers.

On June 4, 2008, the day after San Francisco began offering generous new solar rebates, Ventura, Barahona, and Llorens joined forces to launch a new business. They named it One Block Off the Grid, inelegantly abbreviated 1BOG. The name was, as Llorens sheepishly admitted, "kind of stupid" since they were not doing one block at a time, nor were they taking people off the grid. It was, he explained, a metaphor, the goal being to convert one block's worth of energy usage to solar. Despite the unfortunate moniker, the idea quickly took off. News of 1BOG's pilot campaign spread largely by word of mouth and via farmers markets. Within a few weeks 180 people had signed up as members online. The organization evaluated six installers, eventually choosing one that had a good record in San Francisco and was eligible for the city's maximum $6,000 rebate. Of the 180 members, 70 sent in their data; 42 ended up buying systems. In its second campaign, the organization registered over 1,200 members. This time the chosen installer was SolarCity.

To clear up the misconceptions about solar that pretty much everyone shared, 1BOG provided its members with answers to questions via "solar 101" e-mails and educational webinars. To give potential buyers an upfront expectation of what a system was likely to cost them, 1BOG built an online estimation tool into which they could plug their details. "It's really powerful, extremely accurate," Llorens told me. "It gives you a very solid understanding of the financials of solar in your neighborhood." Thanks to the discount, 1BOG's prices were typically between 15 and 20 percent less than what customers would pay as individuals. For brokering the deal, the organization got a middleman's cut, giving it a revenue stream to underwrite future campaigns. For installers, the advantage of working with 1BOG was a big reduction in the number of fruitless site visits they had to make to close a sale. The website filtered out the unqualified, like renters, and the misinformed, like people who thought that "solar" meant solar hot water. Homes that 1BOG referred to its chosen vendors were more than twice as likely to purchase a system.

From San Francisco, the organization branched out with its no-negotiation, group-purchase model. 1BOG would run a campaign in any city that could persuade a critical mass of people—defined as a hundred interested parties—to sign up on its website. One campaign in San Diego attracted 950 members, out of which 85 ended up purchasing systems. Within California, 1BOG ran campaigns in Los Angeles, Palm Desert, Sacramento, and Sonoma County; outside the state, in Denver, Phoenix, New Orleans, San Antonio, and Philadelphia. Each campaign lasted four months. It began with PR and grassroots outreach, relying on the proven tools of political campaigns, like lawn signs, house parties, and community organizing. In San Diego, 1BOG hired a former Obama campaign staffer. He recruited a corps of passionate young volunteers to canvass their homeowning elders to go solar. Campaigns also made use of the new tools of the Internet, like online advertising, search engine optimization, and viral marketing via social media sites like Facebook and Twitter.

By late 2009 1BOG was launching one new campaign every month. With over 18,000 members it claimed to have been responsible for almost 600 solar installations. But getting a good price was not going to be enough to trigger the hoped-for exponential rise in solar installations. Even with the discount and all the various incentives, a 3-kilowatt system would still leave buyers more than $7,000 out of pocket. Though some members were able to pay this in cash, most needed some form of assistance. To provide it, 1BOG joined forces with partners like SunRun, a company that specialized in providing third-party finance for residential solar customers.

"The real leap forward in the solar industry, perhaps more than the technology itself," Vote Solar's Adam Browning commented, "is what's happening in financial innovation."

The inspiration for SunRun came to Nat Kraemer in 2006 while the thirty-one-year-old navy reservist was on a tour of duty in Afghanistan. "There's nothing like a war to clarify what you want to do with your life," Kraemer said later. He decided that what he wanted to do was to make solar affordable for the masses. That way, Kraemer figured, there would be less need for overseas resource wars. On his return to the US, he hooked up with an old high school buddy, Ed Fenster, a mergers and acquisitions specialist, and Lynn Julich, a venture capitalist who had been Fenster's classmate on the Stanford MBA program. In early 2007 they quit what Julich called their "easy financial jobs" to form SunRun.

For their business model, they adapted a mechanism known as a "power purchase agreement," in which the customer buys electricity from a supplier at a predetermined price. Such agreements had been pioneered for solar several years previously by companies like SunEdison of Baltimore and MMA Renewable Ventures of San Francisco. PPAs had proved popular in the commercial domain, where they funded almost three-quarters of solar installations. Like solar leases, PPAs were based on the recognition that customers were more interested in buying electricity at a price that was certainly more stable and, with a bit of luck, also cheaper than conventional power, than on owning the means by which that electricity was produced. They were happy to pay for solar on a monthly basis, just like they did with their utility bill.

Before SunRun, PPAs had been applied to only a relatively small number of big projects, like the roofs of supermarkets. That way was easy for banks to oversee. But SunRun was proposing to use online marketing to do *thousands* of installations, on individual residences, too many for banks to check on individually. The company had to work hard to persuade potential backers that it was trustworthy. Under the SunRun model customers paid an upfront installation fee of as little as a thousand dollars. The company would own, monitor, and maintain the solar equipment. Installation was done by a specialist partner. Homeowners purchased electricity from SunRun at a rate fixed for the duration of the agreement, typically eighteen years. Over that time, the company claimed, they would save thousands of dollars on utility bills. "This has shifted the question away from, Hey, do you want to buy this really expensive piece of equipment?" Fenster said, "to, Hey, do you like your utility? Would you like a new one? Would you like it to cost less? Would you like it to be green?" It was, he claimed, a predictable and virtually risk-free investment. In Silicon Valley parlance, a no-brainer.

SunRun financed its first system in late 2007. During the next two years the company would pour over $100 million into funding residential solar for a total of more than 10 megawatts. It signed up over 2,700 customers, first in California, then in Arizona, Colorado, and Massachusetts. SolarCity would follow suit, offering power purchase agreements as an alternative to its leases. In late 2009 nearly 90 percent of Californians still bought their systems outright, but leases and PPAs were growing fast. Some observers predicted that, within a couple of years, third-party financing would come to dominate the market.

South of Market Street, downtown San Francisco's main drag, in a utilitarian area that was formerly ground zero for the dot-com boom, a nexus of solar-related outfits had sprung up. SunRun was located just three blocks away from 1BOG's offices in South Park, the advocacy group Vote Solar was even closer, while Clean Power Finance, a start-up that provided a standardized software package to help installers cope with paperwork, was also head-quartered nearby. Like SunRun's principals, many of the youngsters employed by these new firms were Stanford graduates. Llorens reckoned that 80 percent of 1BOG's people were from Stanford. Outside of work they would meet casually with their friends from other firms. "Everybody from Stanford seems to know each other," Llorens said. "That's the whole point, we're bringing some Silicon Valley talent, MBAs, to the industry to fix some of its inefficiencies." Solar, it seemed, had finally outgrown its hippie roots.

Though worthy, these programs were still relatively small scale, measured in hundreds of roofs. California would never meet its ambitious targets for renewables just by adding a few megawatts here, a few megawatts there. At best rooftop solar might ultimately provide 10 percent of the state's electricity. The rest would have to come from much larger installations, like solar power plants. By early 2010 some estimates reckoned there were over a hundred utility-scale solar projects in the US in the planning stage, representing a total of 17 gigawatts of PV. Utility-scale solar is the subject of part 5.

Part 5:
Just the Beginning:

UTILITY-SCALE SOLAR, CHINA,
AND THE SHAPE OF PV TO COME

24

GOOD SOLDIERS

*H*aving spent most of his exceptionally long career as chief executive of large utilities, David Freeman understood the fuddy-duddy mentality that prevailed among power company managers. "Generally speaking," Freeman said, "the folks who run utilities want to do tomorrow what they did yesterday." They resist change, he added, using one of his trademark folksy metaphors, "as much as a little boy resists a bath."

Utilities liked large, central station type plants. Their engineers were familiar with the steam cycle, they identified with the turbines—big hunks of metal that spun round and round. Large rotating masses produced inertia, which was good for maintaining the stability of the grid. Solar, by contrast, power company engineers saw as piddly stuff, more trouble than it was worth. Lacking inertia—like digital computers, solar was either on or off— it threatened to bring their precious grid crashing down. But what California's giant investor-owned utilities especially didn't like about PV was the prospect of losing market share. They feared an unbridled, customer-driven solar market. That was why, as we have seen, they had done everything in their considerable power to discourage its adoption.

In the first decade of the new millennium, however, the power companies came under unprecedented pressure to change their polluting ways. In California the push began in 2002 when Governor Gray Davis adopted a renewable portfolio standard. This obliged utilities to source 20 percent of their electricity from renewables by 2017. Arnold Schwarzenegger upped the RPS ante twice. In 2006 Schwarzenegger brought forward the deadline from 2017 to 2010. Two years later, he increased the ratio of renewables to 33 percent by 2020.

In 2003 California's energy regulators implemented a joint action plan. It stipulated that in meeting new demand, utilities should first adopt energy efficiency measures, then renewables, turning to fossil fuels (meaning natural gas) only as a last resort. In 2006 the state legislature mandated a 25 percent reduction in greenhouse gas emissions by 2020. California's global warming act was informally known as the Pavley Bill, after Fran Pavley, a former middle school teacher turned legislator whose determination to do something about air pollution derived from having seen so many of the kids in her classes suffering from asthma.

In 2007 the California Energy Commission instructed municipal utilities that they were not to ink any new contracts with coal-fired power generators. Hardest hit was Los Angeles Department of Water & Power, the nation's largest muni, with 1.45 million electricity customers. LADWP sourced almost half of its power from two massive coal-fired pants located in Utah and Arizona. In 2009 Los Angeles mayor Antonio Villaraigosa announced that his city would completely replace coal with renewables by 2020. To effect the transition the mayor appointed David Freeman as the utility's interim general manager. "We're moving toward renewables," Freeman pledged. "If we don't, we're going to get fined like hell."

Though utilities were reluctant to embrace change they were also, as Freeman pointed out, like soldiers, used to following orders. If their political masters wanted them to adopt renewables, then so be it. As the end of the decade neared, however, it was obvious that the power companies were going to have to hurry up to meet the 20 percent by 2010 deadline. Which renewables to deploy? Hydro and geothermal were the ones utilities knew best. PG&E's ratio of around 13 percent renewables came mostly from such sources. But hydro was already maxed out: not much chance of building any new dams. Geothermal was proven: California was the site of the world's largest geothermal plants, the Geysers, north of San Francisco. These had been operating since the sixties. But the technology also had downsides, like the toxic sludge it dredged up. As with hydro, the most suitable sites were already long since spoken for. That meant the new renewables would have to be mostly wind or solar. Of the two, because it was more predictable than wind, solar was the option utilities preferred. The output of solar was also a near-perfect match with the peak in demand for electricity that occurred in the early evening, when consumers came home from work and turned on their appliances.

Senior executives at investor-owned utilities were not famous for their visionary leadership. There was, however, one notable exception. In 1980, a few weeks after he became CEO of Southern California Edison, William Gould stunned employees with a letter he circulated within the company. Gould explained that, in the future, the utility's policy would be "to devote our corporate resources to the accelerated development of a wide variety of future electric power sources which are renewable rather than finite."

At the time SCE was 70 percent dependent on fossil fuels. During the eighties under Gould's guidance, the company would derive one-third of its requirements for new power from nontraditional sources. His plan also entailed increasing the purchase of alternative forms of power from third parties, thereby decreasing the company's reliance on large, centralized plants. Engineers at Edison involved in designing coal and nuclear plants were flabbergasted. The prevailing opinion in the technical community was that renewables were a kooky, tree-hugger thing: they could never compete on pure bottom-line economics. The engineers included Gould's son, Bill Gould Jr. He asked his father what in the world had motivated him to come up with such a crazy plan. "Dad patiently explained to me that he had just proclaimed the inevitable future. What he saw before almost anyone else was that public sentiment in favor of renewables would grow to the point where utility executives would have few other choices."

In particular, it was time to deemphasize coal. Within months of Gould's announcement Southern California Edison was sourcing geothermal energy from Brawley (north of the Mexican border), wind power from the San Gorgino Pass (near Palm Springs), and, as we shall see in the next chapter, helping fund a pilot solar thermal facility in the Mojave Desert. SCE was soon, and would remain, the nation's leading clean energy utility, deriving in 2008 16 percent of its electricity from renewable sources. Solar accounted for just 6 percent of the clean energy total, but that was still far more than any other utility, including PG&E, which that same year sourced less than 1 percent of its renewable energy from solar.

Though most SoCal Edison execs were diametrically opposed to Gould, there were one or two who felt he was on the right track. They included Michael Peevey, who became president of the giant utility in the early nineties. After leaving Edison Peevey served as energy advisor to Gray Davis. In December 2002 the governor appointed him president of the California Public Utilities Commission, a move that was criticized by consumer groups, who saw Peevey as an energy industry insider. They were wrong. An economist by training, Peevey had worked for the Department of Labor during the Kennedy and Johnson administrations, then subsequently for the AFL-CIO. Before joining Edison he was president of the California Council for Environmental and Economic Balance. During his tenure at SCE Peevey urged California's utilities to increase their reliance on energy conservation and renewables. He also chaired Calstart, a public-private partnership whose (slightly premature) aim was to spawn an electric vehicle industry based on local aerospace expertise.

As CPUC president Peevey would repeatedly sponsor bold initiatives. On his office desk was a plaque inscribed with the motto "Who dares wins." We have already seen him in action kick-starting the California Solar Initiative and backing the city of Palm Desert in its radical program for the reduction of energy consumption. In principle the CPUC was supposed to be impartial with regard to renewables. In practice Peevey was to be seen at Governor Schwarzenegger's side for the dedication of new solar facilities, waxing lyrical in his remarks about the benefits of such projects. Peevey was an even stronger proponent of energy efficiency. In 2008 he introduced "decoupling plus," an initiative that enabled utilities to earn more money from energy efficiency than from capital investment.

I meet with the commissioner in November 2009 at the CPUC's dilapidated offices in downtown LA. Peevey is jet-lagged, having just returned from a fact-finding mission to China. He is well aware of the antipathy of the state's big utilities toward solar. "They're going to lose market share," Peevey concedes, "and they're not happy about that." Companies like PG&E had made it plain that they would resist efforts to take away their customers. At the same time, however, utilities also had to recognize that solar programs were popular. They could not very well fight everything.

Peevey understands the utilities' concern about market erosion. But he insists that having 85 percent of a growing pie is better than a 100 percent of a declining one. *"The reality is,"* he says, speaking slowly for maximum emphasis, *"there are so many new things to be done."* New things include repairing the state's antiquated grid and building transmission lines and distribution systems. This is after all how utilities have traditionally functioned, undertaking large-scale projects that generate guaranteed financial returns for decades. "This is a capital-intensive business," Peevey explains. "You make your money, not on your sales, but on your capital investment. The name of the game is to make sure your capital expansion is greater than your loss." In his view the state's 33 percent renewables mandate will stimulate a tremendous amount of new capital investment. "Given that mandate, the opportunities for the utilities to grow their business in this coming decade are going to be *gigantic*."

In addition, the utilities also had opportunities for owning solar. "What I told them some time ago is that *they* should be the ones putting in solar panels on people's homes and running them and getting a capital return." That way the power companies get to keep the rebates for themselves and retain control of the market. But utilities had been very slow to take advantage of these new opportunities. "This is not Google we're talking about, or Apple, or Silicon Valley," Peevey sighs. "The utility culture is by nature old-fashioned, conservative, and service-oriented. They like to build big things, not little things." The nature of that culture was changing, but not as fast as some people would like. At least in California it was changing faster than elsewhere. "If you think there's a cultural lag here," he tells me, "this state and its utilities are light-years ahead of most others in the United States."

If its renewable targets were to be met, the state would need a combination of different types of large-scale solar installations. The CPUC had already approved a proposal from Southern California Edison to install 250 megawatts of solar on the roofs of large buildings like warehouses and big box stores. But Peevey reckoned that the biggest share of solar in California by 2015 or 2020 would likely come from large central stations.

When I meet him, Peevey is working on an ambitious new initiative, an idea for a solar park that his old friend David Freeman had recently brought him. Later that day, at Freeman's office at Los Angeles Department of Water & Power, just down the hill from Frank Gehry's futuristic Disney Center, I ask the general manager what the idea entailed. "We want to make solar power a reality," Freeman replies. "By being a reality I mean hundreds and thousands of megawatts. We have to go beyond the rooftops. That's like trying to win a war one house at a time—we'll never get there. We have to go where the sun is hottest and build huge solar arrays. That's what we're planning to do."

Freeman's bold proposal is to construct a new solar station capable of generating 5 gigawatts, equivalent to 10 percent of the state's current energy usage. "If you're trying to do massive solar," he told a reporter, "you need a large flat area that's already under single ownership and you need good sun."

As it happens LADWP owns just such an area at Owens Lake up in the Sierra Nevada. "The Lord is the one that picked that site," Freeman quipped. "I didn't."

Owens Lake is the scene of the infamous water grab by Los Angeles in the early years of the twentieth century. Streams that formerly fed the once-mighty lake were diverted to an aqueduct that channeled the water away to the big city over two hundred miles south. As a result the lake dried up, devastating the local ecosystem. At Owens Lake today, gale-force winds that sweep off the mountains whip up noxious dust storms from salt flats, causing respiratory problems in nearby residents. To keep down the dust, the department has to expend precious water shallow-flooding portions of the lake bed. Conscious of this poisonous legacy, Freeman—"the Green Cowboy" as he was sometimes latterly styled—went up to Owens Lake in person to pitch his plan as a peace offering. "It's a triple hitter," he promised residents. "We'll save water by not having to use so much of it to control dust, we'll cut down on the wind that creates the dust storms with the solar panels, and we'll generate renewable energy at the same time." There was no time to lose. Now aged eighty-three, Freeman was a man in a hurry. "We are in a death struggle with climate change," he warned. "Mother Nature has a time frame and we don't know what it is."

In December 2009 the LADWP board unanimously approved the first phase of his plan. A pilot solar array of up to 10 megawatts would be constructed at Owens Lake at a cost of between $30 million and $40 million, which the department itself would pay. Freeman downplayed the obstacles that such a large-scale facility presented. "It's no different from building a coal or a gas plant," he told me. "It's the bread and butter of building an electric system. You have to find the land, you have to have transmission, you have to pass an environmental review, then you have to find financing. But it can be done, somehow people are making too much of a mystery of it. There are solar plants that have existed in this state, down in the Mojave Desert, for fifteen years." Indeed there are. They and their latter-day successors are the subject of the next chapter.

In April 2010 Freeman stepped down as LADWP chief. His ambitious plan to build a giant solar array has subsequently run into unanticipated problems. Preliminary engineering tests showed solar trackers installed at the southern end of the lake would sink as much as several inches into extremely corrosive mud. Accordingly the utility scaled back its plans from eighty acres to just five. At the same time, questions were being raised about the cost of the proposed solar park. It seemed that whatever transpired, it would likely unfold very slowly.

25

MIRRORS WITHOUT SMOKE

*T*o startled motorists on Interstate 15, the busy freeway that connects Los Angeles with Las Vegas, the ball of glowing white light suspended above the desert looked like one of the mirages for which the Mojave was famous. In fact, the ball consisted of concentrated sunlight. It was the product of almost two thousand "heliostats," giant mirrors mounted on concrete plinths arranged in a disk at Solar One, in its day the world's largest solar power plant. At the center of the disk stood a three-hundred-foot "power tower" made of steel girders. The heliostats focused the sun's rays onto a receiver perched atop the tower. The receiver contained a boiler that turned sunlight into steam that was piped down to ground level to drive a turbine.

Solar One was erected outside the tiny Californian silver mining town of Daggett, about 140 miles northeast of Los Angeles. On November 1, 1982, five hundred invited guests, more than twice Daggett's regular population, gathered in a blue and white striped marquee for the official dedication. In his speech that day William Gould, the visionary chairman and chief executive of Southern California Edison, described the plant as "a giant step toward energy independence. A world desperately dependent upon uncertain oil and gas supplies will learn from Solar One much about the promise of alternative resources, whose implications are enormous."

The experimental $142 million solar thermal plant had been, Gould admitted, mostly funded by federal dollars. The next step, he promised, would be up to private utilities like Edison. His company was considering a new solar facility with an output ten times larger than Solar One's 10 megawatts. Proposals had already been received and "tests performed here will determine for Edison whether construction of an advanced 100-megawatt solar generation station will be a practical option."

Utility-scale solar would prove practical, but the proof would not come from Solar One. During its first years the power tower would fail to live up to expectations. Maintaining a head of steam was hard, passing clouds frequently interrupting operation. Rather, the proof would come from another type of solar thermal plant that employed the less fancy but more reliable technology of parabolic troughs. Giant U-shaped mirrors, these were capable of concentrating sunlight by twenty times or more. Suspended at their focal point was a thin, vacuum-encased tube through which circulated a heat-carrying liquid, typically synthetic oil. The hot oil passed through a vat, turning the water it contained into steam.

The first of what would ultimately be nine solar thermal plants was built in 1984, right next door to the power tower at Daggett, by a feisty upstart called Luz International. Solar Energy Generation System I is a rectangular field containing long rows of shimmering troughs that stand more than twice the height of a man. By 1991 the Luz plants were generating 90 percent of all grid-connected solar energy produced on the planet. Then, in that year, the company went bankrupt.

Most of us as children have used a magnifying glass to focus the sun's rays to burn paper. The idea of concentrating sunlight, of taking a dilute resource and amplifying it, is an ancient one. The first recorded instance of a practical application for "burning glass" dates back to 212 BCE. During the siege of the Sicilian port of Syracuse, the inventor Archimedes—the Eureka! man himself—is said to have deployed reflecting mirrors. He instructed the defending Greek soldiers to use their polished bronze shields to focus sunlight on the wooden triremes of the attacking Roman fleet, setting the warships on fire. Controversy over the credibility of Archimedes's "death ray" raged down the ages until 2005, when resourceful students at MIT replicated the feat. They used 129 hardware-store mirrors to set alight an (admittedly dry) wooden mock-up in less than ten minutes.

At the Paris Exposition of 1878 Augustin Mouchot, an enterprising professor of mathematics, demonstrated that a giant round parabolic reflector trained on an iron cauldron filled with water could produce enough steam to drive a small motor. The demonstration engendered speculation that solar energy might provide unlimited power at almost no cost. However, a commission the French government established to evaluate Mouchot's invention soon concluded France did not have enough sun for solar energy to be applied to industrial ends. The problem of what scientists call "the intermittency of the solar resource"—the inescapable fact that, in most places, the sun is often obscured by clouds—means that concentrator systems must be located somewhere the sun shines more or less all the time. In practice this means in a desert.

Frank Shuman, a hard-headed American engineer, chose to site the world's first solar power plant in Egypt, in the desert south of Cairo. In 1912 Shuman built a station consisting of five fifty-two-meter-long parabolic mirror troughs. They heated steam to drive engines that pumped irrigation water from the Nile into adjacent cotton fields. Though successful, the solar irrigation plant was destroyed during World War I. After the war, oil emerged as the low-cost fuel of choice. Not until more than half a century later, following the oil crises of the seventies, would alternative forms of generating power begin once again to attract serious attention.

In 1979 a small plant based on Shuman-style parabolic trough concentrators for powering irrigation pumps was built outside Coolidge, Arizona. It was a joint effort between the University of Arizona, the Department of Energy, and Acurex, a Californian firm that had developed computer controls to tilt the mirrors so they could track the sun's arc from east to west.

Luz (rhymes with "fuse") means light in Portuguese. Luz is also the location in Genesis where Jacob dreamt of a shining ladder that ascended to heaven. Luz International was founded in 1979 by Arnold Goldman, then just thirty-four years old, an electrical engineer with utopian leanings from Providence, Rhode Island. With his kindly features and chin curtain beard, Goldman somewhat resembled a gnome. His original plan had been to use solar energy to produce steam for industrial purposes, but there was no market for this. Then, serendipitously, he discovered that Acurex, following the success of its Arizona plant, had won a contract from Southern California Edison to supply electricity. Under the Public Utility Regulatory Policies Act of 1978, SCE was obliged to accept power from independent producers.

Acurex could not raise the money to build a plant. Venture capitalists were not interested in funding anything so obviously off-the-wall as solar energy. Goldman stepped into the breach: he persuaded angel investors to stump up the $62 million Luz needed to build a pilot solar thermal plant. The utility was supportive, providing the site at Daggett, some seed funding, and a long-term contract to buy power at a premium price, 14 cents a kilowatt-hour. The price was justified because the plant Luz built would function as a "peaker," its output added on top of base-load power to cope with the spike in demand that occurred during the hottest hours of the day. It was, however, considerably less than the cost of generation, 24 cents.

Luz based its business plan on two premises. First, that the price of fossil fuels—in particular natural gas—would continue to rise. Second, that tax breaks would continue to support investment in new plants. In fact what happened during the eighties was that the price of conventional energy would fall by up to 80 percent, while unsympathetic federal and state governments would progressively eliminate tax breaks.

Under the circumstances what Luz managed to achieve was little short of heroic. Starting in 1984 the company would build at least one solar thermal plant every year. The initial 14-megawatt pilot at Daggett was followed by a second one, twice as big at 30 megawatts, the maximum allowed under the law. Next Luz built five 30-watt plants at Kramer Junction to the east, near Edwards Air Force Base, where supersonic bangs from low-flying jets would sometimes crack the mirrors. Then, having lobbied to get the law changed, the company built two much bigger—80 megawatt—plants at nearby Harper Lake. The nine plants had a total capacity of over 350 megawatts. A further three plants were scheduled for construction when the company went bust.

Economies of scale plus refinements made by the company's engineers to three generations of solar thermal technology kept bringing down the price of the power Luz produced. The new, fourth-generation plants that were to have been constructed in 1994–95 were projected to produce electricity at just six cents a kilowatt-hour, a quarter of the original price, a fourfold reduction in just a decade. But the price of conventional energy had also plummeted, in effect forcing Luz to run as fast as it could just to stay in the same place. William Gould had retired in 1984. His successors at Edison were no longer willing—or required, following changes in the law—to enter into long-term, fixed-price contracts for renewables.

When Luz built its first plant, the company was able to offer investors a 15 percent renewable energy tax credit from the federal government, plus a 10 percent ordinary investment tax credit, and a 25 percent energy tax credit from the state of California. Most of these incentives were axed in 1988. The tax breaks that remained had to be renewed on an annual basis. Delays in their renewal forced Goldman to promise additional returns in order to reassure anxious investors—many of them utilities from other states—which in turn meant that Luz never made any money for itself. The state legislature had also granted Luz an exemption from property taxes. This was controversial, since only one company could benefit. In January 1991, in his last few minutes in office, outgoing Republican governor George Deukmejian vetoed extending the exemption, making it impossible for Luz to finance its tenth plant. The exemption was eventually passed but, by then, the company had run out of money.

Though the firm that built them is no more, the solar thermal plants are still running. Now operated by NextEra Energy, a subsidiary of Florida Power & Light, SEGS I~IX generate, according to Goldman, on the order of $100 million a year in revenue. After more than two decades in operation the plants, it is said, have yet to miss a single hour of output.

Fast-forward fifteen years, to a different world. In 2006 California's investor-owned utilities are hunting for ways to add copious amounts of renewables to their energy mix. Venture capitalists have decided that clean-tech is the next new thing. Arnold Goldman is back in business, with a brand-new company, BrightSource Energy, based in Oakland, flush with $160 million in cash from investors that include Google, Chevron, and Morgan Stanley. He has reassembled many of his old design team to, as he put it, "complete the work we'd started." This time, however, there would be no troughs. Instead, the company planned to use power towers. That way they would be able to up the degree of concentration, from twenty times to a hundred times or more. Higher temperatures meant better steam, hence greater efficiency in driving turbines and much lower cost.

BrightSource could take advantage of recent improvements to the technology. The heliostats it used were much smaller than the mirrors arrayed around Solar One, so less susceptible to wind. Because flat, they were much cheaper to manufacture than the precision-curved parabolic troughs. Better gears and more sophisticated, microprocessor-driven controls meant that the mirrors could track north-south as well as east-west. To showcase its technology the company and its Israel-based subsidiary, Luz II, built a 6-megawatt demonstration plant in the Negev Desert.

The utilities were dishing out big contracts to providers of renewable power. In February 2009 BrightSource announced what it claimed was "the world's largest solar deal," an agreement to sell Southern California Edison 1.3 gigawatts' worth of power. Three months later, the company landed another whopper, this time with PG&E, for 1.31 gigawatt-hours. Some of this would come from three new solar thermal plants whose combined capacity would be 440 megawatts. The plants would be built at Ivanpah, a dry lake near the Nevada border about a hundred miles northeast of Daggett on Interstate 15. The primary attraction of Ivanpah was that, unlike many remote desert sites, it had two major transmission lines running through it.

A few weeks later, in a little-noticed development, personnel from Southern California Edison and a demolition crew strapped explosives to the legs of Solar One. They brought the world's first power tower crashing to the ground. There it was broken up and sold for scrap. *Sic transit gloria mundi.*

Although by far the best credentialed, BrightSource was only one entry in what was rapidly becoming a very crowded field. In addition to domestic solar thermal rivals like Pasadena-based power tower specialist eSolar, contestants included Spanish companies like Acciona and Abengoa. The former had built Nevada Solar One, a 64-megawatt trough plant outside Las Vegas that began operating in mid-2007. The latter had signed a $4 billion agreement to build one 280-megawatt trough plant at Gila Bend, near Phoenix, with US government guarantees of $1.45 billion, and another in the Mojave between Daggett and Kramer Junction.

By 2008 across the southwestern United States an extraordinary solar land rush was under way. That May, having received 125 applications for public land on which to build solar plants, the Bureau of Land Management declared a moratorium on further applications. In July, overwhelmed by a hail of protests, the BLM was forced to reverse its decision. By January 2009 the bureau had received some 223 applications covering 2.3 million acres. For the California Desert District alone, the BLM listed on its website 75 projects representing a grand total of 51.6 gigawatts, more than the state's annual consumption of electricity.

The solar thermal stampede was not confined to the United States. In early 2010 eSolar signed an agreement with a Chinese developer to construct, over the next decade, 2 gigawatts of solar thermal power plants in the Mongolian desert. German companies participated in many of these proposals but only as suppliers of componentry, like precision mirrors and steam turbines. Unlike the US and Spain, Germany has no deserts. Then, in mid-2009, twelve major German firms, including utilities like RWE and E.on, equipment makers like Siemens, and financiers like Deutsche Bank, announced the formation of a consortium called Desertec. The aim of the Munich-based consortium would be to undertake "the biggest solar energy project in the world." Desertec would build a string 100 gigawatts of solar thermal plants stretching across the Sahara. The electricity they generated would be delivered to Europe via undersea cables like the one Siemens was building to supply the Mediterranean island of Majorca from the Spanish mainland. By 2050 Desertec would be capable of delivering up to 20 percent of Europe's energy needs. In total, the project would cost 400 billion euros ($560 billion). The timeline called for contracts to be drawn up by 2012.

It seemed quite logical: Europe needed energy and had cash; Africa needed investment and had sun. But critics of the grandiose plan were quick to point out that building solar power plants in politically unstable countries like Algeria or Libya would saddle Europeans with precisely the same sort of dependency they already faced with oil. The most trenchant criticisms came from Hermann Scheer, who described Desertec as "highly problematic." Scheer objected to the centralized corporate control implicit in the plan. He argued that Desertec would doubtless be subject to the usual problems that dogged mega-projects, like huge cost overruns and missed deadlines. There would also be specific logistical problems, like sandstorms. Above all, the plan would divert investment from projects in Germany itself, most notably putting more solar panels on people's roofs. The price of solar was already on track to reach parity with that of fossil fuels and nuclear. "By the time solar power from North Africa can be supplied for the prices Desertec has promised (which means, not before 2020)," Scheer predicted, "solar power generation will happen at a noticeably lower price here at home."

The financial crisis of 2008/2009 threatened to halt the solar thermal stampede in its tracks. Venture capital had enabled start-ups like BrightSource and eSolar to build their demonstration plants. Building full-scale power plants would be a different story. It wasn't like underwriting whiz-bang software start-ups. "This isn't three guys with a couple of servers who can scale by buying a few Dells," commented Alan Salzman, CEO of Vantage Point Venture Partners, one of BrightSource's original investors. Thousands of tons of steel and glass and concrete and turbines cost more than even Silicon Valley's coffers could cope with.

Where would the $1.37 billion BrightSource needed to build its Ivanpah plants come from? One source was Bechtel Enterprises, the giant construction firm's project development and financing arm, which in September 2009 became an equity investor in the plants. Another was the federal government. In February 2010 BrightSource announced that the Department of Energy had conditionally agreed to provide $1.37 billion in loan guarantees to support the plant's construction. The condition was that the firm needed equity to match the department's loan guarantees.

Finding finance was hard. But the biggest obstacle facing BrightSource, and by extension all the other firms applying to set up power stations in the desert, was passing the environmental review. This was a critical juncture in the permitting process. Prior experience had enabled BrightSource to get the jump on its rivals. Of the 223 applications the Bureau of Land Management had received, the firm's proposal was one of the first to be reviewed. It would be a test case.

At Ivanpah BrightSource had purchased a 4,000-acre site on which it planned to build 3 plants, consisting of a total of 7 power towers and 400,000 heliostats. Before the equipment could be installed, the site had to be cleared and leveled. Rocks and vegetation would have to be removed, then the desert floor scraped. Environmentalists promptly protested. It wasn't that they didn't get the merits of solar power, they just didn't want large plants sited where the mirrors and towers might destroy fragile habitat. The desert was not, as most people supposed, an empty wasteland. It was in fact home to a veritable menagerie of endangered species. They included the Mojave ground squirrel, the western burrowing owl, and the flat-tailed horned lizard. Argument centered on one especially rare critter, *Gopherus agassizii*. Better known as the Mojave Desert tortoise, this just happened to be the official reptile of the state of California. The environmentalists' concerns were lost on Arnold Schwarzenegger. "The Mojave is the best place to have a solar field because it has the most sun," an exasperated governor explained during a December 2008 interview on *60 Minutes*. "But there are some who want to hold it up because they think that it would endanger some animal life. That is going overboard."

BrightStar should have seen it coming. In September 2009 the company was forced to abandon plans to build another plant at Broadwell, east of Daggett, because it was home to a flock of federally protected bighorn sheep. But whereas Broadwell offered one of the most beautiful vistas in the desert—or so the Sierra Club claimed—Ivanpah was not exactly pristine. The site had already been used for livestock grazing. There was also, incongruously, a casino-resort-owned golf club in the vicinity that offered its patrons two eighteen-hole designer courses. Across the highway, Ivanpah dry lake was popular with land sailors. There in March 2009 an intrepid Englishman named Richard Jenkins had broken the world land speed record for a wind-powered vehicle, reaching 126 mph in his yacht, *Greenbird*.

The Sierra Club would not budge. The site had to be preserved because it was home to around two dozen desert tortoises, enough to constitute what environmentalists call an "evolutionary significant unit." BrightSource's original plan had been to hire a tortoise wrangler to round up the reptiles and relocate them to a similar habitat the company would pay for. Unfortunately, a previous attempt at relocating desert tortoises had ended in disaster. The military had wanted to move the reptiles to make way for a new combat training ground in the Mojave. It had planned to helicopter 670 tortoises to new homes, but the relocation was abandoned after 90 of the unfortunate creatures perished. The deracinated tortoises had apparently attempted to crawl back home. Averaging just twenty feet a minute, they were no match for coyotes and cars.

The Sierra Club asked regulators to move the proposed site for the Ivanpah plant closer to the freeway. "The project must not contribute to additional loss of habitat," the club insisted in its government filings. Bright-Source grumbled that such extreme demands and heavy-handed regulation could abort the rebirth of the solar thermal industry, which had languished in suspended animation in California for two decades. In February 2010 the company gave in, submitting a scaled-down alternative design. BrightSource would reduce the footprint of its plant by a quarter, avoiding the area of greatest concern to the environmentalists. The number of 460-foot-tall power towers would be cut back from seven to three; the maximum number of heliostats by 40,000. This would lower the plant's output capacity from 440 megawatts to 392 megawatts. In March 2010 state regulators finally gave BrightSource the nod. The downsized plant began construction that October.

Though only a few of the plans on the drawing board would likely ever get built, solar thermal would undoubtedly make some contribution to the utilities' pressing need for renewable energy. But it would have to compete against solar photovoltaic, a more flexible technology that could be deployed more rapidly, virtually anywhere, and at any scale. The only reason to resort to large central station concentrators had been to save on cost. For most of its history PV had been too expensive for utilities to contemplate. Now, as we shall see in the next two chapters, the price of PV was dropping rapidly in both the traditional crystalline silicon cells and the newer technology of thin film. As a result, utilities were beginning to make their first large-scale deployments of solar PV.

26

BALLAD OF A THIN FILM

*T*he California Public Utilities Commission led by Mike Peevey had made its commitment to solar crystal clear. The utilities responded by backing proposals for the kind of large-scale, remotely located plants they thought made economic sense. Connecting such plants to the grid often meant new transmission lines. But with a BANANA—"build absolutely nothing anywhere near anyone"—mentality rampant among residents along the chosen routes, obtaining permits to construct such lines could take any where from seven to eleven years.

Planners at Southern California Edison scratched their heads. Were there no sites big enough to provide economies of scale, but close enough to customers to not need transmission lines? It turned out there were, in the shape of the roofs of giant warehouses in the regional trucking hub of Inland Empire, just a few miles east of Edison's Los Angeles headquarters. Systems installed on these football field–size roofs would fill a gap in the marketplace, between the small generators covered by the California Solar Initiative and the huge plants envisioned under the state's Renewable Portfolio Standards. You could stick a couple of megawatts of PV on top of a big box. Multiply that by a hundred, the planners mused, and what sort of discount would you get?

"We realized that, as a utility, we could drive a price point change," said Randy Schultz, a project manager at SoCal Edison's project development division. Commercial customers were paying roughly $7 per installed watt of PV, residential customers considerably more. The utility figured that by purchasing in volume, it could knock the price down to just $3.50. "We approached the public utilities commission with an application," Schultz told me. "We said, Let us go out and build generation, and we can do it for approximately half the societal cost of what other people are having to pay." The application envisaged putting 250 megawatts of PV on two square miles of rooftops over five years at an estimated cost of $875 million. It would be the largest rooftop solar project ever undertaken by a utility. With SCE's excellent credit rating, obtaining financing would not be a problem.

Edison launched the project at a warehouse in Fontana, located in the heart of Inland Empire, in March 2008. The giant rooftop provided a wonderful photo-opportunity for Edison International CEO John Bryson, CPUC commissioner Mike Peevey, and California governor Arnold Schwarzenegger. "This proves you can protect the environment and the economy at the same time," a delighted governor told reporters as he flipped a mock switch on the as-yet unbuilt system. Authorization from the CPUC would take another fifteen months to arrive. Meantime, the power company went ahead, paying for the initial phase of the project out of its own pocket. By late 2009 Schultz and his ten-person crew had completed the first, 2-megawatt rooftop in Fontana, a second, 1-megawatt installation in nearby Chino, and were working on the design of a third.

Treading gingerly between the skylights on the roof of the warehouse at Chino, I noticed that the thousands of panels making up the installation were not the traditional mosaic of blue-coated silicon wafers, but a sleek-looking black material. They were in fact made from a compound semiconductor called cadmium-telluride. The panels were supplied by First Solar, a US company which just a few months earlier had leapt ahead of its German and Chinese rivals to become the world's largest maker of PV.

Silicon became the material of choice for solar cells accidentally, as a by-product of Bell Labs' work on transistors. Other materials were much better absorbers of light. There was also the messy business of sawing ingots into wafers, where half your silicon ended up as sawdust. Chemists in particular looked on this primitive method with derision. "Making solar cells from silicon wafers is like making newspapers by chopping logs," as Lord George Porter, a British Nobel chemistry laureate, once put it.

How much more elegant it would be to start with a material whose physical properties matched the distribution of photons in the spectrum of sunlight. Then, instead of gluing the material onto glass, you could simply paint it onto the panel in the form of a thin film. Silicon wafers were hundreds of micrometers thick. Thin films by contrast typically measured just a few micrometers, a huge reduction in raw material costs.

Scientists began chasing the thin-film grail soon after the invention of the solar cell. Several suitable sunlight-absorbing compounds were identified early on, notably copper-indium-selenide (CIS or, with the addition of gallium, CIGS) and cadmium telluride (CdTe, familiarly known as "cad-tell"). During the midseventies, a third thin-film contender emerged from the RCA laboratories at Princeton. Amorphous silicon—a noncrystalline form of the material—quickly became seen by the industry as the likely winner. Cells made from the material worked well under artificial light. In 1980 Sanyo began commercial production of amorphous silicon cells for watches and calculators. By 1984 amorphous accounted for more than 70 percent of the Japanese market.

At ARCO Solar, by 1980 the world's leading maker of solar panels, management was convinced that thin-film amorphous silicon would overtake their conventional, wafer-based crystalline silicon products. Accordingly, the company hired a hundred researchers and assigned them to work on developing amorphous. ARCO also awarded Energy Conversion Devices, a Michigan-based firm, a $25 million R&D contract to make amorphous silicon solar cells. ARCO Solar executives speculated that there would be two markets for the new low-price panels. One was off-grid applications in the third world; the other, "big power utilities that need cheap fuel."

Bill Yerkes, ARCO Solar's outspoken founder, warned the company against putting all its eggs in one basket. Instead of having everyone working on amorphous, Yerkes insisted that it would be more prudent to assign just seventy-five. Twenty of the remaining researchers would work on CIGS, the rest on the outsider, cad-tell. After all, as he pointed out, Kodak had recently made cad-tell cells that were 10 percent efficient at converting sunlight to electricity. That was twice as much as amorphous could manage. Ultimately Yerkes became convinced that cad-tell had the best prospects. In 1985 ARCO Solar unveiled an amorphous silicon module, trademarked Genesis. The company announced that by 1990 it hoped to be producing almost exclusively amorphous cells. A few weeks later Yerkes left ARCO to form his own company, Yerkes Electric Solar—YES for short—to make solar cells from cadmium telluride.

Cadmium made some people uneasy. It was extremely toxic, as the Japanese knew well. In the prewar period, cadmium from Japanese mines had seeped into water used to irrigate rice paddies, accumulating in the tissues of all who ate the rice. It caused *itai-itai byo*—literally, "ouch-ouch disease"—so-called on account of the extreme pain it inflicted in the joints of sufferers. Though there had been no new cases reported since 1946, the Japanese remained leery of the heavy metal. Aware of such concerns but also knowing that in the form of a refractory alloy cadmium was safe, Yerkes refuted the perception in typically head-on fashion. In front of a conference audience, he held up a piece of cad-tell thin film, then proceeded to swallow it. Convincing investors to back his new company was harder. After a year, having failed to find funding, Yerkes gave up and went back to work at Boeing.

For the next twenty years, crystalline silicon would confound predictions. As the efficiency of cells improved and the price of panels dropped, silicon would remain by far the dominant material in the PV market. Then in 2005, with the German market in full spate after the solar feed-in tariffs kicked in, a worldwide shortage of silicon hit. Spot market prices soared to $400 a kilogram. That opened a window of opportunity for cad-tell and its prime exponent, First Solar.

Harold McMaster must have known, as any man must know who starts a company in his sixty-seventh year—past retirement age for ordinary mortals—that the odds he would see his vision to fruition were not good. Indeed, it would take twenty years for McMaster's big idea to reach maturity, by which time he was dead. Before he died, however, McMaster would have the immense satisfaction of knowing that he had been right.

In Toledo, Ohio, known as the Glass City because of its long history of innovation in all facets of that industry, they called McMaster "the glass genius." One of thirteen children, he was born in 1916 into grinding poverty on a tenant farm in nearby Deshler. McMaster went on to earn a master's degree in nuclear physics and more than a hundred patents. A small man, standing barely five feet tall, he liked to think big and move fast.

McMaster is justly famous for his work on tempering glass. The process involves heating followed by rapid cooling. It compresses an otherwise brittle material, giving glass greater tensile strength. Tempered glass is stronger than ordinary glass. It shatters into crumbs rather than sharp-edged shards. This makes it ideal for applications where safety is at a premium, like automobile windshields and skyscraper windows.

In 1971 McMaster started a company, Glasstech, to commercialize a machine he had invented to manufacture tempered glass. People told him he would be lucky to sell five of his machines worldwide. In fact, the firm sold four hundred units in the US alone. Glasstech machines reportedly produce 80 percent of the world's automotive glass and 50 percent of its architectural glass. Small wonder then, that when McMaster founded Glasstech Solar in 1984, fifty-seven local Toledo investors volunteered to back him. In 1987 McMaster sold the parent company for $100 million, retaining only the solar subsidiary. He based it in Colorado to be near the national Solar Energy Research Institute.

The energy crises of the 1970s had prompted McMaster to consider photovoltaics. Transparent glass could be coated with thin layers that changed its color or its ability to pass light: why not with a thin layer of photovoltaic material? McMaster envisioned machines that would crank out solar panels as quickly and cheaply as Glasstech machines made tempered windows. A field of such panels arrayed across two thousand square miles of the desert in Arizona, where he kept a second home, could provide enough energy to power the whole country. It would end America's dependence on imported oil. Since glass is also an amorphous material, it was perhaps natural that McMaster should choose amorphous silicon as a coating for glass panels. Initially, however, the research did not go well. In its first six years Glasstech Solar spent $12 million without producing anything worthy of commercialization. Amorphous cells worked well enough on a small scale in the laboratory, but outside in the real world the panels would lose 20 percent of their efficiency undergoing weathering tests.

While hopes for amorphous faded, cadmium telluride's star was rising. By 1989 US firms like Ametek were exploring low-cost methods for producing cad-tell panels. In 1991 Ting Chu, a researcher at Southern Methodist University in Texas, announced that he had raised cad-tell's conversion efficiency to 15 percent, equivalent to that of crystalline silicon. Though only a laboratory result, this confirmed the material's commercial potential. In 1990 McMaster changed horses. He renamed his company Solar Cells, raised another $15 million—including $2 million of his own money—and relocated with six employees to rented premises on the University of Toledo campus.

In 1997 Solar Cells announced the development of a high-speed deposition process. This could coat a standard two-by-four-foot glass panel with a thin film of cadmium telluride in just thirty seconds. It took rival firms like BP Solar six hours to do the same thing. McMaster, now aged eighty-one, was keen to capitalize on this huge comparative advantage.

In 1997 total global demand for PV was just 125 megawatts. McMaster proposed to build a machine that could produce *500* megawatts' worth of panels a year. He was not interested in the existing market for off-grid solar in developing countries: in his view this was merely a costly distraction. Rather, McMaster believed that his friends in the glass industry would be happy to integrate his panel-coating process into their production lines. High-volume manufacture would drive the price of solar to below a dollar a watt. The primary customer for Solar Cells' panels would be US utilities. This was a high-risk strategy: the glass and power industries are innately conservative. By 1996 Solar Cells had almost run out of cash. McMaster injected more money into the firm, raising his stake from 22 percent to 67 percent. He specified that his investment be used to commercialize the high-speed deposition process. By early 1998 the company was in trouble again, looking for a partner with pockets deep enough to scale up the technology. "Given the kind of money we need," McMaster admitted, "we can't bootstrap our way any longer."

In February 1999 McMaster sold a controlling interest in Solar Cells to True North Partners, a Phoenix, Arizona–based private equity firm. It was run by Michael Ahearn, a lawyer specializing in start-ups, on behalf of John Walton, the second son of Wal-Mart founder Sam Walton and consequently one of the richest men in America. True North's goal was not to earn a targeted return on a cautiously invested portfolio. Rather, it was, as Ahearn boasted, to "go out and build the next great company." They were looking for a technology with the potential to disrupt existing mainstream markets. This was a category into which thin-film cad-tell certainly fitted. To signal their intent, Ahearn and Walton changed the corporate name from the anodyne Solar Cells to the aspirational First Solar. True North invested an initial $16 million in the construction of a plant in Perrysburg, a suburb of Toledo. This, the company boasted, was the largest solar panel factory in the world, capable of producing an unheard-of 100 megawatts of product a year. Predictably, things did not go according to plan. "We had gotten ahead of ourselves," admitted Ahearn, who had by then appointed himself chief executive.

It was 2001 and the company had yet to sell a single panel. Many wrinkles in the manufacturing process remained to be ironed out. Now no longer involved in operations or setting policy, the ever-impatient McMaster was frustrated by the slow pace of progress. "They don't have the right leadership," he grumbled to the *Toledo Blade*. Elsewhere, the news was not good either. In November 2002 BP Solar announced it was shutting down its cadmium telluride plant in Fairfield, California. This had been opened with great fanfare just four years earlier by then vice president Al Gore. But BP's first commercial cad-tell modules failed to live up to expectations. After a few weeks on rooftops, their efficiency slipped from 8 to 6 percent. "That's it," said Dan Shugar, president of leading system integrator PowerLight, "We're not going to use any more of this material." That left First Solar the last firm trying to make cad-tell solar cells.

In 2003 Harold McMaster died, aged eighty-seven. Up until a few days before his death, he was still experimenting with solar energy and talking of forging ahead with ventures that would realize his dream. John Walton would not let him down, pouring $250 million over six years into the fledgling company to get it up and running. Then he too died, in 2005 at the age of fifty-eight, when the experimental plane he was piloting crashed.

First Solar commenced full-scale production of cad-tell panels just in time to catch the wave created by the German feed-in tariffs. In 2005 the company produced 20 megawatts and recorded its first profit. By early 2006 employment at First Solar had doubled to five hundred. Expansions to the Perrysburg plant more than tripled production. Almost all of the 1 million panels the plant produced that year were exported to Germany. In 2006 the German government provided First Solar incentives of $61 million to build a factory at Frankfurt an der Oder—"the other Frankfurt"—in the east of the country, on the Polish border, where unemployment ran around 20 percent. First Solar invested $150 million in the plant, which would ultimately employ four hundred workers. No sooner were its German lines operational than the company announced plans to build an even larger, low-cost production base in Malaysia.

Because of cad-tell's relatively poor efficiency, most of First Solar's panels went to large solar farms, where there were fewer constraints on site size compared with individual roofs. Every few months the company would announce another massive deal to supply such farms. Investors liked what they heard. The company went public in November 2006: within a year its stock had shot from $20 to $220. In March 2009 First Solar announced that its cumulative total production of cad-tell solar modules had reached 1 gigawatt. Getting to the first 500 megawatts had taken more than six years; to the second, just eight months. By the end of that year, the company would muster at its twenty-three lines on three continents a combined annual production capacity of 1 gigawatt, the equivalent of a conventional coal-fired power plant. Harold McMaster's vision of utility-scale PV had been vindicated. Along the way First Solar had also passed an industry milestone, one that had been marked out more than thirty years earlier. Prodigious increases in capacity had been accompanied by economies of scale. Having driven two-thirds of the cost out of manufacturing its cells, First Solar became the first firm to break the dollar-a-watt barrier.

Mike Ahearn gave credit where it was due, to the German feed-in tariff program. "Without forward-looking government programs supporting solar electricity, we would not have been able to invest in the capacity expansion which gives us the scale to bring costs down," Ahearn said. Energy policy, he realized, was the key issue. "It's very likely that we'll look back a few years from now and realize that we're on the cusp of a major transformation in the fundamentals of energy policy," toward a German-style mechanism. Feed-in tariffs provided market certainty. They enabled companies to raise the financing needed to complete large-scale projects. The tariffs also established definite price points and set end dates that gave firms many years in which to recoup their investment—and, more important, make a profit.

Meanwhile, the technology of cad-tell was still far from mature. First Solar expected that its production cost would continue to fall, from 87 cents a watt in 2009 to 70 cents by 2012, while the efficiency of its panels would rise, from 11 percent to 12 percent. By 2014, the cost would be between 52 and 63 cents, while laboratory results indicated that efficiencies of over 16 percent were achievable. "Looking ahead to the next two to four years," Ahearn predicted, "First Solar will be in a position to produce power from the sun *at costs competitive with conventional electricity generated from fossil fuels* [emphasis added]." The company's goal was to increase production by 2012 to more than 2 gigawatts, the equivalent of a large nuclear power plant. Unlike a nuclear plant, which would take at least five years to produce any energy, a well-executed solar power station could be designed, built, and operational in a matter of months. The main problem, according to First Solar president Bruce Sohn, was bureaucracy. In the US, government restrictions could cause a solar project to take three or four years that could be completed in Germany in about six months.

By mid-2009 First Solar had lived up to its name, becoming the world's number one producer of PV panels. In the thin-film section of the market the company was streets ahead of the competition. Though at least fifteen other firms were attempting to make cad-tell solar cells, First Solar accounted for almost all cad-tell production. To allay any remaining concerns about the toxicity of cadmium (and recycle scarce tellurium) the company undertook to dispose of its panels when their working life was over, in twenty years' time.

First Solar's European sales were so strong that until 2007 the company did not even consider the US market. In 2008 First Solar derived 94 percent of its revenues from Europe, Germany alone accounting for 74 percent. In 2009 its order book was so full the firm felt it was not worthwhile exhibiting at US solar industry trade shows.

First Solar was not exclusively concerned with large-scale installations. In 2008 the company invested $25 million in SolarCity. Customers liked the look of the sleek black panels, according to SolarCity CEO Lyndon Rive. Soon, First Solar's products would account for between 60 and 70 percent of the fast-growing installer's residential sales. But the main game was utility-scale projects. In December 2007 First Solar took its first step into the big time, paying $34.4 million for Turner Renewable Energy to acquire the engineering expertise it needed to construct large-scale installations. The deal came with an existing project to build a 7.5 megawatt solar plant outside Blythe, California, in the Sonoma Desert near the border with Arizona. Subsequently upsized to 21 megawatts, this 70-acre, 350,000-panel project was completed in December 2009, with Southern California Edison agreeing to purchase all the power it produced.

In 2008 First Solar supplied 550,000 panels for a 40-megawatt plant on a former military air base outside Leipzig. The following year the company participated in an even larger project, at Lieberose in the state of Brandenburg, on flatlands formerly used by the Soviet Army as a training ground. The 700,000 panels First Solar supplied occupied an area the size of two hundred football fields, enough to host the World Cup several times over. One visitor likened the scene to "a sea of gleaming plasma TV screens." At 53 megawatts, providing enough electricity to power 20,000 homes, it is at the time of writing the largest thin-film plant in the world, the third-largest PV plant overall.

In 2008 OptiSolar, a previously unknown Hayward, California–based maker of amorphous silicon cells, announced a contract with PG&E for the purchase of power. It would come from a giant plant the firm planned to build on nine square miles at Topaz Solar Farm, a sparsely populated swath of land in San Louis Obispo County.* At 550 megawatts the plant would be, by an order of magnitude, the largest photovoltaic project ever built. Opti-Solar had bitten off more than it could chew. Banks weren't lending, certainly not huge sums to a start-up with no track record. In March 2009 First Solar announced it had paid $400 million to acquire OptiSolar's portfolio of projects. In November the company said that, permits permitting, it hoped to begin construction of Topaz Solar Farm in 2010, to start delivering power in 2011, and to complete the project by 2013. Having completed 33 megawatts of utility-scale solar in the US in 2009, First Solar planned to install between 500 and 700 megawatts in North America in 2011 and around 1.5 gigawatts the following year.

There was more. In September 2009 a delegation of senior Chinese business leaders visited First Solar's headquarters in Tempe. The visit was swiftly followed by the announcement that First Solar would build a gargantuan 2-gigawatt PV plant at Ordos, an Inner Mongolian city in China's far north. Ordos is sometimes known as "the empty city," on account of its lavish government projects—fancy apartment blocks and abundant infrastructure—and lack of residents. The plant would be built in four stages, beginning in June 2010 with a 30-megawatt pilot stage, on a piece of land roughly the size of Manhattan. Once completed in 2019 it would produce enough electricity to power 3 million homes. However, the deal was contingent on China enacting a national feed-in tariff that would establish a stable purchase price for solar-generated power. "We're missing a key area: what the price per kilowatt-hour will be," First Solar president Bruce Sohn told a reporter.

As the decade ended, at least sixty other makers of thin-film PV—including sixteen amorphous silicon specialists and no fewer than 22 CIGS start-ups—were attempting to follow the trail that First Solar had blazed. They included Abound Solar, a cad-tell specialist that in July 2010 scored $400 million in government loan guarantees to expand its manufacturing

---

*The site was located on the Carrisa Plains. Few now remembered that this was the location of the ill-fated ARCO Solar plant built back in 1985. The 6-megawatt facility was dismantled in 1995.

turing plant in Loveland, Colorado. But the conventional technology of silicon solar cells still ruled the roost, with a share of around 80 percent. As we shall see in the next two chapters, manufacturers of silicon PV like SunPower and Suntech were also making massive expansions. They would give First Solar a run for its money.

In addition to increased competition, another dark cloud was also looming for First Solar. Silicon is the second most abundant element, making up just over a quarter of the Earth's crust. Tellurium, by contrast, is one of the rarest elements. The silvery-white metal is typically obtained as a byproduct of copper mining. In 2009 First Solar was reportedly already consuming more than 10 percent of the world's supply of tellurium, amounting to just a few hundred tons. Some analysts doubted that the supply could increase fast enough to meet the surging demand.

27

EFFICIENCY, EFFICIENCY, EFFICIENCY

*I*n October 2009, on a former cattle pasture outside the little town (population 6,600) of Arcadia, Florida, Barack Obama commissioned the largest photovoltaic power plant in the US. "It's about time," the shirt-sleeved president said. In fact, for solar-starved residents of the self-styled Sunshine State, it was long overdue.

Since taking over from Jeb Bush in early 2007, Florida's Republican governor Charlie Crist had been doing everything in his power "to protect God's work," as he put it with characteristic piety. Crist looked to Arnold Schwarzenegger as his role model. He had been profoundly influenced by a meeting shortly after being sworn in with his Californian counterpart's environmental point man, Terry Tamminen. Crist had taken the symbolic step of installing solar panels on the roof of the governor's mansion in Tallahassee. "Producing solar energy in the Sunshine State just makes sense," he told the crowd at a global climate change summit he hosted in Miami in mid-2007. With Schwarzenegger and officials from Germany looking on approvingly, the governor signed an executive order calling for Florida to raise the percentage of the energy it derived from renewables from 2 to 20 percent by 2020.

It was not going to be easy. The state's utilities had long claimed that Florida's skies were too cloudy for solar power to be cost-effective. Florida Solar Energy System Incentives, a small-scale state-run rebate program, was popular but forever running out of money. The program was set to expire in 2010. The Sunshine Energy Program, set up by the state's largest utility, Florida Power & Light, charged volunteers $9.75 a month to help "make the construction of solar plants a reality." By June 2008 the program had raked in $11.4 million from 38,929 customers without completing a single large-scale solar project. According to a highly critical report the state's Public Services Commission issued that month, the money had mostly been spent on marketing and administration.

Days later the utility, which the previous year had been refused permission for a huge coal-fired power station next to the iconic Everglades, announced plans to build three new solar plants. They would make Florida number two in the nation for solar energy, boasted FPL chief executive Lewis Hay. In addition to the 25-megawatt plant at Arcadia, the others were a 75-megawatt solar thermal trough plant, and a smaller, 10-megawatt photovoltaic installation at the Kennedy Space Center in Orlando. FPL had decided to go ahead with the plants only after the state legislature had approved a one-off increase in rates that would enable the utility to recover all its costs. Meanwhile, the lawmakers voted down Crist's 20 percent renewable portfolio standard.

The $152 million, 180-acre Arcadia plant consisted of 90,500 black panels. They were mounted on trackers that tilted to follow the sun, "like beachgoers angling for the perfect tanning position," as one observer eloquently put it. The panels were supplied by SunPower, a Silicon Valley–based maker of high-efficiency crystalline silicon solar cells. Arcadia would not be the largest PV plant in the US for long. SunPower had plans to build others that would be ten times larger.

Like First Solar, SunPower's origins dated back a quarter of a century. The company was founded in 1985 by Richard Swanson, then an assistant professor of electrical engineering at Stanford. Swanson had been drawn to Stanford from his native Ohio to study his first love, semiconductors and solid-state physics. His teachers there included Bill Shockley, co-inventor of the transistor. But by the time Swanson graduated in 1969 and moved on to his PhD, the pioneering work had mostly been done. The semiconductor industry was, it seemed to him, maturing. "I was primed to look for something else."

What that something else was became obvious in the wake of the oil crisis of 1973. "That just kind of set the hook," Swanson said. From then on he was "bound and determined to work on PV." It seemed an ideal choice. Swanson was an engineer; here was something where the science was known. You could build a solar cell, the problem was cost. "Solar cells were vastly too expensive at that time to consider for replacing oil." Swanson came up with a concept that would stay with him throughout his career: for solar to be practical, it had to be efficient at converting sunlight to electricity. He embarked on a quest to find ways to improve the efficiency of PV.

In those early days the conventional wisdom held that photovoltaics would be implemented in the form of big solar farms in the desert. The technology that Swanson developed with his students at Stanford targeted utilities. Back then the prevailing view was that wafered silicon would last for maybe another five years before being overtaken by cheaper alternatives. The field split into two camps. The more popular was thin films. The other, much smaller, camp was concentrators.

The attraction of concentrators was that, instead of making the cell thinner, you made it smaller and used a lens or a mirror to cram sunlight onto it, thus increasing the output power. It was a bit like a dairy farmer feeding one cow a hundred times to produce as much milk as a hundred cows. Swanson, with his background in transistors and integrated circuits, knew how to process silicon. He felt more comfortable going the concentrator route. In 1985, frustrated at the glacial pace of change in academia and aware that it was more or less de rigueur at Stanford for professors to start their own firms, Swanson founded a company. Eos Electric Power was based—where else?—in Sunnyvale, part of the Silicon Valley conurbation. Eos was the Greek goddess of dawn who each morning opened the gates of heaven so that the sun god Apollo could sally forth across the sky in his chariot. It was pointed out to Swanson that not many people knew Greek mythology well enough to get the reference. In 1989 the company changed its name to the less poetic but easier to understand SunPower.

As with all start-ups, finding funding was tough. Most venture capitalists think in terms of a five-year horizon. When Swanson was attempting to raise money in 1988, the only way he could make the opportunity look attractive was to extend the financials to 2000, twelve years hence, by which time the company would supposedly be profitable. One seasoned industry veteran told Swanson he had never seen such a long-range business plan. "Do you know how old I'll be in 2000?" the bemused VC demanded. After approaching more than forty venture capital firms, Swanson managed to scrape together enough seed money for the company to continue doing R&D.

Research focused on improving the conversion efficiency of cells. Theory predicted that the maximum efficiency for silicon cells was 28 percent. In practice, the best cells could do was just half that. Why was the performance so poor? Swanson discovered that the nub of the problem was the area where the metal contacts used to extract current from the cell meet the silicon. The bigger this area, the lower the performance. Swanson figured out a way to shrink the contact area to just a point. SunPower would subsequently come up with another innovation, putting the contacts on the back of the cell instead of the front, thus increasing the surface area that sunlight could impact. These improvements were, an admiring Bill Yerkes thought, "the biggest changes since the beginning of the solar cell business."

High-performance concentrator cells were intended for utility-scale applications. That market would take almost twenty years to emerge. Meantime, SunPower was left wandering in the woods attempting to sell products for which there were no buyers. By the early nineties the company was running low on cash. An unexpected commission from Honda saved the day. The Japanese automaker was scouring the world for high-performance cells to power a car the company planned to enter in the 1993 World Solar Challenge. This was a race whose course ran three thousand kilometers across the Australian outback, north to south from Darwin to Adelaide.

SunPower had made a few prototype nonconcentrator cells. To cover the top of a solar racing car would require several thousand such cells. Honda was willing to pay handsomely: SunPower had little choice but to accept the Japanese offer. But with the ink on the contract barely dry Swanson realized, with a sinking feeling, that "we had no clue how to do this." He and his founding team were all former academics; none of them had any experience in mass production. While pondering the problem Swanson happened to bump into an old Stanford friend in a local bakery.

T. J. Rodgers is CEO of Cypress Semiconductor, a Silicon Valley chipmaker he founded in 1982. "We just got this contract from Honda," Swanson told Rodgers, "but we don't know what we're doing." "I've got just the man for you," Rodgers replied, "I know he's available because I just fired him." Mark Allen had a reputation for starting up production lines. "Mark was decompressing from having worked for Cypress for quite a while," Swanson said. "He was taking the summer off, he thought that this would be a fun way to spend his summer, helping us make these solar cells for Honda." In the early nineties the semiconductor industry was undergoing one of its periodic busts. Cypress had just cut one of its shifts. Allen cherry-picked the best people from the shift, including the production manager. "Those guys came over," Swanson told me, "and in the space of a month they converted our sleepy little R&D operation into this beehive of activity running 24/7."

In November 1993 the Honda Dream powered by SunPower cells won the World Solar Challenge in four days two hours, smashing the race record by a full day. SunPower went on to supply cells for Helios, a solar plane developed by NASA. In August 2001 Helios reached an altitude of 96,836 feet, a world record for a nonrocket-powered craft. These record-breaking feats provided excellent publicity, but they didn't bring much money in the door. At least SunPower had proved it could make high-performance cells. Now what Swanson and his team needed to do was strip out some of the complexity and develop standard versions. In March 2000 the dot-com bubble burst. It was not a good time to be chasing new funding.

SunPower would likely have gone bankrupt had not the company been saved, for a second time, by T. J. Rodgers. Hearing that his old friend's outfit was in trouble, Rodgers approached the board at Cypress with a proposal to rescue SunPower. Unfortunately the semiconductor industry was then sliding into one of the worst slumps in its history. "We were getting our ass kicked," Rodgers recalled. Nonetheless, here he was, suggesting that Cypress should invest in, of all things, a solar power company, one that moreover was run by one of his grad school buddies. His board turned him down. So Rodgers wrote a personal check for $750,000 to keep the struggling firm alive.

Thurman John "T. J." Rodgers is one of Silicon Valley's more colorful, larger-than-life characters. An outspoken advocate of free-market capitalism, it seemed odd to some observers that Rodgers should interest himself in a business that throughout its history had depended for its very existence on government subsidies. He was certainly no environmentalist. On one occasion, asked whether he'd gone green, Rodgers pulled a fistful of dollar bills from his wallet. "That's the only green I believe in, baby," he retorted.

It so happened that Rodgers already knew a thing or two about solar. In the immediate aftermath of California's energy crisis, seeking to reduce his company's electricity bills, he had installed a 335-kilowatt solar array on the roof of Cypress's new headquarters in San Jose. The company thus became one of the first in Silicon Valley to adopt renewable energy. "The project was entirely justifiable on economic merits alone," Rodgers insisted. "Commercially sized solar PV is a smart long-term investment, especially during tough economic times." Cypress paid more than $2 million for its system or, as Rodgers put it pithily, "we invested a bunch of bucks to get a bunch of watts." (The company got a bunch of its bucks back, in the form of rebates from the California Public Utilities Commission's Self-Generation Incentive Program.)

The system Cypress bought used conventional silicon solar cells that were 14 percent efficient. That meant they could supply only about a third of the power the building needed. Now here was his buddy Swanson with a design for cells that would be 21 percent efficient, or 50 percent better. Rodgers was canny enough to recognize that if such cells could be made they would give SunPower a huge competitive advantage. He was also humble enough to recognize Swanson's superior talent. "There are few people in my life I have run into who are clearly smarter than me," he said. "At Stanford, there were two: Shockley and Swanson."

Rodgers is a keen competitor. Once he sets his mind to something, he will not give up. Since 1996 Rodgers has poured over $25 million of his own money in what some see as a quixotic attempt—in defiance of climate and geology—to produce the world's best burgundy-style wine, from vineyards he planted in the foothills of the Santa Cruz mountains. Today bottles of his Clos de la Tech pinot noir from Domain du Docteur Rodgers, each proudly bearing around its neck a sealing-wax encased Cypress microchip, sell for upward of $100.

Rodgers would be equally tenacious in his support for solar cells. It took him fifteen months, but eventually he managed to nag his board into backing SunPower. Cypress would ultimately invest $143 million in equipping factories to produce solar cells and modules. By 2004 Cypress had acquired an 85 percent stake in the company. Rodgers himself became SunPower's chairman. His experience, intuition, and determination would prove crucial. "T. J. was the only person that really got it," a grateful Swanson recalled. "He just looked at [our cells] and said, This is gonna happen."

Rodgers's basic insight was that solar cells were semiconductors just like chips. As such they were subject to classic learning curve behavior. In addition to cash, Cypress also provided crucial manufacturing expertise. "Basically what T. J. said was, We've spent the last twenty years honing our manufacturing skills in this intensely competitive global integrated circuit market. You guys could learn it, but it would take you ten years, and you don't have ten years. So what we're gonna do is put your pilot line in our factory in Round Rock, Texas, and you're gonna get inoculated with manufacturing culture."

To be sure, there were some differences between the macroworld of photovoltaics and the microworld of integrated circuits. "At the first meetings we had with Cypress engineers after the merger," Swanson recalled, "they'd ask things like, What's the line width of your circuits? We'd tell them and they'd laugh and say, We don't even know why you're talking with us, this is so easy, we can design your process by lunchtime. Then they'd say, How many wafers are you planning to start? We'd say, We think our first factory's gonna have about a hundred thousand wafers per day. They'd say, You mean a hundred thousand wafers a year, right? No, per day. Then they'd go, Uh-oh, this is gonna be a little harder than we thought."

When Cypress embraced SunPower, PV was still too expensive to realize the old dream of big solar farms in the desert, at least not in the short term. In the interim, a new market driver had emerged in the shape of the German and Japanese government programs. These were based on the idea that solar was of most value on distributed grid-connected rooftops. This market was beginning to explode. In 2001 the entire world annual demand for PV was around 250 megawatts. By 2008 SunPower on its own would be producing that much. Like every other maker of solar cells, SunPower began by shipping most of its products to Germany. For all his dislike of government subsidies, Rodgers had no problem with feed-in tariffs. "I'm real happy to take money from the German government," he said, voicing the common misunderstanding that the German government pays for the feed-in tariff program. (In fact, as we saw in part 3, the incentives are paid by German utilities, who recoup the money from their customers.) At the same time Rodgers conceded that Germany's feed-in tariff program deserved credit for accelerating the acceptance of solar by "something like ten years."

Rodgers recognized early on that if SunPower was to stay competitive with Asian rivals, the company would have to shift its production overseas in order to gain access to low-cost labor. In 2004 SunPower bought a former disk drive factory near Manila and converted it to a plant capable of producing 32 million wafers a year. By 2009 the company was employing five thousand people—up from just fifty in 2003—three-quarters of them in the Philippines. As ever Arnold Schwarzenegger could be counted on to boost anything to do with solar. On a trade mission to China in November 2005, while delivering a speech (written by Terry Tamminen) at Tsinghua University in Beijing, Schwarzenegger pulled a SunPower solar cell from his pocket and held it up. "This is the future, ladies and gentlemen," the governor proclaimed, "the most efficient solar [cell] in the world." SunPower went public later that week. Cypress would be richly rewarded for allowing Rodgers to back his hunch. It would cash in $677 million of SunPower stock and distribute equity worth $2.52 billion to its shareholders.

In 2005 SunPower began shipping products to California. The following year the company reported its first-ever profit. In January 2007 SunPower paid $265 million to acquire PowerLight. It thus gained entrée to the major league: the design, manufacture, and installation of large-scale solar systems. Founded in a garage in Berkeley in 1995, PowerLight rose to preeminence on the basis of PowerGuard, a lightweight solar module that could be installed without having to drill holes in the roof. From the outset company founder Tom Dinwoodie understood that the way to achieve economies of scale was to target commercial rather than residential roofs. PowerLight's first contracts included the 675-kilowatt system on top of San Francisco's Moscone Center and the 335-kilowatt system at Cypress's headquarters in San Jose. PowerLight's big break came in 2004, when the company won a contract to design and build Bavaria Solarpark in Germany. Construction of what was then, at 10 megawatts, the world's largest PV installation was completed in 2005. The $59 million park consists of three separate fields in which 57,600 panels are mounted on single-axis trackers that PowerLight designed.

Solarpark was an enormously significant milestone. It represented incontrovertible proof that solar technology could be applied to generate electricity on a large scale. "Photovoltaic energy is the most intelligent energy technology ever," a triumphant Hermann Scheer asserted at the park's dedication on June 30, 2005. "It establishes a new era of environmentally friendly, sustainable, independent, and cost-effective power for everybody." PowerLight went on to implement even larger projects in Spain and elsewhere in Europe. Prior to 2006 the company had sourced most of its cells from the Japanese firm, Sharp. Following the acquisition SunPower would take over as its prime supplier.

Now, at long last, it was time to demonstrate that the US was also ready to step up to the implementation of large-scale PV. SunPower established a new business unit to focus on direct sales to US utilities. To head the unit the company recruited Ed Smeloff, formerly of the Sacramento Municipal Utility District, the San Francisco Public Utilities Commission, and latterly Sharp. In November 2008, I meet Smeloff in Richmond, across the bay from

San Francisco, at the facility where SunPower assembles its trackers. It is an impressive site, an elongated brick-facade factory that was built in 1931 to house an assembly line for Model A Fords. During World War II tanks rolled off the line straight onto waiting ships. The factory now forms part of the Rosie the Riveter National Park. To me it seemed a potent symbol of the shift from the old, dirty fossil-fuel paradigm (Richmond is blighted with a refinery and associated oil tanks) to the new, clean green one.

I ask Smeloff to explain what has triggered the utilities' sudden interest in solar. He begins by pointing out that, in October 2008, the Bush administration extended the investment tax credit on renewable energy for eight years, simultaneously eliminating a prohibition on utilities obtaining the credit. "Leadership at a handful of utilities—Florida Power & Light, Southern California Edison, PG&E—saw this and said, Wow, this is an opportunity to earn large profits! If we can get our regulators to agree to rate-base solar systems, we can capture the investment tax credit, and have growth earnings above what they were historically."

The first fruit of the SunPower–PowerLight partnership would be the construction of what was at the time the largest PV plant in the US. This is a 14-megawatt system at Nellis Air Force Base—"home of the fighter pilot"—on a disused landfill outside Las Vegas. Completed in December 2007 the plant saves the air force, the federal government's largest consumer of energy, nearly $1 million a year. In May 2009 Barack Obama visited Nellis. "This base serves as a shining example of what's possible when we harness the power of clean, renewable energy to build a new, firmer foundation for economic growth," the president declared.

SunPower differentiated itself from makers of generic crystalline silicon solar by the superior efficiency of its cells. Efficiency turned out to be even more valuable than Swanson had imagined twenty years ago when he was trying to sell the idea to potential investors. Obviously, higher efficiency means that fewer panels are needed for a given application. It turns out that most applications for PV are constrained by area. "This got confirmed in spades when we started working with PowerLight," Swanson told me. "They just loved our panels because it meant they could put more power on a roof."

High efficiency also worked well in a residential context. "If you've ever installed a system on a house," Swanson said, "you realize that there are dormers, trees, chimneys. You've got to put that array in a sweet spot where it gets good sun, but you can only use a relatively small fraction of the roof. We can go in and offer the homeowner enough energy that they can knock their electricity bill down to zero. You couldn't do that with a conventional silicon panel, and with a thin-film panel, I mean, why bother?"

"It's a crystalline world," commented leading PV industry analyst Paula Mints, "and it's because of efficiency, efficiency, efficiency."

SunPower kept pushing the envelope. Its first-generation cells were 21 percent efficient; the second generation, which went into mass production in 2007, 22 percent. In May 2008 the company announced a prototype third-generation cell with an efficiency of 23.4 percent. Expected to be commercially available in 2010, the cell was a key component of the company's plan to halve the cost of a solar system by 2012.* That, T. J. Rodgers believed, would bring SunPower to price parity with PG&E. "Then it's like, we're cheaper, we don't need a subsidy, thank you." Solar, Rodgers reckoned, would become "the biggest market that Silicon Valley has ever seen."

In August 2008 SunPower announced what was its most ambitious project by far. The California Valley Solar Ranch would be a 250-megawatt PV solar plant built in San Luis Obispo County on private land that had not been commercially farmed in over thirty years. PG&E agreed to purchase all the power from the plant. Construction was due to start in 2010, subject to issuance of the appropriate permits. When completed in 2012 the plant would generate enough power for around 90,000 homes. "The California Valley Solar Ranch will generate pollution-free solar power that is competitive with new conventional peak power," announced Howard Wenger, now president of SunPower's utilities and power plants business group.

In March 2010 SunPower won a dramatic victory over archrival First Solar. As we saw in the previous chapter, Southern California Edison had begun installing megawatt-scale PV systems on the roofs of giant warehouses located in Inland Empire east of Los Angeles. In the project's demonstration phase, the utility elected to use thin-film panels made by First Solar. Now Edison announced it was switching to SunPower to supply 200 megawatts—the lion's share—of solar panels to be installed on the roofs of the fifty-odd warehouses it was planning to lease over the next five years.

---

*In June 2010 SunPower announced that it had produced a full-scale solar cell capable of a sunlight-to-electricity conversion efficiency of 24.2 percent, a new world record.

Although SunPower's solar roof tiles were priced higher than rival products, they offered "a significant technological advantage," said Mark Nelson, SCE's director of generation planning and strategy, which made them competitive for large-scale rooftop projects. That advantage was efficiency. The SunPower product was selected because it would allow SCE to produce more power per installation. In early 2010 the company announced that, like First Solar, it would soon muster 1 gigawatt of solar cell production capacity.

SunPower had demonstrated it was possible to succeed at the high-performance end of the PV market. At the low-price end, First Solar dominated. "The only place I wouldn't want to be in this whole thing right now is making generic solar cells," Swanson said, "that's going to be a squeeze." Many of the firms in the generic silicon solar business were Chinese. By far the largest of them was Suntech Power. As we shall see in the next chapter, it too had grand designs for the US utility market.

28

ACCIDENTAL EMPEROR

*I*n 1988 a boyish-looking twenty-five-year-old Chinese named Zhen-grong Shi was selected to pursue post-graduate studies abroad. Expecting to be posted to the US, Shi had been brushing up his English, even working on an American accent. At the last minute his supervisor at the Shanghai Institute of Optics, where Shi had recently obtained a master's degree in laser physics, told him there had been a change of plan. All the slots for the US had been filled; instead, Shi was going to be sent to study electrical engineering in Australia. Shi was stunned. "I didn't even know where Australia was," he confessed.

Like many young graduates, Shi had no clear idea of what he wanted to do. He also lacked confidence in his abilities as a scientist. After a year at the University of New South Wales in Sydney, Australia's top engineering school, he still had not found a vocation. His fellowship had ended, but Shi did not want to return to China. He applied for work as a research assistant in the electrical engineering department but arrived late for the interview. A friend suggested he should go see a professor called Martin Green. "He dropped into my office one afternoon, I think he was looking for a job," Green recalled. "I talked him into enrolling for a PhD with me doing solar cell research." What Shi did not realize was that Green was one of the world's leading photovoltaics researchers. Agreeing to join Green's group was an auspicious decision. It would ultimately lead to Shi becoming one of the wealthiest men in China, the world's first solar billionaire.

Like his peer Richard Swanson of SunPower, Martin Green trained in microelectronics in the late sixties. The young Australian soon became disillusioned with what he saw as the relatively trivial applications in this field. "Making better television sets didn't sound like work for a grown-up person," he said. When photovoltaics became a hot topic following the first oil crisis, Green was able to parlay his expertise in semiconductors into research that seemed to him more meaningful.

In addition to being a first-rate scientist, Green also had the common touch. "What always fascinated me about Martin," his former colleague David Jordan told me, "was his easy familiarity with advanced semiconductor theory and concepts that very few people could comprehend overlaid with a completely down-to-earth approach to the day-to-day world that I and most of the staff at his PV center inhabited."

Green's group was one of the few that managed to survive during the dark days that began during the Reagan era, when research on solar virtually ground to a halt. Maintaining continuity was one of the secrets of their success. "We know stuff that people entering the field now have never even heard of," Green told me. In 1983 Green and his group at UNSW made silicon solar cells that were 18 percent efficient, breaking the world record set by COMSAT. The Australians have kept their place in the record books ever since. In 2008 the group moved the bar to a new high, announcing silicon solar cells that were 25 percent efficient.

The friendly rivalry between Green's group at UNSW and Swanson's at SunPower peaked over the World Solar Challenge race, which is held in Australia every three years. In 1993, as we have seen, Honda chose SunPower cells for its winning entry. Three years later Green had his revenge, when Honda selected UNSW cells. The Japanese won the race for a second time, their car averaging almost ninety kilometers an hour over the three-thousand-kilometer course.

For his pioneering work Green has received many honors, most notably the 2002 Right Livelihood Award, the "Alternative Nobel Prize" that Hermann Scheer had won three years before. (Scheer nominated Green for the prize.) Being based in Australia, a country where powerful coal and mining interests dictate energy policy, sometimes made things difficult for him. Especially when it came to the ultimate responsibility a professor has toward his students: finding them jobs.

When the solar industry surged following the passage of Germany's Renewable Energies Law in 2000, entrepreneurial Chinese led by Zhengrong Shi rushed to establish start-ups to manufacture solar cells and modules. In addition to Shi, former UNSW students would become top executives at five other leading Chinese PV makers—JA Solar, Solarfun, Trina, Sunergy, and Yingli Green Energy. In 2007 the combined market capitalization of these firms was more than $10 billion. The next year China overtook Germany to become the world's largest manufacturer of solar cells. By 2009 Chinese firms accounted for 43 percent of the world's solar panels. Green reckoned that 70 percent of the Chinese output came from companies headed by his graduates.

Zhengrong Shi was born, the younger of twins, in 1963 to a family of poor farmers on Yangzhong, an island in the Yangtze River about 150 miles northwest of Shanghai. It was a time of widespread famine caused by the Great Leap Forward, Mao Zedong's disastrous experiment in forced agricultural collectivization. With two older children to support, Shi's parents were forced to give him up for adoption. Though rural, Yangzhong had a good school system and Shi proved a dedicated student. In 1978, aged just sixteen, he went to university. That year Chinese leader Deng Xiaoping made the strategic decision to send 3,000 students abroad each year for further education. The anti-intellectual Cultural Revolution had devastated China's scientific and engineering community. Deng realized that if China was to recover economically, the community had to be rebuilt. Shi was one of the beneficiaries of this foreign studies program.

At UNSW working under Green on thin-film silicon solar cells, for the first time in his life Shi found himself enjoying doing research. Making breakthroughs at the cutting edge was exciting. Green reckoned that Shi was one of the brighter graduate students he had mentored. The young Chinese had a good work ethic: he was always able to make progress where others had stalled. Shi earned his PhD in record time, in just two and a half years. In addition to authoring scientific papers, during his years in Australia he would receive eleven patents. In 1995 Green appointed Shi deputy director of research at Pacific Solar, a spin-off from his group. Pacific Solar's aim was to commercialize crystalline silicon on glass, a second-generation thin-film solar technology. Gradually, however, Shi became bored. He needed a new challenge.

Shi had maintained links with his homeland. During the nineties he and Green had visited Chinese manufacturers with the idea of persuading them to license UNSW's technology. But China's solar cell makers lacked initiative: they were not interested in doing anything new and different. In 2000 Shi was approached by local government representatives from Wuxi, an upcoming industrial hub seventy miles west of Shanghai, not far from his birthplace. They offered him $6 million to set up a factory to make conventional silicon solar cells. Initially Shi was skeptical. Corruption in China was rife. He was unsure that such a venture could succeed. He and his wife had taken out Australian citizenship, bought a house in Sydney, had two Australian-born sons. They had no plans to abandon their well-established, comfortable life and return to China. A two-week trip to the mainland changed his mind. The infrastructure was improving rapidly. The official attitude had turned positive.

China's leaders had realized that there was a flaw in their clever strategy of sending the nation's best and brightest overseas to study. Having completed their education, the scholars were supposed to come back and do their bit for the motherland. As of 1997, however, less than a third of the 293,000 students who had gone to the West had returned. To reverse this brain-drain, Beijing introduced new incentives to lure back its world-class talent. Entrepreneurs in particular were encouraged by offers of cheap land at science parks. There, local officials would expedite paperwork and limit regulatory constraints. For the exiles there was also an intrinsic appeal to the idea of setting up shop at home. Working for Western firms, hindered among other things by their poor English, Chinese engineers could only go so far. As an old Chinese saying has it, "Better to be the head of a chicken than the tail of an ox." Meaning that even if you only have a small business, at least you are in charge.

People from Wuxi are known for their shrewdness. "We want sons like you to come back and be bosses here," the investment committee told Shi. He accepted their offer. After fourteen years in Australia it was time to go home. His country needed him. "I realized that if I came back I could really do something, because I had so much more advanced experience and knowledge than people here," he said. In 2001 Shi started up Suntech Power with twenty employees. He soon discovered he had a talent for business as well as science. In part this was out of necessity: money was tight. In addition to the $6 million in seed funding from the local government, Suntech also managed to garner $5 million in research grants. Shi was able to pick up secondhand equipment cheaply from AstroPower, a bankrupt US solar cell manufacturer. His workforce of technical college graduates cost him just $250 a month each. Even so, at one point in the early years he could not afford to pay his employees. But Shi showed guts and an unrelenting determination. "Zhengrong is a hugely talented guy," said David Hogg, Shi's former boss at

Pacific Solar. "The fact that he's been successful doesn't surprise me one little bit, the guy's completely driven."

In September 2002 Suntech began production on a line that had a capacity of 10 megawatts, four times larger than China's entire output of PV the previous year. By 2003 Shi was selling panels at $2.80 a watt, well below his competitors' prices, while maintaining a profit margin of 20 percent. Having demonstrated that the PV business could be profitable, Shi wanted to expand rapidly, building huge new factories to capitalize on the apparently insatiable demand. His chairman baulked. Realizing that a company could only have one boss, in 2005 Shi borrowed $100 million from US investors, including Goldman Sachs, to buy out his local shareholders. That December Suntech floated on the New York Stock Exchange, the first private Chinese company to do so, raising $400 million and making Shi, at least on paper, a billionaire.

Following the enactment of the revised solar feed-in tariff in 2004, the German market took off. When supply from local firms like Q-Cells and SolarWorld could not meet the surge in demand, German photovoltaic installers turned to China. By 2006 Suntech was exporting more than 90 percent of its products. Eighty percent of sales went to Europe, primarily Germany. China leapt to fourth place in solar manufacture, with Suntech accounting for more than half of Chinese output. The following year, the company's workforce tripled to 3,500 at four giant factories. Suntech passed Sharp as the world's number one producer of silicon solar modules, second overall only to First Solar.

In 2008 China overtook Germany as the world's largest manufacturer of solar cells. The Chinese share of the market leapt to 30 percent, up from 15 percent in 2006. In early 2009 Suntech announced that its production capacity had reached 1 gigawatt. The company replaced Q-Cells as the world's largest producer of silicon solar cells. Then the global financial crisis hit. A massive oversupply of panels caused prices to plunge. Suntech was forced to throttle back production and to cut its workforce by 10 percent. The big question was, could China's domestic market expand to pick up the slack?

The Chinese were in one sense already the world's largest consumers of solar energy. China is the biggest producer of solar water heaters, accounting for around 60 percent of solar hot water capacity. An estimated 30 million Chinese households mount solar water heaters on their roofs.

In January 2006 the Chinese government passed a law mandating that 15 percent of China's energy would be sourced from renewable sources by 2020. The country's hydroelectric program had culminated in the giant Three Gorges Dam. That meant further increases in capacity would have to come mostly from wind and solar. In 2008 China installed about 50 megawatts of PV, twice as much as the previous year, but still a relatively tiny amount. But with environmental disaster looming, the Chinese government began to move away from smog-causing coal toward renewable sources. In China, once the government makes a decision, things can happen quickly. New targets were announced. The percentage of renewables was increased from 15 to 20 percent by 2020. The amount of that electricity to be sourced from solar was revised upward from 1.8 gigawatts to at least 10 gigawatts, double that amount according to some reports.

Frequent visitors to China, like the CEO of Silicon Valley–based solar start-up Innovalight, Conrad Burke, were awestruck by what they saw there. "The sheer scale of some of these solar factories is quite scary, just the amount of investment that the Chinese have made in the gigawatt level, and that they can scale even further if they put their minds to it," Burke told me. "It's shockingly fast."

In June 2009 local government in Jiangsu province, home to more than 160 Chinese PV companies including Suntech, announced a German-style cost-covering feed-in tariff. The following month the central government in Beijing launched its Golden Sun Program. This would fund the construction of 294 solar projects with a combined capacity of 642 megawatts. The projects, most of which would be installed at remote industrial and commercial sites, were expected to be completed within three years. China's ministry of finance would underwrite between 50 and 70 percent of the $3 billion program cost. Suntech alone submitted applications for Golden Sun projects totaling 179 megawatts. In July 2007 the company announced that it had signed letters of intent to build 1.8 gigawatts of solar projects for provincial and city governments in China. However, these were contingent on Beijing establishing a national feed-in tariff scheme for utility-scale solar plants.

When the California Solar Initiative got under way in 2007, just 2 percent of the modules used were Chinese. By the fourth quarter of 2009, the ratio had risen to 46 percent. In the first half of 2010 the state-controlled China Development Bank extended a whopping $24 billion in loans to Chinese manufacturers, including the "big three," Suntech, Trina, and Yingli Green Energy, to build new factories. The loans were enough to increase world solar cell and module capacity by 100 percent, according to New Energy Finance, a London-based offshoot of Bloomberg. One observer predicted that two-thirds of all new growth in solar production would come from Chinese manufacturers. Analysts speculated as to whether in the coming decade the world's largest market for solar power would be the US or China. Suntech, it was clear, planned to play a leading role in both.

In October 2008 Suntech purchased EI Solutions, a Pasadena-based solar integrator and installer. The company also teamed up with MMA Renewable Ventures of San Francisco, a specialist in large-scale solar project finance, to form a joint venture called Gemini Solar Development. In March 2009 Gemini announced that it would build a 30-megawatt solar power plant for Austin Energy, a municipal utility based in the Texan capital. This would address a shortage of daytime peak power that Austin Energy faced. According to Shi, Suntech was involved in bids with Gemini and other developers for over 2 gigawatts of projects. In November 2009 the company announced that it would begin assembling modules at a plant in Phoenix, becoming the first Chinese company to export jobs to the US. It was, said Shi, "the first step in a long-term strategic investment in the North American market." Already in the crucial California market, Suntech claimed to have won a share of 25 percent, second only to rival SunPower.

Though the Americans still held the advantage in terms of efficiency, the Chinese were beginning to nip at their heels. This was in large part thanks to ongoing intellectual input from Australia. In late 2009 Suntech began shipping modules containing its new high-performance Pluto cells. The single crystalline versions were 19 percent efficient, the cheaper polycrystalline versions almost 16 percent, a world record for such cells. The underlying technology was licensed from the University of New South Wales. The relationship extended well beyond licensing. UNSW provided training for Suntech staff while Suntech gave students from the university hands-on access to its facilities. Most significantly, Stuart Wenham, Martin Green's colleague and fellow UNSW professor, served as Suntech's chief technology officer. Reflecting the closeness of the connection, Shi flew an Australian flag outside his headquarters in Wuxi.

These days when he came to New York, Shi would lunch, he told people, with the likes of Henry Kissinger. What the architect of the open China policy and arguably the greatest beneficiary of that policy discussed over their meal is not recorded. No doubt they would have touched on long-term strategy, on how the US and China would have to work together. Certainly Shi had grand ambitions for the future. "In twenty years' time," he said, "we are thinking of Suntech not just as a producer of solar panels, but as an energy company, like BP or Shell."

What would the solar industry look like in twenty years? With residential, commercial, and utility-scale markets all over the world continuing to grow rapidly, the change would likely be dramatic. "The solar industry is very dynamic," Shi said. "We're very young, far from mature yet." Indeed, as we shall see in the final chapter, PV still had plenty of room for improvement.

NANO SOLAR

*F*or photovoltaics Germany's Renewable Energies Law had been the Big Bang. In the ten years since the Bundestag passed its epoch-making act in 2000, the solar industry had blossomed from a tiny niche, expanding at a compound annual rate of 40 percent. If that rate were to continue for the next twenty years, then solar would take over *all* electricity generation. That was not going to happen, but there could be little doubt that by 2030, photovoltaics would account for a significant portion of the world's energy mix, up from a fraction of 1 percent today.

Almost all the signs were positive. By the end of 2009 the young solar industry had installed a cumulative total of 20 gigawatts of PV, the equivalent of forty medium-size coal-fired power stations. A quarter of that wattage had been installed in the previous twelve months. Though buffeted by the global financial crisis, the resilient solar industry had quickly bounced back. It was on track to resume its growth trajectory, with 4 gigawatts of panels slated for shipment in the second quarter of 2010 alone. The following year it expected to pass the 10-gigawatt milestone. Hundreds of new projects of all shapes and sizes filled the pipeline. Prices were falling by 5 or 6 percent a year. In some places—like Hawaii—solar electricity from commercial installations had already achieved parity with conventional grid power.

Europe continued to dominate the solar market, with a share of around 75 percent. Germany still towered head and shoulders above the rest, but promising new markets were also opening up. They included Italy, an ideal prospect because of its combination of high electricity prices and lots of sun. In 2009 the Italians installed over half a gigawatt of photovoltaic capacity. In March 2010 SunEdison, a US firm, announced plans to build near Venice what would be, at 72 megawatts, the world's largest solar farm. Pundits predicted that the Italian solar market would be the first not to need subsidies to break even.

Though still relatively slow, the pace of growth of solar in the US was accelerating. In 2010 US installations were expected to top 1 gigawatt for the first time. Some analysts predicted that in a few years the US would become the world's largest market for solar. Spurred by its new feed-in tariff, Japan was picking up steam. Though yet to transform rhetoric into reality, China with its horrendous environmental problems and lack of conventional infrastructure was potentially the biggest market of all.

In the space of a single decade the dynamic young solar industry had achieved significant scale. The industry mustered at least five firms that boasted an annual production capacity of a gigawatt plus. In addition to well-established names like Sharp and Sanyo, the roster included a slew of aggressive young start-ups, like Suntech and SolarWorld, which had shot from out of nowhere to preeminence. The solar industry was also profitable for the first time in its history, with revenues of $40 billion in 2009. It had generated tens of thousands of jobs and created enormous wealth for its shareholders. Healthy revenues meant that firms had money available for research and development. By 2008 a single company, SunPower, was spending more on R&D on PV than the US Department of Energy. As a result the rate of technological development was accelerating.

Every aspect of the manufacturing process was being optimized. The larger the volume of production, the more raw material costs predominate. Efforts therefore focused on reducing those costs. At the bulk level REC Silicon, a Norwegian renewable energy firm, had built a new $1.7 billion plant at Moses Lake, Washington. Its aim was to strip 40 percent out of the cost of producing silicon. The firm had abandoned the traditional, stop-start batch method in favor of a much more energy-efficient, continuous method. Meanwhile, start-ups like Hillsboro, Oregon–based Solaicx were developing innovative lower-cost ways to grow silicon ingots.

At the cell level First Solar had already proved it was possible to produce PV for less than a dollar per watt. How much cheaper could solar cells get? Could the industry get where it needed to go—generating electricity that was cheaper than that generated by coal—by incremental improvements alone? Or were revolutionary breakthroughs still needed? A good place to find answers to these questions was Silicon Valley. With its unique concentration of talent and an environment that rewarded risk-taking, the high-tech mecca had spawned a slew of start-ups, most of them with names that seemed to include the letters S-O-L. "I believe that the valley will be the place from which the next wave of solar and other alternative energy technologies will emerge," SunPower chairman T. J. Rodgers predicted.

First Solar was a maker of thin-film cells. As they attempted to replicate First Solar's success in cutting costs, a major concern for makers of crystalline silicon cells was how to reduce the quantity of bulk material they needed to make their wafers. Wafers are produced using wire saws, extraordinary pieces of equipment that resemble an egg-slicer. They deploy wires that are hundreds of miles long to produce thousands of wafers at a time. The problem is that by its nature this process is wasteful, grinding almost half the starting silicon into dust. Pure material is hard to recover from the resultant slurry.

Using a combination of thinner wires and better abrasives, SunPower had managed to slim the thickness of its wafers from 300 micrometers down to just 165 micrometers. The result was a huge increase in the number of wafers that could be sliced from the same ingot. Others, like San Jose–based start-up Silicon Genesis, were working on a cleaving process that did away with wires altogether. It produced flexible "foils" of material that were just 20 microns thick, more like thin films than wafers.

In addition to reducing the cost of raw materials, makers also worked on improving the performance of their cells. In particular, of the crystalline silicon cells that still accounted for 80 percent of the solar market. For all the success of cadmium telluride, many observers believed that silicon would remain the material of choice for PV. With huge new plants like REC's coming online there were now few constraints on supply and silicon was, after all, along with steel and cement, the world's most studied material. "Silicon has a reliability record which is unmatched by any other material," said T. J. Rodgers. "God put silicon on this planet for a reason," joked Conrad Burke, CEO of Sunnyvale–based start-up Innovalight. "I'm telling you, silicon will win this war."

Innovalight had developed a new formulation of the material it called "silicon ink." This was a dark brown liquid that resembled coffee. It contained nanoscale particles of silicon. The ink is printed onto conventional silicon wafers by adding an extra step to the production process. The ink layer improves the sunlight-to-electricity conversion efficiency of the resultant cells by at least 1 percent. That may not sound like much, but squeezing out even a small improvement enables makers to charge a higher premium for their products, hence boosts their profits.

Innovalight has demonstrated cells with efficiencies of close to 19 percent. Burke believes that the technology has the potential to go "way over 20 percent." The company's target customers are manufacturers of generic crystalline silicon solar cells with efficiencies of around 15 or 16 percent. Though some are huge firms, they lack a technological edge that would allow them to differentiate their products. Innovalight claims to have sold its silicon ink to ten companies. Thus far three, the leading Chinese firms JA Solar, Yingli Green Energy, and Solarfun, had publicly announced plans to use the liquid to enhance their products. Burke estimated that by the end of 2010 Innovalight would have the capacity to produce a gigawatt of silicon nanocrystals. Performance-enhancing additives like silicon ink were just one way of improving cells. "There's a whole realm of other things in the nano regime where you can multiply the efficiencies," Burke told me.

Such "other things" were the specialty of the Photovoltaics Centre of Excellence at the University of New South Wales in Sydney. There, for over twenty years, as we saw in the previous chapter, Martin Green and his group had been setting the pace in PV research, designing and building the world's highest-performance crystalline silicon solar cells. Wafer-based cells represented the first generation of PV technology; thin-film cells, the second. Now Green saw the need for a third generation that would combine the high performance of the former with the low cost of the latter. To create third-generation cells would require fundamentally different underlying concepts. In 2002 Green challenged the young researchers in his group to go back to the drawing board and reinvent the solar cell.

Kylie Catchpole was one of the youngsters who picked up the gauntlet Green had thrown down. She investigated the strange optical properties of metals. Light striking tiny particles of metal causes them to give off waves of electrons known as "plasmons." The waves ripple outward, scattering the light. Catchpole realized that this property could be exploited to solve the low-efficiency problem that bedevils thin films. Because only a few microns thick, such cells are typically too thin to absorb the full spectrum of sunlight. In particular, the longer infrared wavelengths pass straight through them. Catchpole found that by sprinkling nanoparticles of silver on the surface of a thin-film solar cell, she could deflect the light off the metal so that it bounced back and forth within the cell. The plasmons enhanced the amount of light the cell absorbed by up to 30 percent.

In 2007 Catchpole published a paper on her work. Judging by the number of times it has subsequently been cited by other scientists, the paper caused a big splash. Among the many researchers who followed up on her discovery were colleagues at the California Institute of Technology in Pasadena and the Institute for Molecular and Atomic Physics in Amsterdam. They have since demonstrated a plasmonic solar cell that may be more compatible with standard manufacturing processes. It takes advantage of a new technology called nano-imprinting to fabricate light reflectors on the back of the cell. The concept has commercial potential. "We're at the stage where we have numbers where we can go to companies and say, Hey, we can prove this works, so it should work in your process as well," Albert Polman, who directs plasmonics research in Amsterdam, told me.

One firm, Suntech Power, is sponsoring research on plasmonic solar cells at Swinburne University of Technology in Melbourne. The group is led by Min Gu, another graduate of Green's group, and a friend of Suntech CEO Zhengrong Shi. "Plasmonic solar cells will be twice as efficient as the current generation of cells, and will also cost significantly less to run," Gu said in April 2009. "By working with Suntech in the development phase we can ensure the technology can be transferred to the production line. This should allow us to have cells ready for manufacture within five years." "Plasmonic technology has the potential to take solar cells to the next level," Shi agreed.

Plasmonics was just one of many approaches researchers were taking to the development of third-generation solar cells. At Caltech Harry Atwater, the researcher who coined the name "plasmonics," and his team were also working on silicon nanowires. These were arrays of vertically aligned crystals of silicon. Under the electron microscope, they looked like blades of grass. The arrays were able to absorb light between twenty and fifty times better than ordinary cells. To make such cells took only 1 percent as much material as conventional crystalline cells. Even better, instead of expensive ultra-high-purity material, the nanowires could be made out of cheaper lower-grade "metallurgical" silicon. Grown on a flexible plastic sheet, the arrays could be peeled off and used in a variety of new applications, such as the curved roof of a car. "Integrated photovoltaics" was how Atwater described them.

"For all the hoopla that we've celebrated for the last two or three years in solar, I still think that we haven't really started yet," Conrad Burke told me. "The best is yet to come." There was a good chance, Burke thought, that many as-yet-unknown technologies would emerge. "It's still rather early in the cycle to put your finger on it and say, Here's what it's gonna end up looking like." There could be more than one winner. For solar it was just the beginning.

Though the final outcome was still unclear, there could be no doubting that the process of switching to solar was well under way. The snowball was rolling down the mountainside, gathering momentum and volume as it went. But for all the progress being made in photovoltaics on the technology front, supportive policy was still crucial if the momentum was to be maintained. In Germany, the birthplace of the solar feed-in tariff and still by far the world's largest solar market, significant developments were taking place. The epilogue examines these developments.

EPILOGUE

NO SWITCHING BACK

*O*n September 27, 2009, following what was described as "an exceptionally boring campaign," Germans elected a new, center-right government on a platform of budget cuts. Chancellor Angela Merkel's conservative Christian Democratic Union formed a coalition with the pro-business Free Democrats. The Greens increased their share of the vote. The big loser was the Social Democratic Party. Soon, rumors were rife that the new coalition was planning to slash solar feed-in tariffs by up to 30 percent. "The discussions these days are frightening," a clearly worried Hans-Josef Fell e-mailed me.

Growth in the German solar market had far exceeded the corridor defined by the Renewable Energies Law. The threshold was 1.5 gigawatts; sales in 2009 topped 2.3 gigawatts. This triggered an automatic degression: from January 1, 2010, tariffs would be reduced by 9 percent for rooftops and 11 percent for ground-mounted systems instead of the regular reductions of 8 percent and 10 percent. The problem was that system prices had declined dramatically, by 40 percent or more. Senior politicians within the new government argued that subsidies should be cut proportionately. The PV industry retorted that system prices were not the same as production costs. In the industry's view, a reduction of 5 percent would be more appropriate.

Critics argued that overly generous solar feed-in tariffs were becoming unaffordable. Just for panels installed in 2009 electricity consumers would pay around $18.5 billion over twenty years. Even Greens like Fell conceded that cuts needed to be made. His old comrade-in-arms Hermann Scheer likewise admitted that the tariffs could not continue at their current rate. Subsequent media reports suggested that any reductions would probably be moderate, to prevent damaging the fast-growing solar sector. The tariffs would be tweaked, not completely revamped. Then, in mid-January 2010, environment minister Norbert Roettgen dropped a bombshell, announcing additional cuts of between 16 and 17 percent on new roof sites and 25 percent on open-field solar parks. The revised tariffs would be introduced on April 1.

The announcement of these tougher-than-expected measures drew howls of protest, in particular from the eight-hundred-member German Solar Industry Association. Such severe reductions would cause the industry irreparable harm. Companies would be driven into bankruptcy, thousands of jobs would be lost, production would disappear offshore. "These excessive cuts are threatening one of our country's most important job and economic engines," warned the association's president, Günther Cramer, CEO of SMA Solar Technology. In 2009 the German solar power industry generated sales of more than $10 billion. A total of 60,000 jobs were now directly dependent on photovoltaics. The association held a protest rally outside Merkel's CDU headquarters in Berlin. It set up a website to gather signatures for a petition.

"Such cuts would put too much pressure on the German corporations and the advantage of the Chinese manufacturers (exchange rates, subsidies, low wages, etc.) would be even greater," Fell wrote me. "For this reason we are advocating a smaller reduction in reimbursements." Coming from the Greens this argument was not surprising, but some other participants in the debate took unexpected positions. For example, the Free Democrats had long been implacably opposed to the Renewable Energies Law. Now the junior partners in the new coalition reversed their stance, arguing that the proposed cuts were too severe, they threatened one of Germany's most successful industries. "We can't take an axe to [the solar sector]," said Michael Kauch, the FDP's environmental policy spokesperson. Merkel's conservative Bavarian allies, the Christian Social Union, were likewise opposed to swingeing reductions. The southern state is the location of about 40 percent of Germany's solar installations. "The plans . . . will have an unacceptable impact on Bavaria," complained state premier Horst Seehofer.

Conversely, the editors of *Photon* magazine, the solar industry's leading in-house booster, argued in favor of significant reductions in subsidies. "Otherwise, an unfettered expansion of capacity would drive up electricity bills and could trigger consumer protests," explained Anne Kreutzmann, the magazine's managing director. Installing excessive amounts of solar might mean that customers would have to pay too much, thereby destroying support for solar among the German public. Especially since the majority of PV panels were now being made by Chinese, not German manufacturers. By 2010 Germany's share of the world solar market had slipped to 15 percent and was continuing to decline.

The solar industry association countered by quoting a recent poll. This found that 71 percent of Germans favored *increasing* the surcharge on their electricity bills, to 5 percent from the current 3 percent, in order to support a greater deployment of photovoltaics. "There's really strong backing in the population for solar energy generation," claimed industry spokesperson David Wedepohl.

Electricity prices in Germany were rising faster than in other European countries. The industry argued that government-imposed feed-in tariffs were largely to blame. But a report commissioned by the Greens found that the hikes stemmed from Germany's giant utilities leveraging their near monopoly on generation and distribution. The pro-tariff backlash caused the government to tone down its plans. The cuts in subsidies for solar parks would be reduced from 25 percent to 15 percent. The introduction of the cuts would be delayed by three months, from April to July. But tariffs for rooftop systems would still be slashed by 16 percent.

In Germany's parliamentary system the upper house, the Bundesrat, has to sign off on new legislation. In early June 2010, as approval ratings for the Merkel government slumped to record lows, the Bundesrat voted against the proposed reductions in feed-in tariffs. A mediation committee was established to hammer out a compromise. The committee took several weeks to settle the dispute. The tariffs would be reduced by the proposed amounts, but the industry would be given a further three months' grace to adjust to the new conditions.

Even with the cuts, the market was not expected to collapse. To the contrary, the solar industry was booming as never before, with supply struggling to keep pace with apparently insatiable demand for panels. By mid-2010 Germany was on track to install 6.5 gigawatts, a 70 percent increase over 2009. Full-throttle expansion was expected to continue in 2011, with orders for a massive 9.5 gigawatts in the pipeline, equivalent to the cumulative total installed capacity at the end of 2009.

The government stressed it expected solar to continue growing, reaching 66 gigawatts by 2030. By then, insiders predicted, solar would have long since become cheaper than conventional methods of energy generation, rendering further discussions over feed-in tariffs academic. Already, on some days, the country was generating as much as 10 percent of its total electricity supply from the sun. For Germany, there would be no switching back.

Appendix

MAGIC CRYSTALS: A NONSCIENTIST'S GUIDE TO HOW SOLAR CELLS WORK

Solar cells are alchemical devices. They transform something free—sunlight—into something valuable—electricity. This seemingly miraculous metamorphosis occurs by manipulating the material at the atomic level. In effect, you persuade the electrons it contains to dance to your tune.

To make a cell, you start with some magic crystals. Solar cells can be made from almost anything. (In the early days, scientists reportedly made some cells out of spinach.) The material of choice—from which more than 80 percent of all solar cells are made—is silicon. Mostly, this means either silver-colored, single-crystal stuff that comes in salami-shaped ingots. Or sparkly, metal-flake-sheen, multi-crystalline stuff that comes in square blocks. Either way, your first step is to slice up your starting material into wafers. You do this using a wire, just like on a cheese cutter. Since ultra-pure silicon is expensive, the trick is to make your wafers as thin as possible. An ingot sliced with a wire saw can yield several thousand wafers, each about twice the width of a human hair.

Now comes the clever part. The idea is to harness the photons—energy-laden particles of sunlight—that are continually smashing into the wafer, using them to jolt some electrons out of their cosy orbits around the atomic nucleus. Thus simultaneously orphaned and energized, the electrons can then be rounded up and moved out in the form of an electric current. You contrive this by doctoring your silicon, peppering either side of the wafer with traces of two elements. The first element contains too many electrons, making one side electrically negative (or minus [–], just like on a battery terminal). The second element contains too few electrons, making the other side electrically positive (+). The result is a silicon sandwich with a bottom layer, which has a lack of electrons, and a much thinner top layer—"the skin of the wafer," as they say in the industry—which has an abundance of them. Where the two layers meet, their opposite charges create an electric field. The field acts as a barrier through which electrons from the top layer cannot pass. The only way for the sunlight-boosted particles to go is up, to the surface. There, they are captured by a grid of metal contacts.

Electric current flows out from the contacts to do work—like run your air conditioner—then back, via a second set of contacts on the base of the cell. This completes the circuit. Since no electrons are used up, the solar cell (or at least the silicon part of it) never wears out. Unlike all other electricity-generating technologies, solar cells have no moving parts. That accounts for their remarkable longevity left outside in the open air, exposed to all weathers. Makers routinely guarantee their solar panels for twenty years. You won't get that long a warranty on many products.

Untreated silicon is shiny and reflects the sun, which is not what you want. To prevent the sunlight bouncing off, you coat your cell with an anti-reflective material, often a lustrous blue in color. Then you wire the cell up to other cells and glue them, using a transparent adhesive, onto the back of a glass panel. *Et voilà*—PV. Elegant, eh?

One other thing you should know about solar cells is that they produce continuous, direct current (DC). This doesn't matter if you are using your solar system off-grid, in isolation, to drive appliances that can run on batteries. But the electric grid is based on alternating current (AC). So if you are going to hook up your solar system to feed power into the grid, you need a gizmo for converting DC to AC. By rights this should be called a converter, but for some reason (don't ask) it is known as an inverter.

ACKNOWLEDGMENTS

Writing a book may be the closest a man can get to having a baby. To be sure the origin is asexual, the gestation period longer, and the birth pangs less painful, but the feelings of pride (and relief) at having brought something new into the world are perhaps not entirely dissimilar. One big difference, however, is that producing a book like this one would not have been possible without generous help from many individuals during its four-year passage to parturition.

First and foremost, I want to thank the eighty-plus scientists, engineers, entrepreneurs, politicians, activists, analysts, and others who were gracious enough to spare me the time for an interview. In particular, those who were prepared to undergo the ordeal multiple times. Especially Bill Yerkes and Harry Lehmann, each of whom I interviewed on four occasions. Their input and insights were absolutely invaluable. Don Osborn and Tom Starrs also allowed me to interrogate them at greater-than-usual length.

From the outset Paula Mints, doyenne of PV analysts, has been my guiding light. With patience and good humor she drew on her unparalleled knowledge of the industry to point me in the right direction and disabuse me of misconceptions. Paula was always prepared to answer my queries, no matter how trivial. She also volunteered to read through and comment on my first draft. David Hochschild was likewise an invaluable volunteer reader. His early enthusiasm for my work provided much-needed reassurance. Needless to say, remaining errors in the text are due to my own inadequacies.

In almost thirty years as a journalist covering the electronics industry I had never been to Germany. It would have been impossible for me to organize the interviews in that country without the unstinting help from Christian Bertsch, formerly of the State of Victoria's European Office in Frankfurt, now of Greentech Management. In Melbourne Christian subsequently became my guide on all things German, as well as a good friend. I am deeply grateful to him.

I thank my good friends Jonathan Friedland and Victoria Godfrey for their gracious hospitality in Los Angeles; and likewise in Marin County,

Maria and Kyle Thayer. For looking after me so thoughtfully in Hammelburg, *Vielen Dank* to Annemarie Fell. And for their kindness in taking a stranger into their home in Santa Barbara, thanks also to Sara and Bill Yerkes.

I doff my cap to John Perlin, author of *From Space to Earth: The Story of Solar Electricity*, which is as good an introduction to the subject of photovoltaics as you could wish for. I drew extensively on this exemplary work for some of my early chapters.

To Alison Daams, ace librarian at the Moonee Valley Library, my gratitude and admiration once again for her tenacity in tracking down documents, no matter how ancient or arcane. And to Victor McElheny, my undying gratitude for his constant inspiration and support.

Carol Riordan deserves a round of applause for her thoroughness in compiling the daily National Center for Photovoltaics (NCPV) Hotline, an indispensable resource in keeping me up to date with developments in the fast-moving world of photovoltaics. And for allowing an apparently uncredentialed writer from the antipodes to attend Solar Power '09, the Solar Energy Industry Association's magnificent conference and exhibition in Anaheim in 2009, my sincere thanks to Monique Hanis.

Thanks to my agent, Mike Hamilburg, for finding a good home for this book.

Melbourne, December 2010

SOURCES

Introduction

Keefe, Bob. "Ted Turner Says He's Serious about Solar Power." *Atlanta Journal-Constitution*, September 26, 2007.

Obama, Barack. Remarks to Joint Session of Congress. February 24, 2009, www.whitehouse.gov/the_press_office/remarks-of-president-barack-obama-address-to-joint-session-of-congress/ (accessed November 11, 2010).

Olson, Syanne. "McKinsey Study Finds Electricity from Renewable Energy Is a Feasible Goal by 2050." PV-tech.org, April 29, 2010.

Perez, Richard. "Comparing the World's Energy Resources." www.asrc.cestm .albany.edu/perez/ (accessed November 11, 2010).

SEIA/SCHOTT Solar. "94% of Americans Say Solar Energy Development Is Important, Renewable Energy World." June 10, 2008.

Shaw, George Bernard. *Maxims for Revolutionists*, *Man and Superman*. 1903.

"Solon—Solar Energy Commerial." www.youtube.com/watch?v=A2naWGkKE 0&feature=related (accessed November 11, 2010).

Wesoff, Eric. "Milestone: 10 Gigawatts of Solar Panels in 2010." *Greentech Media*, October 7, 2010.

1

Bell Telephone Laboratories. Press release, April 25, 1954.

Bradley, Robert L. *Capitalism at Work*. Salem, MA: M&M Scrivener Press, 2009, p. 249.

Broad, William J."US Has Plans to Again Make Own Plutonium." *New York Times*, June 27, 2005.

Carter, Jimmy. Digest of Other White House Announcements. October 21, 1977, www.presidency.ucsb.edu/ws/index.php?pid=6826 (accessed November 11, 2010).

———. "The Moral Equivalent of War." Reprinted in *Time*, October 18, 1982, www .time.com/time/magazine/article/0,9171,949597,00.html (accessed November 11, 2010).

————. The President's Proposed Energy Policy. April 18, 1977, www.pbs.org/wgbh/amex/carter/filmmore/ps_energy.html (accessed November 11, 2010).

————. Proclamation 4558—Sun Day 1978. www.presidency.ucsb.edu/ws/index.php?pid=30559 (accessed November 11, 2010).

————. Remarks at the Solar Energy Research Institute. May 3, 1978, www.presidency.ucsb.edu/ws/index.php?pid=30746 (accessed November 11, 2010).

————. The Road Not Taken (Trailer). Footage from June 1977, www.youtube.com/watch?v=v9VD6MdEt0U (accessed November 11, 2010).

————.Solar Energy Message to the Congress. June 20, 1979, www.presidency.ucsb.edu/ws/index.php?pid=32503 (November 11, 2010).

Clarke, Arthur C. "Extra-Terrestrial Relays." *Wireless World*, October 1945, www.lakdiva.org/clarke/1945ww/1945ww_oct_305-308.html (November 11, 2010).

Ford Foundation. *A Time to Choose*, by S. David Freeman et al. 1974, p. 319.

Freeman, S. David. Interview with author, November 6, 2009.

————. *Winning Our Energy Independence*. New York: Gibbs Smith, 2007, p. 1.

Grossman, Karl. "Nukes in Space in Wake of *Columbia* Tragedy." www.21stcentury radio.com/articles/03/0224176.html (accessed November 11, 2010).

Kazmerski, Larry. Interview with author, July 5, 2007.

Kreiser, Harry. S. David Freeman interview, Institute of International Studies, UC–Berkeley, September 29, 2003, globetrotter.berkeley.edu/people3/Freeman/freeman-con0.html (accessed November 11, 2010).

Lovins, Amory B. "The Road Not Taken." *Foreign Affairs*, October 1976.

Manchester, Harland. "The Prospects for Solar Power." *Reader's Digest*, July 1955, p. 57.

Newton, James. *Uncommon Friends*. San Diego: Harcourt Brace Jovanovich, 1987, p. 31.

Perlin, John. *From Space to Earth: The Story of Solar Electricity*. Cambridge, MA: Harvard University Press, 2002, pp. 25–31, 41–46. The early history of the solar cell is taken from a variety of sources, notably Perlin.

Perlin, John, et al. "It Still Works!" *Solar Today*, January/February 2004, p. 27.

"Silicon Genesis Europe: Round Table Interviews." November 16, 2006, www.silicongenesis.stanford.edu/transcripts/electronicaeng.htm (accessed November 11, 2010).

Strong, Steven J. *The Solar Electric House*. Emmaus, PA: Rodale, 1987, p. 7.

"Sun Electricity." *Time*, July 4, 1955, www.time.com/time/magazine/article/0,9171,807289,00.html (accessed November 11, 2010).

"Vast Power of the Sun Is Tapped by Battery Using Sand Ingredient." *New York Times*, April 26, 1954, p. 1.

"White House Solar Panel Goes on Display at Carter Library." Jimmy Carter Library and Museum, news release, March 27, 2007.

2

Berger, John J. *Charging Ahead*. Berkeley: University of California Press, 1998, pp. 75–96, 110–21.

"COMSAT Collection Ms. 459." Sheridan Libraries, Johns Hopkins University, www.library.jhu.edu/collections/specialcollections/manuscripts/msregisters/ms 459.html (accessed November 11, 2010).

Gay, Charlie. Interview with author, June 10, 2007.

Linden, Lawrence H., et al. "Solar Photovoltaic Industry." MIT Energy Laboratory Report, December 1977, p. 40, n. 1.

McKenzie, Peter. Interview with author, March 21, 2007.

Varadi, Peter. Honorary Lecture by the John C. Bonda Prize Winner, Peter F. Varadi, Nineteenth European Photovoltaic Solar Energy Conference, Paris, June 7–11, 2004.

———. Interview with author, June 14, 2007.

Wald, Matthew L. "ARCO to Sell Siemens Its Solar Energy Unit." *New York Times*, August 3, 1989.

Yerkes, Bill. Interviews with author, October 10, 17, 2006; November 7, 2006; June 5, 2009.

3

Berger, John J. *Charging Ahead*. Berkeley: University of California Press, 1998, pp. 75–96.

Berman, Daniel M., and John T. O'Connor. *Who Owns the Sun?* White River Junction, VT: Chelsea Green, 1996.

Berman, Elliot. Interview with author, April 5, 2007.

Bruton, Tim. Interview with author, June 14, 2007.

Obituaries of Robert O. Anderson: *New York Times*, December 6, 2007; *Guardian*, December 7, 2007; *Times* (UK), December 10, 2007.

Perlin, John. *From Space to Earth: The Story of Solar Electricity*. Cambridge, MA: Harvard University Press, 2002, pp. 52–67.

Reece, Ray. *The Sun Betrayed*. Boston: South End Press, 1979.

Spaeth, Anthony. "ARCO's Solar Chill." *Forbes*, November 12, 1979, p. 84.

Yerkes, Bill. Interviews with author, October 10, 17, 2006; November 7, 2006; June 5, 2009.

4

Gerstenzang, James. "Oil Executive Breaks with Industry." *LA Times*, May 21, 1997.

Jester, Terry. Interview with author, May 9, 2007.

McKenzie, Peter. Interview with author, March 21, 2007.

Mrohs, Mark. Interview with author, December 2, 2008.

Robertson, Wayne. Interview with author, August 13, 2008.

Yerkes, Bill. Interviews with author, October 10, 17, 2006; November 7, 2006; June 5, 2009.

5

Aitken, Don. Interview with author, August 31, 2007.

Clark, Bill. Interview with author, August 13, 2007.

Firor, Kay. Interview with author, August 16, 2007.

Kimura, Osamu, and Tatsujiro Suzuki. "30 Years of Solar Energy Development in Japan." Paper presented at the Conference on the Human Dimensions of Global Environmental Change, Berlin, 2006.

"The Public Utility Regulatory Policies Act." www.americanhistory.si.edu/powering/past/h4main.htm (accessed November 11, 2010).

Schewe, Phillip F. *The Grid*. Washington, DC: Joseph Henry Press, 2007, pp. 164–73.

Shugar, Dan. Interview with author, November 17, 2008.

Smeloff, Ed, and Peter Asmus. *Reinventing Electric Utilities*. Washington, DC: Island Press, 1997, pp. 19, 22.

Weinberg, Carl. Interview with author, August 8, 2007.

Weinberg, Carl, and Donald E. Osborn. "Key Utility Contribution to the US Grid–Connected PV Market: The PG&E and SMUD Solar Experience." International Solar Energy Society, Solar World Congress, Florida, 2005.

Yerkes, Bill. Interviews with author, October 10, 17, 2006; November 7, 2006; June 5, 2009.

6

Aitken, Don. Interview with author, August 31, 2007.

Allen, Arthur. Denis Hayes quoted in "Prodigal Sun." *Mother Jones*, March 1, 2000, www.motherjones.com/politics/2000/03/prodigal-sun (accessed November 11, 2010).

Beattie, Donald, ed. *History & Overview of Solar Heat Technologies*. Cambridge, MA: MIT Press, 1997, p. 34.

Berger, John J. *Charging Ahead*. Berkeley: University of California Press, 1998, p. 124.

Carpenter, Paul R., and Richard D. Tabors. "A Uniform Economic Valuation Methodology for Solar Photovoltaic Applications Competing in a Utility Environment." MIT Energy Laboratory report, June 1978.

Carter, Jimmy. Solar Photovoltaic Energy Research, Development, and Demonstration Act of 1978 Statement on Signing HR 12874 into Law. November 4, 1978, www.presidency.ucsb.edu/ws/index.php?pid=30122 (accessed November 11, 2010).

Commoner, Barry. *The Politics of Energy*. New York: Knopf, 1979.

Egan, Michael. *Barry Commoner and the Science of Survival*. Cambridge, MA: MIT Press, 2007, p. 163.

Freeman, S. David. *Winning Our Energy Independence*. New York: Gibbs Smith, 2007, p. 96.

Friedman, Thomas L. "Revolutionary Changes for Solar Field." *New York Times*, August 18, 1981.

Hammond, Allen L., and William D. Metz. "Solar Energy Research: Making Solar after the Nuclear Model?" *Science*, July 15, 1977, p. 241.

Hayes, Denis. *Rays of Hope*. New York: Norton, 1977, p. 168.

"How Solar Technologies Will Work." *BusinessWeek*, October 9, 1978, p. 96.

Kidder, Tracy. "The Future of the Photovoltaic Cell." *Atlantic Monthly*, June 1980, p. 69.

Koff, Stephen. "Was Jimmy Carter Right?" *Cleveland Plain Dealer*, October 1, 2005.

Landsberg, Mitchell. "State Has Lost Global Lead in 'Green' Power." *LA Times*, May 16, 2001.

Maycock, Paul D., ed. *Photovoltaic News*, February 1987.

Reagan, Ronald. First Inaugural Address. January 20, 1981, www.reagan2020.us/speeches/First_Inaugural.asp (accessed November 11, 2010).

———. Official Announcement of Candidacy for President. November 13, 1979, www.reagan2020.us/speeches/candidacy_announcement.asp (accessed November 11, 2010).

Revkin, Andrew C., and Matthew L. Wald. "Solar Power Wins Enthusiasts but Not Money." *New York Times*, July 16, 2007.

Schwartz, William A., and Stephen A. Herzenberg. "Federal Report Says Solar Electricity Will Be Unimportant Till Next Century." *Harvard Crimson*, February 3, 1979.

Strong, Steven J. *The Solar Electric House*. Emmaus, PA: Rodale, 1987, p. 4.

Teller, Edward. Interview in *Playboy*, August 1979.

———. "I Was the Only Victim of Three Mile Island." Op-ed, *Wall Street Journal*, July 31, 1979.

Wald, Matthew L. "ARCO to Sell Siemens Its Solar Energy Unit." *New York Times*, August 3, 1989.

———. "US Companies Losing Interest in Solar Energy." *New York Times*, March 7, 1989.

Williams, Neville. *Chasing the Sun*. Gabriola Island, BC: New Society Publishers, 2005, p. 89.

7

Behrman, Daniel. *Solar Energy: The Awakening Science*. Boston: Little, Brown, 1976, pp. 70–89.

Berger, John J. *Charging Ahead*. Berkeley: University of California Press, 1998, pp. 122–34.

Berman, Daniel M., and John T. O'Connor. *Who Owns the Sun?* White River Junction, VT: Chelsea Green, 1996, pp. 183–86.

Böer, Karl. Interviews with author, July 17, 21, 2007.

Butti, Ken, and John Perlin. *Golden Thread*. Palo Alto: Cheshire Books, 1980, pp. 200–16.

Daugherty, John. "Sundown for Solar." *Phoenix New Times*, June 30, 1993.

Fitzgerald, Mark C. "Solar at the White House." *Solar Today*, May/June 2003.

"Future Large-Scale Terrestrial Uses of Solar Energy." Proceedings of the 25th Annual Power Source Conference, 1972, pp. 145–48.

Hallowell, Christopher. "Why Create Energy When You Can Save It?" *Time*, March 15, 1999.

Kelly, Charles. "Valley Philanthropist John F. Long Dies at 87." *Arizona Republic*, March 1, 2008.

Kidder, Tracy. "The Future of the Photovoltaic Cell." *Atlantic Monthly*, June 1980.

Lim, Bill B. P. *Solar Energy in the Tropics*. Dordrecht: Reidel, 1983, p. 234.

Strong, Steven J. *The Solar Electric House*. Emmaus, PA: Rodale, 1987, pp. 1–4.

8

Lindsey, Robert. "Physicist's Solar Airplane Set to Challenge the English Channel." *New York Times*, June 9, 1981.

Muntwyler, Urs. E-mail correspondence, June 27, 2009.

Perlin, John. *From Space to Earth: The Story of Solar Electricity.* Cambridge, MA: Harvard University Press, 2002, pp. 148–52.

"The Quiet Achiever." www.snooksmotorsport.com.au/solartrek/Solar_Trek/bp_solar_car_crossing_of_austral.htm (accessed November 11, 2010).

Real, Markus. Interview with author, November 10, 2007.

Real, Markus, and Hans Lüdi. "Project Megawatt: Experience with Photovoltaics in Switzerland." Conference Record of the Twenty-second Photovoltaic Specialists Conference, IEEE, October 7–11, 1991, pp. 574–75.

"Sunrayce '93." US DoE, www.osti.gov/bridge/purl.cover.jsp;jsessionid=D1D010. D8DA96A45690A7983513B97A64?purl=/10159587-cvRG3B/webviewable/ (accessed November 11, 2010).

Warner, Cecile. Interview with author, July 13, 2007.

9

Aitken, Donald, et al. "SMUD PV Program Review." Final report, December 30, 2000, p. 6.

Akst, Daniel. "All Charged Up in Sacramento." *LA Times*, November 17, 1992.

Collier, Dave. Interview with author, July 28, 2007.

Freeman, S. David. *Winning Our Energy Independence.* New York: Gibbs Smith, 2007, pp. 3, 28.

Hughes, Wade. Interview with author, March 25, 2009.

Kreiser, Harry. S. David Freeman interview, Institute of International Studies, UC–Berkeley, September 29, 2003, www.globetrotter.berkeley.edu/people3/Freeman/freeman-con0.html (accessed November 11, 2010).

Osborn, Don. Interviews with author, July 26 and August 9, 2007.

Smeloff, Ed. Interview with author, November 18, 2008.

Smeloff, Ed, and Peter Asmus. *Reinventing Electric Utilities.* Washington, DC: Island Press, 1997, pp. 27, 29, 43, 47.

10

Nelson, Les. Interview with author, September 12, 2007.

Proceedings of American Solar Energy Society Annual Conference. June 25–30, 1994, San Jose, California.

Redgate, Pat. Interview with author, September 14, 2007.

Sindelar, Allan. "Preparing for a PV Future." 1993, www.ibiblio.org/london/alternative-energy/homepower-magazine/archives/35/35p82.txt (accessed November 11, 2010).

Starrs, Tom. Interviews with author, August 28 and September 5, 2007.

11

Cryanoski, David. "Japan Goes for the Sun." *Nature*, April 30, 2009, p. 1084.

Doe, Paula. "Explosive Growth Reshuffles Top 10 Solar Ranking." *Renewable Energy World*, September 12, 2008.

Fackler, Martin. "Japan Sees a Chance to Promote Its Energy-Frugal Ways." *New York Times*, July 4, 2008.

Friedland, Jonathan, and Bob Johnstone. "Samurai Sorcerer." *Far Eastern Economic Review*, June 3, 1993, pp. 60–65.

Green, Martin. *Power to the People*. Sydney: UNSW Press, 2000, p. 57.

Japan: An Illustrated Encyclopedia. Tokyo: Kodansha, 1993, p. 341.

"Japan May Force Utilities to Buy Surplus Domestic Solar Power." AFP, February 24, 2009.

"Japan's Solar Module Shipments Double in 2009/2010." Reuters, May 28, 2010.

Johnstone, Bob. *We Were Burning*. New York: Basic Books, 1999, pp. 158–69.

Kimura, Osamu, and Tatsujiro Suzuki. "30 Years of Solar Energy Development in Japan." Paper presented at the Conference on the Human Dimensions of Global Environmental Change, Berlin, 2006.

"Kondo Set to Resign over Solar Blunder." *Japan Times*, October 25, 2000.

Kuwano, Yukinori. "Present Status & Future Prospects for Solar Cells." 1990.

Maeda, Risa. "Japan Solar Subsidies Lure Fewer Users Than Planned." Reuters, April 8, 2009.

Maycock, Paul D., ed. *Photovoltaic News*, February 1987, p. 5.

Ristau, Oliver. "The Photovoltaic Market in Japan: Unquestioned Leadership of World Market." September 15, 2001, http://www.solarserver.com/solarmagazin/artikelseptember2001-e.html (accessed November 11, 2010).

Sandtner, Walter. Interview with author, January 17, 2008.

"Sanyo Develops HIT Solar Cells with World's Highest Energy Conversion Efficiency of 23%." Sanyo press release, May 21, 2009.

Sharp Corporation. *1912–1992: Eighty Years of Sincerity and Creativity.* 1992.

"Sharp Solar Gets New Boss." *Photon International,* June 2007.

The Sunshine Project: For Establishing Clean, New Energy Technology. MITI, March 1986

12

Hockenos, Paul. *Joschka Fischer and the Making of the Berlin Republic: An Alternative History of Postwar Germany.* Oxford: Oxford University Press, 2008, pp. 155, 210, 212.

Jacobsson, Staffan. E-mail communication, April 7, 2008.

Jacobsson, Staffan, and Volkmar Lauber. *The Politics and Policy of Energy System Transformation—Explaining the German Diffusion of Renewable Energy Technology,* Energy Policy 34. Elsevier, 2006, pp. 256–76.

Lehmann, Harry. Interviews with author, April 8 and 21, 2008.

SFV website. www.sfv.de (accessed November 11, 2010).

Teske, Sven. Interview with author, March 26, 2008.

Von Fabeck, Wolf. Interview with author, November 3, 2008.

Uphues, Andreas. "Tension in Aachen." *Focus* 22 (1993), www.focus.de/politik/deutschland/solarenergie-spannung-in-aachen_aid_140598.html (accessed November 11, 2010).

13

Cramer, Günther. Interview with author, November 7, 2008.

Fleck, Olaf. Interview with author, November 6, 2008.

"Heinz Reisenhuber." en.wikipedia.org/wiki/Heinz_Riesenhuber (accessed November 11, 2010).

Hoffmann, Winfried. Interview with author, November 13, 2007.

Jacobsson, Staffan, and Volkmar Lauber. *The Politics and Policy of Energy System Transformation—Explaining the German Diffusion of Renewable Energy Technology,* Energy Policy 34. Elsevier, 2006.

Lehmann, Harry. Interview with author, January 18, 2008.

Moehrstedt, Udo. Interview with author, June 10, 2008.

Räuber, Armin. Interview with author, November 12, 2007.

Sandtner, Walter. Interview with author, January 11, 2008.

Sietman, Richard. "Nuclear Energy: FRG's Burdens." *Physics World* (March 1989), www.physicsworldarchive.iop.org/index.cfm?action=summary&doc=2%2F3% 2Fphwv2i3a4%40pwa-xml&qt= (accessed November 11, 2010).

Teske, Sven. Interview with author, March 26, 2008.

14

Böer, Karl. Interview with author, July 21, 2007.

Eurosolar Journal, September 1989.

Fleck, Olaf. Interview with author, November 6, 2008.

Johnson, Bob. Interview with author, March 19, 2008.

Lehmann, Harry. Interview with author, April 8, 2008.

Mints, Paula. Interview with author, September 15, 2007.

Pearce, Fred. "Bring on the Solar Revolution." *New Scientist*, May 24, 2008.

Sandtner, Walter. Interview with author, January 11, 2008.

Scheer, Hermann. *Energy Autonomy*. London: Earthscan, 2007, p. 52.

———. Interview with author, November 14, 2007.

———. Homepage, www.hermannscheer.de/en/ (accessed November 11, 2010).

———. *The Solar Economy*. London: Earthscan, 2004.

———. *A Solar Manifesto*. 2nd ed. London: James & James, 2001, pp. 8, 9, 242.

Williams, Neville. *Chasing the Sun*. Gabriola Island, BC: New Society Publishers, 2005, p. 269.

15

NB. The chronology for this chapter is anchored by articles by Bob Johnson in *Solar Flare*, a PV industry newsletter he edited while working for market researcher Strategies Unlimited.

Fell, Hans-Josef. Interview with author, November 8, 2007.

Häberlin, Heinrich. Interview with author, February 7, 2008.

Johnson, Bob. Interview with author, March 19, 2008.

Kreutzmann, Anne. "A New Path to Self-Sustaining Markets for Photovoltaic." *EuroSun* 1996.

Lehmann, Harry. Interviews with author, January 18; April 8, 21; November 10, 2008.

Schulte-Janson, Dieter. Interview with author, February 24, 2009.

Solar Flare, September 1997.
Von Fabeck, Wolf. Interview with author, November 3, 2008.

16

EWS website. www.ews-schoenau.de (accessed November 12, 2010).
Fell, Hans-Josef. Interviews with author, November 8, 2007; November 8, 2008. See also www.gruene-bundestag.de/cms/members/dok/204/204403.hansjosef_fell.html (accessed November 12, 2010).
Isenson, Nancy. "Germany's 'Electricity Rebels.'" *Deutsche Welle*, April 18, 2005.
Johnson, Hugh. *World Atlas of Wine*. London: Mitchell Beazley, 1977, p. 158.
Kulish, Nicholas. "Munich Redux: Germany's Hot Spot of the Moment." *New York Times*, April 13, 2008.
Suzuki, David, and Holly Dressel. *Good News for a Change*. Vancouver: Greystone Books, 2003, p. 312.

17

Blue, Laura. "Lessons from Germany." *Time*, April 23, 2008.
BSW. "The German Solar Industry Has Come of Age." en.solarwirtschaft.de/home/photovoltaic-market/german-market.htm (accessed November 12, 2010). For facts and figures about the German PV market, see inter alia.
Burgermeister, Jane. "South Korea Taps Germany to Help Grow Its Solar Industry." *Renewable Energy World*, April 29, 2009.
Collier, Robert. "Germany Shines a Beam on the Future of Energy." *San Francisco Chronicle*, December 20, 2004.
Fell, Hans-Josef. Interviews with author, November 8, 2007; November 8, 2008.
Gaertner, Reiner. "Germany Embraces the Sun." *Wired*, July 9, 2001.
"The German Solar Energy PV Market." *Solar Plaza*, July 16, 2009.
"Germany's Largest Solar Power Project Nears Completion." *Solarbuzz*, August 20, 2009.
Goetzberger, Adolf, and Volker Hoffmann. *Photovoltaic Solar Energy Generation*. Berlin: Springer, 2005, p. 171.
Hockenos, Paul. *Joschka Fischer and the Making of the Berlin Republic: An Alternative History of Postwar Germany*. Oxford: Oxford University Press, 2008, p. 282.
Jacobsson, Staffan, and Volkmar Lauber. *The Politics and Policy of Energy System Transformation—Explaining the German Diffusion of Renewable Energy Technology*, Energy Policy 34. Elsevier, 2006.

Kant, Immanuel. *Prolegomena*. Cambridge: Cambridge University Press, 2004, p. 132.
Landler, Mark. "Germany Debates Subsidies for Solar Industry." *New York Times*, May 16, 2008.
———. "Hot German July Doesn't Faze Farmer Who Reaps the Sun." *New York Times*, July 26, 2006.
Morris, Craig. "Much Ado about Germany?" *Solar Today*, February 2007.
"Phase One for World's Largest Solar PV Plant." *Renewable Energy World*, June 3, 2005.
"Photovoltaik-Weltmeister Bayern: PHOTON, DGS und Solar-Initiativen feiern 'Ein-Prozent-Party.'" February 4, 2007, www.solarserver.de/news/news-6727.html (accessed November 12, 2010). About the 1 Percent party.
Renewable Energy Journal, no. 10 (June 2000).
Scheer, Hermann. Interview with author, November 14, 2007.
Stein, Christof. The Experience of Germany on Photovoltaic Incentives, Federal Ministry for the Environment, Nature Conservation and Nuclear Safety. Presentation, March 8, 2004, Rome, www.wind-works.org/FeedLaws/Germany/ChristofSteinBMFUSolarExperience.pdf (accessed November 12, 2010).
"Successful Renewable Energy Law." Planning Association, Energy and Environment. November 2, 2003, www.planungsgemeinschaft.de/index2.html?/cgi/enews.cgi?tabelle=gb&no=14&generate=0&id=155&p_lang=2~Planungsgeme inschaft (accessed November 12, 2010). Jürgen Trittin quote.
Teske, Sven, and Volker Hoffmann. *A History of Support for Solar Photovoltaics in Germany*. N.p., 2006.
Voosen, Paul. "Spain's Solar Market Offers a Cautionary Tale about Feed-In Tariffs." *New York Times*, August 18, 2009.
Wang, Ucilia. "Solar Prices Set in Germany." *Greentech Media*, June 6, 2008.
Weber, Eicke. Interview with author, November 9, 2007.

18

"Celebrities in Switzerland—Brenninkmeijer." www.switzerland.isyours.com/e/celebrities/bios/151.html (accessed November 12, 2010). On Marcel Brenninkmeijer.
Cramer, Günther. Interview with author, November 7, 2008.
Fisher, Marc. "East Germany's Bitterfeld: Grimiest Town in Dirtiest Country." *Washington Post*, April 16, 1990.
Hogg, David. Interview with author, November 15, 2007.

Milner, Anton. Interview with author, February 11, 2008.

Mints, Paula. "The PV Industry 2009: In Search of Stability and Sustainability." *Renewable Energy World*, August 31, 2009.

Osborne, Mark. "Q-Cells Highlights Module Business Model as CIGS Thin-Film Ramps." *PV-tech.org*, May 13, 2010, http://www.pv-tech.org/news/_a/q-cells _highlights_module_business_model_as_cigs_thin_film_ramps/ (accessed November 12, 2010).

Reiner Lemoine Foundation. www.reiner-lemoine-stiftung.de/eng/reinerle.htm (accessed November 12, 2010).

Rust, Achim. "Zur Sonne!" *Welt Online*, May 14, 2004, www.welt.de/print-welt/ article313624/Zur_Sonne.html (accessed November 12, 2010).

"Schmierig statt haarig." *Manager Magazin*, September 23, 2003, www.manager -magazin.de/koepfe/artikel/0,2828,266816,00.html (accessed November 12, 2010). On Immo Ströher.

Steitz, Christoph. "German Solar Firms Hit by Price War." Reuters, August 13, 2009.

Tuohy, William. "Pollution a Bitter Fact of Life in Germany." *Chicago Sun-Times*, March 11, 1990.

Wenzel, Von Kirsten. "Reiner Lemoine." *Tagesspiegel*, March 16, 2007, http://www .tagesspiegel.de/wirtschaft/unternehmen/reiner-lemoine/823408.html (accessed November 12, 2010).

Williamson, Hugh. "The Sun Shines Bright on Q-Cells." *Financial Times*, December 18, 2007.

19

"Adam Browning, Vote Solar Initiative." *Grist*, January 30, 2004, www.grist.org/ article/vote/P5 (accessed November 12, 2010). On Bonnie Raitt concert.

Adelman, Ken. Solar Warrior Homepage. www.solarwarrior.com (accessed November 12, 2010).

Aguirre, Abby. "The Outsider." *Nation*, March 3, 2005.

Asmus, Peter. "Bad Energy." *Metro*, August 22–28, 2002.

Browning, Adam. Interview with author, February 12, 2009.

"California Electricity Crisis." en.wikipedia.org/wiki/California_electricity_crisis (accessed November 12, 2010).

Capretz, Nicole. Interview with author, February 26, 2009.

Carney, James, and John Dickerson. "The Rocky Rollout of Cheney's Energy Plan." *Time*, May 19, 2001.

Hochschild, David. Interview with author, February 18, 2009.

Huard, Ray. "San Diego Panel OKs Plan for More Renewable Energy." *San Diego Union-Tribune*, September 5, 2003.

"Interview: Governor Gray Davis." www.pbs.org/wgbh/pages/frontline/shows/blackout/interviews/davis.html (accessed November 12, 2010).

Kay, Jane. "SF Could Be Leader on Solar Energy." *SF Gate*, October 27, 2001.

Lee, Mike. "San Diego Leads California in Solar Installations." *San Diego Union-Tribune*, July 16, 2009.

Mayfield, Kendra. "Fog City Catches a Few Rays." *Wired*, January 7, 2003.

Mieszkowski, Katherine. "You Gotta Fight for Your Right to Go Solar." *Salon*, June 3, 2004.

Mullman, Jeremy. "Power Politics." *SF Weekly*, August 1, 2001.

Raitt, Bonnie, et al. "Don't Let the Utilities Turn Out the Lights on Solar Energy." www.envirolink.org/resource.html?itemid=20020809125752581715&catid=8 (accessed November 12, 2010).

Saldaña, Lori, and Donna Frye. "The Argument Against: Region Should Not Rely on Imported Power." *San Diego Union-Tribune*, December 14, 2007.

"San Francisco Voters Back Solar Power." Associated Press, November 7, 2001.

Schrag, Peter. "Blackout." *American Prospect*, February 26, 2001.

"Self-Generation Incentive Program." October 4, 2010, www.dsireusa.org/incentives/incentive.cfm?Incentive_Code=CA23F&re=1&ee=1 (November 12, 2010).

Starr, Kevin. *Coast of Dreams*. New York: Knopf, 2004, pp. 594, 596.

Testimony of David Freeman before the Subcommittee on Consumer Affairs. Foreign Commerce and Tourism of the Senate Committee on Commerce, Science and Transportation, May 15, 2002.

"The Vote Solar Initiative." Homepage, www.votesolar.org (accessed November 12, 2010).

20

Adelson, Andrea. "Making a Difference: From the Utility Business to US Energy Secretary?" *New York Times*, November 22, 1992.

Aitken, Don. Interview with author, August 31, 2007.

"Arnold Details Specifics of Environmental Action Plan." www.schwarzenegger.com/en/news/uptotheminute/news_upto_en_env_action.asp?sec=news&subsec=uptotheminute (accessed November 12, 2010).

Barringer, Felicity. "With Push Toward Renewable Energy, California Sets Pace for Solar Power." *New York Times*, July 16, 2009.

Beslau, Karen. "The Green Giant." *Newsweek*, April 16, 2007.

"The Best Energy Bill Corporations Could *Buy*." *Public Citizen*, August 8, 2005, www.citizen.org/cmep/energy_enviro_nuclear/electricity/energybill/2005/articles .cfm?ID=13980 (accessed November 12, 2010).

Bohan, Drew. Interview with author, April 8, 2009.

Broehl, Jesse. "Rebirth of the California Million Solar Roof Plan." *Renewable Energy World*, February 28, 2005.

Browning, Adam. Interview with author, February 12, 2009.

Bush, George W. President's State of the Union Address. January 31, 2006, www.washingtonpost.com/wp-dyn/content/article/2006/01/31/AR200 6013101468.html (accessed November 12, 2010).

"Bush Pushes for Renewables." *Environmental News Service*, October 12, 2006.

Del Chiaro, Bernadette. Interview with author, January 18, 2010.

Dicum, Gregory. "Plugging Into the Sun." *New York Times*, January 4, 2007.

"DoE Releases Details of Solar America Initiative." *Renewable Energy Today*, February 10, 2006.

Douglas, Elizabeth. "Solar Subsidy Plan Is Passed." *LA Times*, January 13, 2006.

Environment California. "Million Solar Roofs Bill (SB 1) Signed into Law." Press release, August 21, 2006.

———. "Summer Report 2006." Vol. 4, no. 1.

Fine, Howard. "Utility Player." *Los Angeles Business Journal*, April 26, 2004.

Fletcher, Ed. "Soaking Up the Sun." *Sacramento Bee*, May 16, 2003.

Hertsgaard, Mark. "Terry Firma: Interview with Tamminen." *Grist*, August 19, 2004.

Hochschild, David. "Where We Go from Here." *SF Gate*, September 16, 2005.

Jiménez, Viviana. "World Sales of Solar Cells Jump 32 Percent." *Earth Policy Institute*, January 1, 2004.

Kamen, Al. "Culture Shock." *Washington Post*, May 23, 1997.

Lifsher, Marc. "Governor's Solar Plan Is Generating Opposition." *LA Times*, June 27, 2005.

———. "Rebate Rule Chills Sales of Solar." *LA Times*, May 8, 2007.

Little, Amanda. "Schwarzenegger's Solar-Roof Plan Could Get Sidelined by Partisan Squabbling." *Grist*, June 23, 2005.

"Million Solar Roof Initiative Sent to State Legislature." *Renewable Energy World*, March 14, 2005.

Molloy, Tim. "Schwarzenegger Won't Apologize for Girlie Men Remark." Associated Press, July 19, 2004.

Mulligan, Helen. Interview with author, April 30, 2009.

Office of the California Governor. "Gov. Schwarzenegger Announces One Million Solar Roofs by 2018." Press release, February 28, 2005.

————. "Gov. Schwarzenegger Participates in Launch of New Solar Energy Facility." Press release, October 23, 2008.

————. "Gov. Schwarzenegger Signs Legislation to Complete Million Solar Roofs Plan." Press release, August 21, 2006.

"President Bush Leads Energy Panel Discussion." NREL Newsroom, February 21, 2006.

Redgate, Pat. Interview with author, September 14, 2007.

Roberts, David. "One Nation, under Terry." *Grist*, January 4, 2007.

Rose, Craig. "Solar Energy's Day Is Dawning." *San Diego Union-Tribune*, December 10, 2006.

"Sunny California Flirts with Million Solar Homes Proposal." *Environmental News Service*, August 6, 2004.

Tamminen, Terry. Interview with author, February 26, 2009.

Thompson, Don. "Homes Urged to Go Solar." Associated Press, August 4, 2004.

"Time to Turn to Solar Energy." *Oakland Tribune*, April 25, 2005.

US DoE. Million Solar Roofs Initiative. Final report, October 2006, www.nrel.gov/docs/fy07osti/40483.pdf (accessed November 12, 2010).

21

Altman, Alex. "Incentives, Tax Credits, Loan Programs and Tariffs Create a Perfect Storm for Development of Valley's Solar Industry." *Public Record*, November 11, 2008.

Barringer, Felicity. "Berkeley Approves City-Backed Loans for Solar Panels." *New York Times*, September 18, 2008.

Conlon, Pat. Interview with author, March 14, 2009.

DeVries, Cisco. Interview with author, February 13, 2009.

Fehrenbacher, Katie. "Q&A with Berkeley Mayor Tom Bates on Solar Financing Plan." November 6, 2007, http://gigaom.com/cleantech/qa-with-berkeley-mayor-tom-bates-on-solar-financing-plan (accessed November 12, 2010).

Ferguson, Jim. Interviews with author, March 24, 2009; December 3, 2009.

Gammon, Robert. "Will Berkeley's Solar Plan Go Viral?" *East Bay Express*, November 19, 2008.

Jones, Carolyn. "Biden to Model Solar Finance Plan on Berkeley's." *SF Gate*, October 20, 2009.

————. "Bright Future as Berkeley Starts Solar Program." *San Francisco Chronicle*, February 28, 2009.

Kaufman, Leslie. "Harnessing the Sun, with Help from Cities." *New York Times*, March 15, 2009.

Kaufmann, K. "Hundreds Sign Up for Palm Desert Energy Financing." *Desert Sun*, August 18, 2008.

———. "No Excess Solar Payments This Year." *Desert Sun*, April 30, 2009.

———. "Palm Desert Energy Loans for Solar Sell Out in 20 Minutes." *Desert Sun*, July 6, 2009.

———. "Palm Desert Helps Secure Solar Tax Credit." *Desert Sun*, January 28, 2009.

Koot, Edwin. "Incredible Growth in Spanish Solar Energy Market Spells Good and Bad News for PV Industry." *Solar Plaza*, January 21, 2009.

La Ganga, Maria. "Recycling Is a Religion for Berkeley Mayor Tom Bates." *LA Times*, August 8, 2009.

Lee, Mike. "San Diego Leads California in Solar Installations." *San Diego Union-Tribune*, July 16, 2009.

Office of Energy and Sustainable Development. "Berkeley Climate Action Plan." www.ci.berkeley.ca.us/ContentDisplay.aspx?id=19668 (accessed November 12, 2010).

"Palm Desert to Provide Loans for Greenovating Homes." *Business Wire*, July 27, 2008.

Roosevelt, Margot. "California Adopts Innovative Solar Loan Law." *LA Times*, July 22, 2008.

Rosenberg, Martin. "Making the Hard Changes." *EnergyBiz*, June 2008. Mike Peevey on energy efficiency.

Schwartz, Naoki. "Poster Child for Efficient Energy Use." Associated Press, July 29, 2007.

St. John, Jeff. "San Francisco Sets $20M–$30M Goal for Berkeley-Style Solar Loan Program." *Greentech Media*, March 24, 2009.

Strassman, Marc. Interview with Jim Ferguson. *Utopia News*, February 28, 2009.

Timiraos, Nick. "Fannie and Freddie Resist Loans for Energy Efficiency." *Wall Street Journal*, March 25, 2010.

22

Barringer, Felicity. "Feed-In Tariffs for Solar Continue to Spread." *New York Times*, July 22, 2009.

Brinkman, Rob. "New $550m Coal Plant? Citizens Group Urges Cleaner Alternatives." *Gainesville Iguana*, May/June 2005, www.afn.org/~iguana/archives/2005 _05/20050501.html (accessed November 12, 2010).

CEC. "2007 Integrated Energy Policy Report (Adopted December 5, 2007)." www .energy.ca.gov/2007_energypolicy/index.html (accessed November 12, 2001).

Cinnamon, Barry. Interview with author, November 3, 2009.

———. A New Paradigm for Interconnection, presentation for Solar Power International 2009, October 28, 2009, Anaheim, CA.

Cory, Karlyn, Toby Couture, and Claire Kreycik. "Feed-In Tariff Policy: Design, Implementation, and RPS Policy Interactions." *NREL Technical Report*, March 2009.

CPUC. "CPUC Approves Feed-In Tariffs to Support Development of Onsite Renewable Generation." Press release, February 14, 2008.

Ehrlich, David. "Who's Lining Up for California's Feed-In Tariffs." *Cleantech Group*, February 20, 2008.

El Gammal, Adel. Feed-In Tariffs, presentation for Solar Power International 2009. October 28, 2009, Anaheim, CA.

Galbraith, Kate. "Europe's Way of Encouraging Solar Power Arrives in the US." *New York Times*, March 13, 2009.

Geesman, John. Feed-In Tariffs, presentation for Solar Power International 2009, October 29, 2009, Anaheim, CA.

Gipe, Paul. "California Energy Commission to Recommend Renewable Energy Feed-In Tariffs." Wind-Works.org, December 4, 2007, www.wind-works.org/FeedLaws/USA/CECConsidersFeed-inTariffs.html (accessed November 12, 2010).

Gonzalez, Angel, and Keith Johnson. "Spain's Solar Power Collapse Dims Subsidy Model." *Wall Street Journal*, September 3, 2009.

"Green Growth: Are Feed-In Tariffs the Answer?" *Washington Monthly*, www.newamerica.net/events/2009/green_growth_are_feed_tariffs_answer (accessed November 12, 2010).

Hull, Dana. "Barry Cinnamon, CEO of Akeena Solar, Inc." *Mercury News*, October 31, 2009.

"Interview with John Bertolino." *Etopia News*, July 20, 2009.

Kachan, Dallas. "Solar Insiders Lament Lack of US Feed-In Tariff." *Inside Greentech*, March 13, 2007.

Klebensberger, Boris. Interview with author, November 20, 2009.

Lacy, Stephen. "Building a FIT Renewable Energy Market in the US." *Renewable Energy World*, March 10, 2008.

Loder, Asjylin. "Gainesville Utility Places Premium on Solar Power." *St. Petersburg Times*, January 1, 2009.

Marshall, Christa. "Some Utility Executives Think German Approach Won't Work Nationally in US." *Earth News*, July 28, 2008.

Martin, Dan. "Feed-In Tariffs Have Earned a Role in US Energy Policy." *Renewable Energy World*, August 31, 2009.

"A New Kind of 'Titletown.'" *Gainsville Today*, March 2009.

Regan, Ed. Interview with author, December 18, 2009.

Renewable and Appropriate Energy Laboratory. "UC Berkeley Study Touts Economic Benefits of a Feed-In Tariff." Press release, July 7, 2010.

Schwarzenegger, Arnold. Letter to members of California State Senate. gov.ca.gov/ pdf/press/2009bills/SB32_NegreteMcLeod_Signing_Message.pdf (accessed November 12, 2010).

"Solar Advocates Applaud PG&E Commitment to Expand Net Metering Program." *Business Wire*, October 29, 2009.

"Solar Fact Finding Mission to Germany for Utility Decision Makers." *Summary Report*, June 8–13, 2008.

Steffen, Sarah. "Drastic Cuts to Subsidies for Solar Energy Remain Contentious." *Deutsche Welle*, November 19, 2009.

Strassman, Marc. "Interview with Ed Regan." *Utopia News*, February 12, 2009.

Sweet, Cassandra. "Calif. Governor Signs Solar-Power Feed-In Tariff Bill into Law." *Dow Jones Newswires*, October 14, 2009.

Voosen, Paul. "Spain's Solar Market Crash Offers a Cautionary Tale about Feed-In Tariffs." *New York Times*, August 18, 2009.

23

Alsever, Jennifer. "The Last Green Mile." Fast Company, June 1, 2007, http://www.fast company.com/magazine/116/next-innovation.html (accessed November 12, 2010).

Awn, Rich. Interview with Dave Llorens, cofounder of 1BOG.org. Green Radio, March 31, 2009.

Baron, Rachel. "Power Purchases to the People." *Greentech Media*, September 26, 2007.

Chafkin, Max. "Entrepreneur of the Year: Elon Musk." *Inc. Magazine*, December 2007.

Daniels, Cora. "Solar Powerhouse: Lyndon Rive Likes the Unusual. He Sells the Sun and Plays Hockey Underwater." *Men's Fitness*, September 2008.

Dolan, Kerry. "Low-Cost Solar for the Masses." *Forbes,* May 26, 2010.

Fenster, Ed. The Solar House That Jack Built: Residential Solar Project Financing, presentation for Solar Power International 2009, October 28, 2009, Anaheim, CA.

Gage, Deborah. "Startup Sells Solar Panels Online." *SF Gate*, May 18, 2008.

Greenpeace International. "San Francisco Voters Back Solar Measures Championed by Greenpeace." Press release, November 11, 2001, www.greenpeace. org/international/press/releasessan-francisco-voters-back-sola (accessed November 12, 2010).

Herring, Garrett. "SunRun Generation Seeks to Make PV Affordable with Residential PPAs." *Photon International*, January 2008.

Hymas, Lisa. "We Built This Solar City." *Grist*, April 11, 2008.

Johnston, Marsha. "Residential Solar PPAs Continue to Drive Solar Market Growth." *Renewable Energy World*, January 20, 2009.

Jurich, Lynn. Breaking Down Market Barriers in a Growing Industry, presentation for Solar Power International 2009, October 28, 2009, Anaheim, CA.

Kanellos, Michael. "Cutting Down Solar Costs with Satellite Imagery." *Green Tech*, April 20, 2008.

Karney, Bruce. Interview with author, January 15, 2010.

Kennedy, Danny. Interview with author, November 25, 2008.

Kilduff, Paul. "Shining a Light on SF Solar Options." *SF Gate*, October 11, 2008.

Llorens, Dave. Emergence of Service Providers: Breaking Down Market Barriers in a Growing Industry, presentation for Solar Power International 2009, October 28, 2009, Anaheim, CA.

———. Interview with author, January 22, 2010.

Marshall, Matt. "Musk Leads $10M Investment in SolarCity—to Provide Solar for All." *VentureBeat*, September 15, 2006.

Matz, Michael. "Solar Block Party." *Photon*, November 2009.

Mitra, Sramana. "Driving Solar Growth through Financing: SolarCity CEO Lyndon Rive." July 22, 2009, www.sramanamitra.com/2009/07/22/driving-solar-growth-through-financing-solar-city-ceo-lyndon-rive-part-1 (accessed November 12, 2010).

———. "From Greenpeace to Green Power: Sungevity CEO Danny Kennedy, on Strategy." February 25, 2009, www.sramanamitra.com/2009/02/25/from-greenpeace-to-green-power-sungevity-ceo-danny-kennedy-part-1/ (accessed November 12, 2010).

Montgomery, Jim. "Five Predictions for PV Solar." *Photovoltaic World*, January 12, 2010.

Neukermans, Armand. Interview with author, December 20, 2009.

Riddell, Lindsay. "Sungevity Activates White House Campaign." *San Francisco Business Times*, June 15, 2010.

Rive, Lyndon. How the US Can Become *THE* World's Largest Solar Market, presentation for Solor Power International 2009, October 28, 2009, Anaheim, CA.

Schmit, Julie. "SolarCity Aims to Make Solar Power More Affordable." *USA Today*, November 9, 2009.

Smith, Rebecca. "Lightening the Load." *Wall Street Journal*, October 6, 2008.

"SolarCity Collective Power Program Successfully Concludes in Portola Valley." *PR Newswire*, December 19, 2006.

"SolarMan." November 11, 2009, www.youtube.com/watch?v=D7W1W5hCnZY (accessed November 12, 2010).

Soto, Onell. "Power to the People." *San Diego Union-Tribune*, May 31, 2009.

SunRun. "1BOG and SunRun Sweeten the Deal for Home Solar Electricity." Press release, March 19, 2009.

Wesoff, Eric. "SEIA on Utility-Scale Solar." *Greentech Media*, March 18, 2010.

Woody, Todd. "The Dell of Solar Energy." *Fortune*, April 18, 2008.

24

Adelson, Andrea. "Making a Difference." *New York Times*, November 22, 1992.

Bodine, Mike. "DWP Head to Float Owens Lake Solar Panel Proposal." *Inyo Register*, January 9, 2010.

Eilperin, Juliet. "California Tightens Rules on Emissions." *Washington Post*, September 1, 2006.

Fine, Howard. "Utility Player." *Los Angeles Business Journal*, April 26, 2004.

Freeman, David. Interview with author, November 6, 2009.

Freeman, S. David. *Winning Our Energy Independence*. New York: Gibbs Smith, 2007, p. 170.

Gould, Bill. Personal communication with author, February 5, 2010.

Grasseschi, Wendilyn. "Owens Lake Solar Project Could Fuel 10 Percent of State's Power Needs." *Mammoth Times*, January 8, 2010.

Institute for Local Self-Reliance. "Renewable Portfolio Standards—California, New Rules Project." www.newrules.org/energy/rules/renewable-portfolio-standards/renewable-portfolio-standards-california (accessed November 12, 2010).

Isensee, Laura. "Los Angeles Eyes Owens Lake for Huge Solar Project." Reuters, February 11, 2010.

"Leading from the Front." *Economist*, February 21, 1998.

Office of California Governor. "Gov. Schwarzenegger Signs Landmark Legislation to Reduce Greenhouse Gas Emissions." Press release, September 27, 2006.

Peevey, Michael. Interview with author, November 6, 2009.

PG&E. "2008 Electric Power Mix Delivered to Retail Customers." www.pge.com/myhome/edusafety/systemworks/electric/energymix/index.shtml (accessed November 12, 2010).

Roosevelt, Margot. "State Acts to Limit Use of Coal Power." *LA Times*, May 24, 2007.

Sahagun, Louis, and Phil Willon. "DWP Scales Back Its Owens Lake Solar Test." *LA Times*, July 6, 2010.

SCE. "2008 Renewables Summary." www.sce.com/PowerandEnvironment/Renewables/ (accessed November 12, 2010).

———. "Former SCE Chairman and CEO William R. Gould Dies." Press release, March 14, 2006.

"Schwarzenegger's Executive Order Raises California RPS to 33% by 2020." *Renewable Energy World*, September 17, 2009.

Shugar, Dan. Photovoltaics in the Utility Distribution System. Photovoltaic Specialists Conference, 1990.

State of California Energy Action Plan. September 21, 2005, www.energy.ca.gov/energy _action_plan/2005-09-21_EAP2_FINAL.PDF (accessed November 12, 2010).

Willon, Phil. "Owens Lake as Solar Power Plant?" *LA Times*, December 2, 2009.

Zahniser, David, and Phil Willon. "LA Utility Shelves Plans for Solar Farm Near Salton Sea." *LA Times*, November 16, 2009.

———."Villaraigosa Unveils Solar Plan for Los Angeles." *LA Times*, November 25, 2008.

25

"Archimedes' Death Ray: Idea Feasibility Testing, MIT 2.009 Produce Engineering Processes." October 2005, web.mit.edu/2.009/www/experiments/deathray/10 _ArchimedesResult.html (accessed November 2010).

"Augustin Mouchot." wikipedia.org/wiki/Augustin_Mouchot (accessed November 12, 2010).

Baker, David. "PG&E Expands Solar Power Plans." *San Francisco Chronicle*, May 14, 2009.

Berger, John J. *Charging Ahead*. Berkeley: University of California Press, 1998, pp. 23–47. See for an overview of Luz International.

Blood, Michael. "Endangered Tortoises Snarl Solar Energy Plans." Associated Press, January 2, 2010.

BrightSource Energy. "BrightSource Energy Proposes Reduced Footprint Alternative Mitigation for Ivanpah Solar Electric Generating System." Press release, February 11, 2010.

Butti, Ken, and John Perlin. *Golden Thread*. Palo Alto: Cheshire Books, 1980, pp. 101–11.

Cejnar, Jessica. "Edison Dismantles Daggett Solar Project." *Desert Dispatch*, July 27, 2009.

"Daggett, California." en.wikipedia.org/wiki/Daggett,_California (accessed November 12, 2010).

Eurosolar. "Desertec: Why Roam Far and Wide, as the Good Is Found Close to Home." Press release, June 17, 2009.

Ewers, Justin. "Obama's Stimulus Keeps the Solar Power Dream Alive for Start-Ups." *U.S. News & World Report*, March 9, 2009.

Garg, H. P. *Advances in Solar Energy Technology*. Berlin: Reidel, 1987, p. 184.

Genesis 28: 12, 19.

Goldman, Arnold. Interview. *Beyond Zero Emissions*, May 22, 2008.

Johnson, J. T. "The Hot Path to Solar Electricity." *Popular Science*, May 1990, p. 82.

Kanellos, Michael. "Ausra on the Block: Get Used to Acquisitions in Solar." *Greentech Media*, November 16, 2009.

———. "Solar Thermal: Which Technology Is Best?" *Greentech Media*, April 28, 2009.

Locke, Robert. "World's Largest Solar Power Plant Dedicated." Associated Press, November 3, 1982.

Lotker, Michael. "Barriers to Commercialization of Large-Scale Solar Electricity: Lessons Learned from the Luz Experience." SAND91-7014, November 1991.

Maloney, Peter. "Solar Projects Draw New Opposition." *New York Times*, September 24, 2008.

Mieszkowski, Katherine. "The Tortoise and the Sun." *Salon*, January 22, 2009.

Sahagun, Louis. "Environmental Concerns Delay Solar Projects in California Desert." *LA Times*, October 18, 2009.

———. "Solar Energy Firm Drops Plan for Project in Mojave Desert." *LA Times*, September 18, 2009.

"Sandia Labs Shares Major Solar Success with Industrial Consortium." Press release, June 5, 1996.

SCE. "Southern California Edison and BrightSource Energy Sign World's Largest Solar Deal Agreement for 1,300 Megawatts of Clean and Reliable Solar Power." Press release, February 11, 2009.

"Schwarzenegger's Green Challenge." *60 Minutes*, December 21, 2008, www.cbsnews.com/stories/2008/12/19/60minutes/main4677334.shtml (accessed November 12, 2010).

"Science: Archimedes' Weapon." *Time*, November 26, 1973.

Streater, Scott. "California Desert on Pace to Become World's Solar Capital." *New York Times*, August 13, 2010.

Taub, Eric. "Reclaiming His Place in the Sun." *New York Times*, September 24, 2008.

Wald, Matthew. "Turning Glare into Watts." *New York Times*, March 6, 2008.

Wang, Ucilia. "The Rush for Gigawatts in the Desert Explodes." *Green Techmedia*, January 9, 2009.

Wasserman, Harvey. "Sunrise on Alternative Power." *LA Times*, March 9, 1991.

Weiss, Michael. "Everybody Loves Solar Energy, but . . ." *New York Times*, September 24, 1989.

Woody, Todd. "Big Solar's Day in the Sun." *Business 2.0*, June 5, 2007.
———. "China Powers the Global Green Tech Revolution." *Grist*, January 11, 2010.
———. "The Southwest Desert's Real Estate Boom." CNN Money, July 11, 2008.
Zeller, Tom. "Europe Looks to Africa for Solar Power." *New York Times*, June 22, 2009.

26

"Analyst Predicts First Solar to Become Largest Solar Module Manufacturer in '09." PV-tech.org, June 17, 2009.
Appleyard, David. "Utility-Scale Thin-Film: Three New Plants in Germany Total Almost 50 MW." *Renewable Energy World*, March 11, 2009.
"ARCO's Big Bet." *Time*, January 28, 1980.
Berkman, Leslie. "From Rooftop to Shining Roof." *Press Enterprise*, March 27, 2008.
"Cadmium Telluride Photovoltaics." en.wikipedia.org/wiki/Cadmium_telluride _photovoltaics (accessed November 12, 2010). See for a history of cadmium telluride solar cells.
Daily, Matt. "First Solar Sees Contract for China Power Supply." Reuters, February 23, 2010.
Dickerson, Marla. "Rooftop Solar Project Powers Up." *LA Times*, December 2, 2008.
Duce, John, and Indira Lakshmanan. "First Solar to Build World's Largest Solar Power Plant in China." Bloomberg, September 9, 2009.
Ehrlich, David. "California Dreamin' of Miles of Solar." Cleantech.com, March 28, 2008.
———. "First Solar Buys Renewable Energy." Cleantech.com, December 3, 2007.
Fairley, Peter. "BP Solar Ditches Thin-Film Photovoltaics." *IEEE Spectrum*, January 10, 2003.
First Solar. "First Solar Announces Two Solar Projects with Southern California Edison." Press release, July 16, 2008.
———. "First Solar Welcomes EU Endorsement of Euro 45 Million Aid for New Solar Module Plant in Germany." Press release, April 27, 2006.
"First Solar & NRG Open 21 MW Blythe Project." *Renewable Energy World*, December 22, 2009.
"First Solar Hosts Chinese National Leadership Delegation Seeking Renewable Energy Solution." *Business Wire*, September 8, 2009.

"First Solar Produces 1 Gigawatt of Clean Solar Electricity." *Business Wire*, March 20, 2009.

"First Solar Secures Financing for 53 Megawatt Solar Power Plant in Germany." *Business Wire*, April 21, 2009.

"First Solar Sees Costs Down by a Third in 5 Years." Reuters, June 24, 2009.

Gelsi, Steve. "For First Solar's Michael Ahearn, a Year in the Sun." *Market Watch*, December 6, 2007.

"Glasstech Builds on Its Pioneering Ideas." *Glasstech World*, 2008.

Hand, Aron. "First Solar's Sohn Makes Case for Solar Cost Perspective." *PV Society*, July 23, 2009.

"Inventor Became Philanthropist." Obituary of Harold McMaster, *Toledo Blade*, August 26, 2003.

Iverson, Wesley. "Amorphous Silicon Wins Place in the Sun." *Electronics*, April 5, 1984.

Johnson, Nicholas. "Abengoa, Abound to Get $1.85 Billion in US Solar Project Loan Guarantees." Bloomberg, July 3, 2010.

Kanellos, Michael. "First Solar's Stock: From $20 to $220 in a Year." *Greentech Media*, November 8, 2007.

Kanter, James. "First Solar Claims $1-a-Watt 'Industry Milestone,' Green Inc." *New York Times* blog, February 24, 2009.

Lifton, Jack. "The Tellurium Supply Conjecture." *Resource Investor*, July 9, 2009.

Marcial, Gene. "First Solar's Bright Future." *BusinessWeek*, August 9, 2009.

Moore, Matt, and Paul Burkhardt. "German Green Technology Forges Ahead." Associated Press, July 19, 2007.

Pakulski, Gary. "First Solar Renews Its Push for Viability." *Toledo Blade*, February 16, 2003.

———. "Future Looks Sunny as Plant Adds Capacity for Solar Panels." *Toledo Blade*, April 11, 2006.

———. "Insiders at Perrysburg's First Solar Disagree on Firm's Prospects." *Toledo Blade*, September 23, 2001.

Popular Science, May 1989, p. 156.

Porter, Lord George. Comment on making solar cells from silicon heard by author at a conference in Tokyo in the mid-1980s.

Quinn, James. "Maverick Using New Technology." *LA Times*, April 2, 1985.

Randazzo, Ryan. "Solar Technology Fuels Phoenix Company's Spectacular Growth." *Arizona Republic*, November 24, 2007.

Schultz, Randy. Interview with author, October 26, 2009.

"Second Look: Michael Ahearn of First Solar, Inc. Says Europe's Energy Policies Leaving U.S. Behind." Knowledge W. P. Carey, July 29, 2009, http://knowledge .wpcarey.asu.edu/article.cfm?articleid=1799 (accessed November 12, 2010).

Stone, Daniel. "Not in Anyone's Backyard." *Newsweek*, January 13, 2010.

Stuart, Becky. "Utility Scale Solar Set to Take Off." *PV Magazine*, July 20, 2010.

Veiga, Alan. "SoCal Edison to Build $875 Million Solar Energy Installation." *Mercury News*, March 27, 2008.

Wald, Matthew. "Two Large Solar Plants Planned in California." *New York Times*, August 15, 2008.

"Waldpolenz Energy Park: Thin-Film Photovoltaic on the Way to the World's Biggest Solar Electric Power Park." *Solarserver*, September 11, 2008.

Wang, Ucilia. "Armed for Battle: First Solar vs. SunPower vs. Suntech." *Greentech Media*, March 4, 2009.

———. "First Solar Panels Big with SolarCity Customers." *Greentech Media*, April 17, 2009.

———. "Strong Demand, High Pricing for Tellurium, Indium." *Greentech Media*, December 2, 2009.

Wells, Edward. "Going for Broke." *Inc. Magazine*, June 1998.

Yerkes, Bill. Interview with author, October 29, 2009.

27

Adams, David. "FPL Goal: Three Solar Plants in 2009." *St. Petersburg Times*, June 25, 2008.

Anderson, Zac. "Solar Plant Set to Open, Even as Shadows Loom." *Herald Tribune*, October 14, 2009.

Ball, Molly. "Nellis Tour Gives Obama Close Look at Solar Future." *Las Vegas Review-Journal*, May 28, 2009.

"Briefly Noted . . . Solar Energy." *Renewable Energy World*, December 4, 2001.

California Valley Solar Ranch. Homepage. www.californiavalleysolarranch.com (accessed November 12, 2010).

"Cypress Semiconductor to Deploy Clean, Onsite Solar Power." *Business Wire*, November 14, 2001.

"Cypress Startups: History, Theory of Funding, Lessons." September 3, 2009, www.cypress.com/?rID=36694 (accessed November 12, 2010).

Dinwoodie, Tom. Interview with author, December 11, 2008.

"Governator to China: Go Green." *Red Herring*, November 14, 2005.

Gunther, Marc. "For Solar Power, the Future Looks Bright." *Fortune*, October 4, 2007.

Holahan, Catherine. "T. J. Rogers' Startup Strategy." *BusinessWeek*, December 3, 2007.

Jones, James. "Arcadia Shines for Obama." *Bradenton Herald*, October 28, 2009.

Kho, Jennifer. "SunPower Buys PowerLight: $265M." *Red Herring*, November 15, 2006.

Kling, Jim. "Academic Spinoff: Sunpower." *Science Careers*, November 12, 2004.

Little, Amanda. "Charlie Crist Superstar." *Grist*, January 14, 2008.

Loder, Asjylyn, and Craig Pitman. "FPL Unveils Plans for a Solar Plant." *St. Petersburg Times*, September 27, 2007.

Marinucci, Carla. "Governor Suggests China to Invest in California's Clean Technology." *San Francisco Chronicle*, November 15, 2005.

PG&E. "PG&E Signs Historic 800MW Photovoltaic Solar Power Agreements with OptiSolar and SunPower." Press release, August 14, 2008.

"President Obama Delivers Remarks at Solar Energy Center." *Washington Post* transcript, October 27, 2009.

"The Rebel Vintner." *Newsweek*, February 20, 2006.

Robinson, Peter. "John McCain, Meet T. J. Rodgers." *Corner*, February 15, 2008.

Smeloff, Ed. Interview with author, November 18, 2008.

Smith, Stephen. "President Obama Visits Florida Solar Facility—A Trip Filled with Irony." *Institute for Southern Studies*, November 2, 2009.

"Southern California Edison Orders 200 Megawatts of SunPower Panels for Large Utility Solar Project." *Business Wire*, March 10, 2010.

Stuart, Becky. "New World Record for Solar Cell Efficiency Announced." *PV Magazine*, June 24, 2010.

"SunPower Adds 40-Megawatt Capacity to California Valley Solar Ranch." *Renewable Energy World*, May 3, 2010.

"SunPower Announces World-Record Solar Cell Efficiency." *PR Newswire*, May 12, 2008.

Swanson, Richard. Interview with author, November 19, 2008.

Sweet, Cassandra. "Silicon Substitute." *Wall Street Journal*, February 8, 2009.

Wirbel, Loring. "Cypress' Solar Power Gambit Bears Fruit in SunPower IPO." *EE Times*, January 16, 2006.

"World's Largest Solar Photovoltaic System Commissioned." *Renewable Energy World*, July 11, 2005.

Yerkes, Bill. Interview with author, November 7, 2006.

Zachary, Pascal. "Silicon Valley Starts to Turn Its Face to the Sun." *New York Times*, February 17, 2008.

28

Ariel, Yotam. "Incentives, Falling Cost and Rising Demand in China's PV Market." *Renewable Energy World*, November 13, 2009.

"Australian Scientist Stresses Solar Energy in Meeting Demand." Xinhua, September 13, 2007.

Bai, Jim, and Chen Aizhu. "China Selects 294 Solar Power Plants for Subsidy." Reuters, November 15, 2009.

Biggs, Stuart, and John Duce. "Yingli $5.3 Billion Loan May Boost Solar Panel Supply." *BusinessWeek*, July 12, 2010.

"China Will Have Solar Feed-In Tariff in Place in 2009—Suntech." *Renewable Energy World*, August 21, 2009.

"Chinese Scientist Finds Wealth in Solar." Associated Press, July 23, 2007.

"Chinese Solar Cells Swamping Subsidized German Market." *Earth Times*, March 28, 2008.

Fisher, Richard, and Phil McKenna. "The Sun King." *New Scientist*, November 10, 2007.

Flannery, Russell. "Sun King." *Forbes*, March 27, 2006.

"Gemini to Build 30-MW Plant for Austin Energy." *Renewable Energy World*, March 12, 2009.

Green, Martin. Interview with author, October 23, 2006.

———. "Shi Zhengrong." *Time*, October 10, 2007.

Groom, Nichola, and Leanora Walet. "China, US Offer No Quick Boost to Solar Slump." Reuters, August 4, 2009.

"Growing Pains: The Rise of Big Solar." *Economist*, April 15, 2010.

"Highest Silicon Solar Cell Efficiency Ever Reached." *Science Daily*, October 24, 2008.

Hogg, David. Interview with author, November 15, 2007.

Hug, Rolf, and Martin Schachinger. "Chinese Solar Modules Penetrating the German Market." *Solarserver*, November 10, 2006.

Jordan, David. Interview with author, July 17, 2007.

Osborne, Mark. "Suntech Grabs 25% Share in California: US Plant Plans Imminent." PV-tech.org, August 21, 2009.

Powell, Bill. "China's New King of Solar." *Fortune*, February 11, 2009.

Shi, Zhengrong. Presentation at Solar Power International 2009, October 28, 2009, Anaheim, CA.

Smith, Deborah. "Arise the Sun King." *Sydney Morning Herald*, September 12, 2006.

"Solar—Two World Records." *Australian Energy News* 7, March 1998.

"Solar Water Heating." en.wikipedia.org/wiki/Solar_water_heating (accessed November 12, 2010).

Stuart, Becky. "China to Be World's Largest PV Market." *PV Magazine*, August 4, 2010.

"Suntech Achieves 1GW Solar Cell and Module Production Capacity." *PR Newswire*, January 8, 2009.

"Suntech CTO Stuart Wenham Receives William R. Cherry Award for Outstanding Achievements in PV Technology." *Solar Plaza*, June 11, 2009.

"Suntech Pushes Pluto Tech to 19% Efficiency on Monocrystalline Solar Cells, 17% on Multi Cells." PV-tech.org, March 27, 2009.

Wang, Ucilia. "Here Comes China's $3B 'Golden Sun' Projects." *Greentech Media*, November 16, 2009.

———. "Suntech Buys EI Solutions, Teams Up with MMA." *Greentech Media*, October 2, 2009.

———. "Suntech Claims New World Record in Silicon Panel Efficiency." *Greentech Media*, August 19, 2009.

———. "Suntech Laid Off 10%, Factories Running at 50%~60% Capacity." *Greentech Media*, January 12, 2009.

———. "Suntech Vies for Big Share of China Subsidy Program." *Greentech Media*, May 21, 2009.

———. "True or False: China Will Be a Bigger Solar Market Than the US." *Greentech Media*, July 17, 2009.

Watts, Jonathan. "Bright Future for China's Solar Billionaire." *Guardian*, July 25, 2008.

———. "Energy in China." *Guardian*, July 25, 2008.

Yuankai, Tang. "Solar Power Shines." *Beijing Review*, September 2006.

Zweig, David, and Stanley Rosen. "How China Trained a New Generation Abroad." Science and Development Network, May 22, 2003.

29

Burke, Conrad. Interview with author, November 3, 2009.

Caltech. "Caltech Researchers Create Highly Absorbing, Flexible Solar Cells with Silicon Wire Arrays." Press release, February 16, 2010.

Catchpole, Kylie. Interview with author, March 3, 2010.

Green, Martin. *Third Generation Photovoltaics*. Berlin: Springer, 2003, pp. 1–3.

Johnson, Colin. "Silicon Ingot Maker Solaicx Customizes Gear for Solar." *EE Times*, March 26, 2009.

Kanellos, Michael. "Innovalight Signs with Yingli for Second Chinese Solar Deal." *Greentech Media*, July 26, 2010.

———. "Solar's Big Question: What Happens in the Second Half?" *Greentech Media*, March 30, 2010.

Krupp, Fred, and Miriam Horn. *Earth: The Sequel*. New York: Norton, 2009, p. 23.

McNichol, Tom, and Michael Copeland. "Here Comes the Sun." *Business 2.0*, March 2, 2007.

Mints, Paula. "10 Years in the Sun: The Most Profitable Decade in PV History Draws to a Close." *Renewable Energy World*, March 5, 2010.

Podewils, Christoph. "Sawing, Peeling and Blasting." *Photon International*, August 2008.

Polman, Albert. Interview with author, March 8, 2010.

Sibley, Lisa. "JA Solar to Use Innovalight's Silicon Ink Technology." Cleantech Group, September 15, 2009.

"Silicon vs. CIGS: With Solar Energy, the Issue Is Material." *ZD Net*, October 2, 2006.

"Slashed Subsidies Send Shivers through European Solar Industry." *New York Times*, March 31, 2010.

Swanson, Richard. Interview with author, November 19, 2008.

"Swinburne University of Technology and Suntech Team to Develop Next Gen Solar." *PR Newswire*, April 6, 2009.

"The View from Moses Lake, Part 1: REC Silicon Makes a Big Solar-Poly Bet on a Fluidized Bed." *PV Tech*, March 22, 2010.

Wesoff, Eric. "A Lifetime in the Solar Industry." *Greentech Media*, March 30, 2010.

Epilogue

Blau, John. "Germany Takes a Close Look at Its FIT." *Renewable Energy World*, December 15, 2009.

———. "How Much PV Capacity Is Actually Installed in Germany?" *Renewable Energy World*, December 21, 2010.

"Boost for Solar Industry as the Bundesrat Rejects FIT Reductions." *Recharge News*, June 7, 2010.

"BSW: German Solar Demand Still Booming in 2010." *Renewable Energy World*, May 27, 2010.

Dohmen, Frank, et al. "Solar Industry Fights to Save Subsidies." *Spiegel*, April 22, 2010.

Enkhard, Sandra, et al. "Bundestag and Bundesrat Remain Divided Over FITs Issue." *PV Magazine*, June 17, 2010.

"German Federal Election 2009." en.wikipedia.org/wiki/German_federal_election , _2009 (accessed November 12, 2010).

"German FIT Reduction Deal Reached." *Renewable Energy World*, July 7, 2010.

"German Politicians Settle Cuts in Solar Power Tariffs." Earthtimes.org, February 23, 2010.

"Germany's Solar Industry Worried about Future in Wake of Subsidy Cuts." *Deutsche Welle*, January 21, 2010.

"iSuppli Details Impact on German Solar Market When Feed-In Tariff Changes Hit." *PV-tech.org*, May 20, 2010.

Kirschbaum, Erik. "German FDP Opposes Solar Incentive Cuts." Reuters, February 2, 2010.

————. "German State Premier Seehofer Objects to Solar Cuts." Reuters, March 12, 2010.

————. "Germany May Trim Solar Power Incentives." Reuters, December 17, 2010.

Kuhlmann, Karl. Interview with SAG Solarstrom. *Solar Plaza*, May 12, 2010.

Levitz, David. "Greens Say Utilities Have Overcharged for Electricity by One Billion." Reuters, August 3, 2010.

Parkin, Brian, and Nick Comfort. "German Solar-Park Aid Cut Less Than Expected in Draft." Bloomberg, February 23, 2010.

Paulsson, Lars, "Solar Doubling, Gas Glut Drive Down German Power Prices." Bloomberg, September 22, 2010.

"Slashed Subsidies Send Shivers through European Solar Industry." *New York Times*, March 31, 2010.

Stratman, Klaus. "Solar Power Is Unaffordable." *Handelsblatt*, June 21, 2010.

Van Loon, Jeremy. "German Solar Market May Extend Growth in 2011, Phoenix Says." Bloomberg Business Week, August 4, 2010.

Wicht, Henning. Interview with *Solar PV.tv*, March 19, 2010.

INDEX

371